THE LIBRARY
ST. MARY'S COLLEGE OF MARYLAND
ST. MARY'S CITY, MARYLAND 20686

079116

The Hungry Fly

This volume is published as part of a long-standing cooperative program between Harvard University Press and the Commonwealth Fund, a philanthropic foundation, to encourage the publication of significant scholarly books in medicine and health.

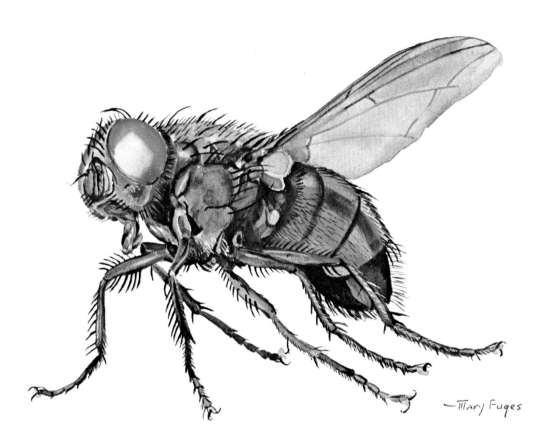

The Hungry Fly

A Physiological Study
of the Behavior
Associated with Feeding

V. G. Dethier

A Commonwealth Fund Book

Harvard University Press
Cambridge, Massachusetts
and London, England
1976

Copyright © 1976 by the President and Fellows of Harvard College
All rights reserved
Printed in the United States of America

Library of Congress Cataloging in Publication Data

Dethier, Vincent Gaston, 1915–
    The hungry fly.

    "A Commonwealth Fund Book."
    Bibliography: p.
    Includes index.
    1. Phormia regina.   2. Insects—Physiology.
3. Insects—Behavior.   4. Animals, Food habits of.
I. Title.
QL537.C24D47   595.7'74   75-6643
ISBN   0-674-42710-6

The selection from *The Creatures' Choir* by Carmen Bernos de Gasztold, translated by Rumer Godden, which appears on p. 479, is reprinted with the kind permission of The Viking Press, Inc. English text: Copyright © 1965 by Rumer Godden. All rights reserved.

Dedicated to my former students who taught me more than they realized: Yousif Mohammed Arab, William Belzer, and late David R. Evans, David Falk, Jack Gans, Alan Gelperin, Nancy Goldrich, Casmir T. Grabowski, George W. Green, Frank E. Hanson, Edward S. Hodgson, Mary Caroline McCutchan, Margaret Nelson, Elizabeth Omand, Gerald Pollack, Marion V. Rhoades, Arnold Slonim, Thomas Smyth, and Myron L. Wolbarsht.

Preface

Those of my colleagues who raise flies do so for the purpose of feeding lizards, fish, birds, or mantids—which may be the real raison d'être of flies; flies raise flies to beget more flies, and so on ad infinitum; our laboratory raises flies in order to study them. Why, you might well ask, study flies and devote an entire book to the hungry fly?

Living things have everything important in common—the carbon atom as the structural corner stone, RNA and DNA, the cell, and, in most animals, the neuron and muscle. The blowfly (*Phormia regina* Meigen) as one variation on a theme has been spectacularly successful in coping with the problems of life. It must successfully derive energy from the environment. In this pursuit it exhibits the phenomena that we describe in ourselves as hunger, satiation, motivation, learning, et cetera. Its basic problems are the universal ones. An inquiry into its methods is revealing.

This inquiry, however, is not the sole intent of this book. It is written to show how ideas develop, to demonstrate that advances in knowledge are like blind pseudopodia, which make tentative explorations in one direction, withdraw from dead ends, extend anew in others. It is also intended to show that science is not truth graven in imperishable stone; that, instead, today's truth is most often tomorrow's error. The chapter on chemoreception, for example, may already be out of date insofar as truth is concerned, but in perusing it the reader will, it is to be hoped, detect the stumbling about, the searching among the trees for the forest.

It is for these reasons that a goodly balance of old work is recounted, that some antiquated experimental procedures are described in detail, and that the names of people occupy a prominent place in discussions. It is fashionable in this generation to relegate old work to the storage stacks of libraries, on the grounds that the enormous increase in publications makes vigorous pruning of references mandatory or that much of the older work is incorrect anyway. Old work is given space here not out of misplaced reverence, nor to reemphasize old themes nor even to perpetuate error, but to trace the development of ideas, to place newly acquired facts in a proper perspective, to reveal the error of looking at the status quo and accepting it as gospel truth, and above all to extricate the thread of continuity which in the final analysis directs the pattern of our efforts. Similar motives underlie the emphasis on names. People are as important as facts, especially since facts do not always speak for themselves. Ideas, the development of which one wishes to trace, are the products of people, and knowing which sets had common origins sometimes helps us to understand the interrelations of these ideas.

Experimental techniques, even old ones, are described in some detail not because anyone might ever wish to repeat the experiments, but because techniques frequently reveal in startling clarity the intellectual approach to problems, how the development of ideas is shaped by the technology of the time, and that many experiments were never done not because of lack of conceptualization but because of lack of technical wherewithal.

And finally a word about what this book is not about. It is not about chemoreception or neurophysiology. It may justifiably claim to serve as a source book of work done and a stimulus for work to be done. It is a book to be enjoyed. In these respects it is a farrago, but a farrago whose theme is a description of behavior in terms of underlying physiological mechanisms.

Since this book represents endeavor extended over a considerable span of time, the assistance of numerous individuals and institutions is most gratefully acknowledged. I thank the Commonwealth Fund for a grant-in-aid to assist in the preparation of this monograph; the John Simon Guggenheim Memorial Foundation, which on two occasions awarded me a Fellowship; the National Science Foundation, which supported the research; and Sussex University, Brighton, England, the London School of Hygiene and Tropical Medicine, and the Landbouwhogeschool, Wageningen, The Netherlands, all of whom were unstinting in their hospitality.

V.G.D.

Sussex
England

Contents

		Page
1	Introduction	1
2	The Search: Appetitive Behavior	4
3	To Each His Taste: Discrimination and Preference	34
4	The Etiquette of Eating: Mechanisms and Control of Ingestion	67
5	Detecting the Unpalatable and the Dangerous	119
6	The Sweet Tooth and the Freshness of Water	160
7	The Flavor of Things: Codes and Information	186
8	Avoiding the Temptation of Gluttony	228
9	Food For the Next Generaton: Specific Hungers	299
10	Thirst	336
11	Winter: Diapause	353
12	Microscopic Brains	363
13	Profiting By Experience	410
14	The Fly and the Concept of Motivation	460
15	Epilogue	477
	Index	481

The Fly

Busy, curious, thirsty fly,
Drink with me, and drink as I;
Freely welcome to my cup,
Could'st thou sip, and sip it up;
Make the most of life you may,
Life is short and wears away.

Just alike, both mine and thine,
Hasten quick to their decline;
Thine's a summer, mine no more,
Though repeated to threescore;
Threescore summers when they're gone,
Will appear as short as one.

William Oldys, *On a Fly Drinking Out of a Cup of Ale*

1
Introduction

'I'll not hurt thee,' says my Uncle Toby rising from his
chair, with the fly in his hand, . . . *this world is
surely wide enough to hold both thee and me.'*
Sterne, *Tristram Shandy*

Appetite for food and appetite for sex are two of the most important, if not the most important, driving forces underlying the design for living. These appetites of mice and men, of flies and all dioecious animals have this in common: they are developed in a manner that ensures the survival of an individual for a critical span of time, and through him the survival of the species. While the individual exists, its structure and behavior are geared to the economy of staying alive long enough to mate and begin the production of others of its kind. In the course of progressing toward this destiny, the individual may fall prey to any number of other forms of life and serve to fulfill their destinies instead of its own. Or it may succeed in attaining its goal, die, and return its elements to the earth. As long as it lasts, however, it requires energy and matter.

The blowfly (*Phormia regina*) like all species of life, is a temporary form through which flows energy and matter, the matter becoming, for a while, fly and then passing on. The fly is just another way to reverse entropy on this planet, to defy, apparently, the Second Law of Thermodynamics. It is another way to build orderly complexity in a system characterized by increasing disorder and randomness.

We could study the fly for itself. Then we would know all about the fly. Were that the sole goal of this research, we would have learned little. The fly is one of more than two million forms of life. We would know all about one two-millionth of life. On the other hand, if we were to view the fly as one successful solution to the problem of maintaining a working interface between the cosmos of life and the greater world, as one solution to the basic problems that are common to all, we might, just might, gain some insight to multiple fundamental phenomena.

Animals are improbable accretions of matter. Real species are every bit as improbable as are the figments of diseased minds or of the imagination of a Hieronymus Bosch or Dali or Escher. The fly is a most improbable beast. From our point of view, the fly is a small animal. From an amoeba's point of view it is a giant. In the general scheme of things, it must be classed as small. On a purely descriptive level, what is it that meets the eye? We see an organism averaging eight millimeters in length and weighing about thirty milligrams. Its wingspread is about twice the length of its body. If we are able to overlook the fly's scatological way of life, we see a thing of beauty, a jet jewel, from whose polished surface the sun coaxes an iridescent display to rival that of the soap bubble, a jewel whose diaphanous wings bear it aloft with consummate skill, the curvature of whose eyes flows in smoothest arc, whose faceted design rivals the honeycomb in hexagonal perfection, whose hairs curve in marvelously fluted columns rivaling the best of gothic architecture. And privately within, its softer self is laced with the exquisite silver filigree of its air-filled tracheae. There is perfection in its parts and gracefulness in all its movements.

The analytic but less poetic eye observes a miniature articulated beast encased in armor. A semirigid nonliving armor constructed of lipid-impregnated tanned protein intermingled with an extraordinarily resistant compound, chitin, and further protected on the outside surface with a thin layer of cement and of wax. Although this cuticle together with the underlying hypodermis and associated glands may truly be considered a versatile functional organ, from a physical point of view, the fly "lives" in an impervious box. This box, the exoskeleton, provides not only protection against the vicissitudes of an inimical environment but also bodily support and points of attachment for muscles. Structurally speaking, the fly differs from a vertebrate in much the same fashion as a small frame house differs from a skyscraper. In both cases the smaller unit can be adequately supported by external framing while the larger requires internal trussing.

The fly is designed in the form of three imperfect spheres arranged in line: head, thorax, and abdomen. These constitute a more extreme division of labor than occurs in the vertebrate body, each third being highly developed for one major function. It is surprising what a large percentage of the body has been allocated to the production of energy for present and future generations, that is, to digestion and reproduction. The abdomen houses the major portion of the digestive tract, the portion that stores and processes food. It contains also the fat body, the analogue of the vertebrate liver, and the organs of excretion and reproduction. It is the least rigid section of the body. The exoskeleton of the abdomen is constructed of rigid plates that are joined by extensive, soft, folded intersegmental membranes. Flexible coupling allows the abdomen to undergo enormous distention when a maximum volume of food is ingested or when large numbers of eggs are produced.

The thorax is given over to locomotion. Contained therein are the muscles to drive the six legs and the enormous flight muscles. The thorax is more rigid than the abdomen but still sufficiently elastic that the muscles can deform it to drive the wings indirectly. There is scant room for any other organs within it. The head, the most rigid of the three compartments, is primarily neural. Containing the brain and the majority of sense organs, it is the center of the receipt and processing of information. There is no single specialized compartment for respiration comparable to the thoracic cavity of vertebrates. The transport of gases is accomplished by an intricately ramifying system of tubes and sacs, the tracheal system, which delivers oxygen directly to all cells of the body and removes carbon dioxide as directly. The blood does not serve a primary respiratory function as it does in vertebrates.

Of the several structural differences between the fly and the vertebrate, one of the most striking from a functional point of view is the arrangement of fluid compartments. The principle compartments in the vertebrate are: the cardiovascular, the cerebrospinal, the lymphatic, the coelom; and intercellular spaces. In the fly there are, aside from the heart and the vessel leading from it, no vessels carrying circulating fluid. The heart is an open tube. Blood leaving it bathes all organs of the body directly. The body cavity of the fly is therefore filled with blood. Space that is not filled with organs or blood is occupied by air sacs connected to tracheae.

Although the fly and the vertebrate differ in their organization, their goals are the same. Living things are circumscribed cosmoses, designs for utopias where constancy is the goal. The machinery of life operates best under constant conditions. The limits of tolerance are narrow. Existing in a greater world where change is the rule, the minimum constant challenge is to maintain the status quo in the midst of change; the maximum challenge is to exercise autonomous change independently of or even in opposition to change in the wider outside world. These islands of balancing constancy that are organisms, man, the fly in his chitinous box, win their utopias only at cost. Energy is required to maintain temperature stability, to keep water balance, to drive locomotion (which is necessary to exploit new sources of energy when the immediate source is depleted), and to find other flies to make more flies. And energy is required to run the machinery designed to process and utilize energy. The cycle closes upon itself. The primary immediate goal is eating. To this end are so many of the ambitions of men and the activities of flies directed.

2
The Search: Appetitive Behavior

> *What . . . spirit can it be that prompts*
> *The gilded summer-flies to mix and weave*
> *Their sports together in the solar beam*
> *Or in the gloom of twilight hum their joy?*
> Wordsworth, "Despondency Corrected"

For most animals eating begins with search, with a restless kind of locomotory behavior that tends to increase the probability of an individual's encountering food. Some animals, like the cockroach *Nauphoeta cinerea*, decrease their activity as the length of deprivation increases (Reynierse, Manning, and Cafferty, 1972); but most, as the period of time since the last meal lengthens, become more and more active and drain more and more of their energy reserves. Instead of conserving dwindling supplies by remaining inactive, they gamble on finding food before collapsing. The falcon and the kestrel fly in increasingly active patterns, the wolf and the lion forage vast expanses of hunting territory, a hydra extends its tentacles to their fullest, waving them back and forth, a hungry human urbanite searches for a restaurant. When no food is forthcoming in the immediate vicinity, or existing supplies have been completely exhausted, individuals and populations move. Plague locusts move on after they have devastated the land, human populations emigrate from the country to the cities in times of famine, nomadic people decamp when an area can no longer support them or their flocks—all the historic human emigrations of the past have had their basis ultimately in lack of food, irrespective of whether the immediate causes might have been economic, social or epidemiological. When organisms are hungry (to use the word in its widest common-sense connotation), they move. From an analytical point of view it is possible to equate hunger with movement. Satiation is the antithesis of hunger. It is associated with lack of movement.

To place the search for food in a proper perspective it is helpful to give some consideration to remote antecedent events. An investigation of the broader aspects of individual activity patterns provides a useful base. Obviously no animal has the stamina to be continuously and uniformly active. Although the behavior of individual cells or organs might seem to contradict this statement, even they alternate periods of greater activity with periods of lesser activity. Not even that most active of organs, the mammalian heart, maintains a constant beat; it slows during sleep. Conversely, no animal, however sluggish, is continously inactive. Were this the case, that animal would never reproduce; and it would survive only so long as its energy reserves remained or the food immediately adjacent to it could diffuse passively into it. A resistant egg, an encysted amoeba, a protozoan spore, or the larva of the midge *Polypedilum vanderplanki* that withstands one and one half years of desiccation (Hinton, 1951), although they can remain alive for months, years, decades, and possibly centuries, are, from the point of view of accomplishment, less actively alive than potentially alive. Thus it is that organisms alternate periods of activity with periods of rest.

The frequencies of different cyclic biological processes vary from milliseconds to years. Of the several rhythms of activity that occur, the most common and fundamental is approximately twenty-four-hourly, hence the designation circadial (Danilevsky, Goryshin, and Tyshchenko, 1970). This rhythm is controlled by an internal biological clock, the location and identity of which remain a mystery. The clock measures off time with fair accuracy, and, like our more familiar mechanical clocks, can be set for any time. Of the several environmental factors (variously called Zeitgeber, entraining agents, synchronizers) that set clocks, photoperiod is the most important. Biological clocks, assumed to be internal oscillators, are set to initiate and terminate activity in some relation to solar time. The vast majority of animals are either diurnal or nocturnal, although other patterns of activity can also occur.

It was realized early that a start toward understanding appetitive behavior of the blowfly and the relation between feeding and activity could be achieved only by first becoming acquainted with whatever endogenous rhythms the fly might possess. One of the early attempts to measure the activity of flies involved the ingenious use of large cardboard boxes (Barton Browne and Evans, 1960). Four boxes, approximately 0.5 meter square, were placed in line and connected with stemless glass funnels (Fig. 1). Both ends of the funnels were flush with the boxes they connected. Two hundred flies were released into the first box, where they milled about according to their state of activity. Their locomotion was essentially random, and sooner or later, sooner if they were very active, they chanced upon the funnel and passed through to the next box where the process was repeated. In a sense they were diffusing from one box to the next, and the rate was

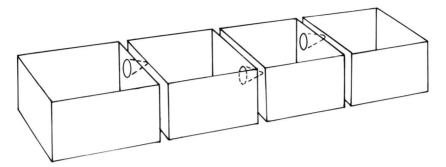

Fig. 1. An actograph for populations of flies (from Barton Browne and Evans, 1960).

related to their state of activity. Any number of boxes could be connected in a series. If one chooses to leave the last box in line open, the effect is tantamount to building an infinite series. The funnels biased diffusion in one direction, and the bias could be accentuated by placing a light at the far end of the funnel. At intervals the flies in each box were counted. A measure of activity was derived from "funnel passages." Funnel passages were expressed as a percentage of all of the transits that would occur if every fly moved from the box it was in at the beginning of the counting period to the last box.

This simple device had the merit of providing an estimate of both walking and flying activity. A more sensitive technique, but one that did not take flying into account because the flies were given only limited opportunity to fly, was devised by Green (1964a). This actograph (Fig. 2) provided measures of the activity of individuals. Each fly was housed in a tubular cage of nylon mesh approximately 5 cm long, 1.8 cm in diameter, mounted on thin balsa wood frames lined with shim brass on one side, and pivoted on two needle points soldered to the brass liners at the center points of the long arms of the frames. An actograph unit consisted of three nylon chambers set in a partitioned box so that the pivot points of each chamber rested in mercury-filled conical depressions in a common brass bar connected to one side of a 24-volt d.c. electrical circuit. A fine wire from each chamber touched an adjustable mercury contact connected to the opposite side of the circuit whenever the cage was tipped in one direction. The opposite end of the cage rested on a stop adjusted to limit the angle of tipping. As the fly jumped about or moved from one end of the cage to the other, the circuit was opened or closed depending on what direction the fly jumped. Locomotor activity was recorded as the number of tilts per unit time.

It is clearly impossible to separate an animal completely from environmental stimulation, so that true spontaneous activity is difficult to assess. To approach the ideal as closely as possible, temperature and relative humidity were rigorously controlled, a fan outside operated continuously to provide white noise and mask possible auditory cues, and the chambers were separated from one another by partitions to ensure privacy for each fly. Each fly could be fed measured volumes of any desired solution—feeding did not require opening the apparatus. A 0.98-mm tubing projected through the floor of each cage and connected externally to a microburette. The basic experiment consisted of continuously monitoring the activity of an individual in constant darkness from the instant of emergence from the puparium until the instant of death from starvation. A puparium was inserted into a slit in the end of the cage with its anterior end projecting inwards so that the emerging fly would step directly into the cage.

The first results showed that every fly behaves in essentially the same way (Fig. 3). There is a brief flurry of activity at emergence at 1400 hours of the first day of adult life. Activity during the remainder of the day and following night is low. From then until about 24 hours before death activity follows a regular ±24-hour cycle. The total

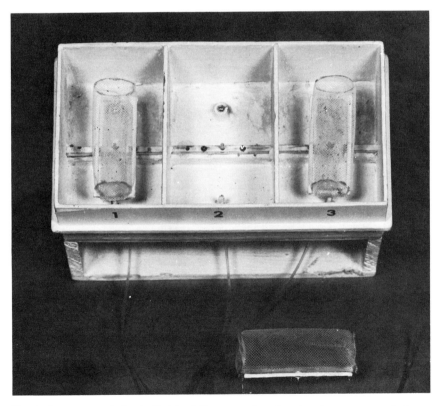

Fig. 2. An actograph for individual flies; one chamber has been removed. A capillary feeding tube attaches to each chamber (from Green, 1964a).

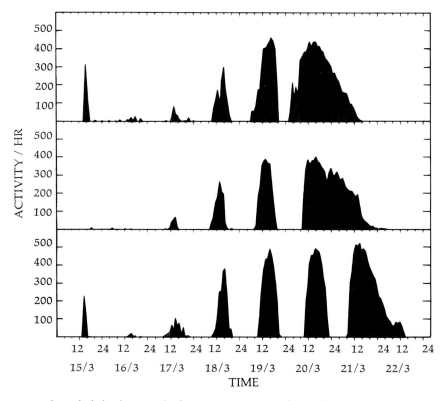

Fig. 3. Circadial rhythms in the locomotor activity of three females of *Phormia regina* from emergence (first indication of activity) to death from starvation in total darkness (from Green, 1964a).

amount of activity increases daily until death approaches, whereupon activity declines. This cycle is correlated with the light/dark cycle of the rearing schedule. The biological clock is set by light even before the fly has emerged from its puparial skin and begun adult life. From that moment on its activity is locked into the solar day. However, it can be reversed. Flies that have been exposed to a reversed day/night cycle for six days and then placed in constant darkness faithfully keep time with the reversed day (Fig. 4). Light is such a powerful incentive to activity that the endogenous circadial cycle is damped in constant lighting, and the flies are active throughout their lives (Fig. 5). The increase of activity with time and its decline with approaching death persist. All subsequent studies of the relation between feeding and locomotion were conducted under constant lighting.

During the daytime many environmental factors modify activity. Studies of the behavior of the closely related fly *Phormia terraenovae* in Finland have shown that flying activity is strictly dependent on temperature and cloud cover (Nuorteva, 1966). North of the arctic circle, in

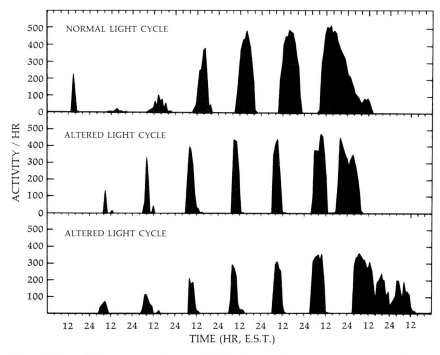

Fig. 4. Phase differences in the circadial rhythms of locomotor activity of flies exposed to a normal light-dark cycle (light 0520–1850 hr.) and an altered light-dark cycle (light 2400–1200 hr.) prior to tests in constant darkness (from Green, 1964a).

continuous sunlight, the flies retain their circadial rhythm. No activity occurs during the solar night because the temperature at ground level never reaches the 12°C required for the initiation of activity. At excessively high temperatures also the flies are inactive. This happens farther south at noon when the temperature exceeds 28°C. The activity curve of flies in these regions is clearly bimodal. Bimodality may also be a function of season. Many flies have two peaks of activity, morning and evening, in the summertime, but only one in spring and autumn (Norris, 1966; Sytshevskaya, 1962).

Although the bimodal cycle may be dictated by ambient conditions, a strictly endogenous origin cannot be ruled out of consideration. Experiments with the tsetse fly *Glossina morsitans* Westw. comparable to Green's experiments with *Phormia* have shown that the tsetse fly has a diel rhythm even in constant darkness and that this rhythm is more or less swamped in constant light. Under normal light-dark schedules there is a bimodal rhythm that corresponds to behavior in the field. Since relative humidity, temperature, and other ambient variables were controlled in the laboratory, it was concluded that the bimodal activity in this fly is not derived from environmental influences

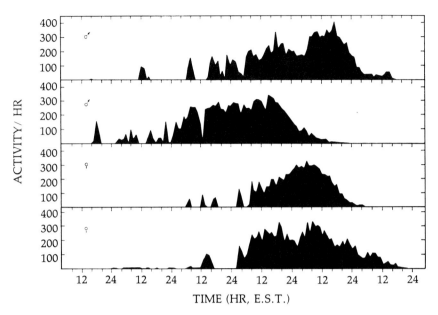

Fig. 5. Locomotor activity patterns of flies exposed to constant light from emergence to death from starvation (from Green, 1964a).

(Brady, 1972, 1974). Another blood-sucking dipteran of the tropics also exhibits a bimodal pattern of activity. This is the malarial mosquito *Anopheles gambiae.* Although bimodality is undoubtedly circadial, its existence does not mean that the underlying endogenous rhythm has two peaks (Jones, Cubbin, and Marsh, 1972). It is quite conceivable that there are factors suppressing the middle of a sustained circadial activity so that the net overt behavior is bimodal.

Other things being equal, the activity of flies is not identical from one day to the next. Days of intense flying activity are frequently followed by days in which the activity is less than would have been predicted from meterological conditions. Nuorteva suggested that a decrease in endogenous activity following feeding to repletion is responsible for the lack of absolute correlation with meteorological conditions.

At this point a description of the beginning of a summer day in the life of a young fly serves to link our discussion of locomotion with feeding behavior. The first rays of the rising sun have warmed the surface of a maple leaf on the underside of which clings a fly. The fly has spent the night there immobilized by low temperature, the absence of light to stimulate it, and the cycling of its internal clock, which had ordained a period of inactivity. With the advent of daylight all of these conditions are reversed. With warmth the fly's metabolism accelerates. With light its central nervous system is barraged by incoming signals from the huge compound eyes and hidden heat receptors. With the onset of a new day the mysterious internal clock triggers activity.

The fly begins grooming, making little scrubbing motions with its forelegs as it removes minute motes of detritus. It mutually cleans its proboscis and forelegs by massaging the former with the latter. It cleans its head. It rubs its hind legs together with precision. It smooths down its wings with the same pair of legs. Then, elevating the wings enough to insert the hind legs under them, it polishes its iridescent abdomen. So it continues its ministrations as the sun pursues its ascent. Its toilet finally completed, the fly jumps downward from the underside of the leaf, turns a half roll, and flies out into the sunlight.

It spends most of the day alternately flying about randomly and landing to rest and explore surfaces. All of these activities, if continued for long, require more energy than the fly possesses. During the night, even though resting, it has depleted some reserves, and as it flies it rapidly consumes others. When the fly had emerged from its puparium, it was obese with fat inherited from its larval days. This was stored in the cells of its fat body, loose aggregations of cells scattered throughout most of the body. By age four days, however, almost all of this had been used and, as a result of feeding, had been replaced by glycogen, which was stored not only in the fat body but also in the halteres (the greatly modified hindwings, which operate as gyroscopes) and the massive flight muscles. In old age, four to five weeks, these storage depots would be meager regardless of the amount of food eaten. During the prime of life, however, the depots would be depleted in proportion to the energy expended.

During the night when the fly was resting, when it groomed, and whenever it walked, it would draw upon residual fat reserves as well as glycogen; however, flying demands enormous amounts of energy—and rapidly. The average *Phormia* weighing 25 mg in the prime of life beats its wings 12,000 strokes a minute on a pleasant summer day (25°C). Each stroke utilizes the equivalent of 1.24×10^{-7} grams of glycogen per gram of muscle. To convert this to work, 1.11×10^{-4} cm^3 of oxygen per gram of muscle are required. The total energy expenditure is 0.521×10^{-3} calories per gram of muscle. Looked at another way, the blowfly is utilizing the equivalent of 35.5 mg/g of glycogen per hour and expending a total of 152 cal/g of energy (Hudson, 1958). A fly that has been fed 1.8 mg of a 1 M glucose solution could fly nonstop for a maximum of three hours, at the end of which time it would have used up to 95% of its available energy supplies. Younger and older flies are not capable of such a sustained effort. Smaller species of flies have less endurance. *Drosophila melanogaster,* weighing 0.5 mg in its prime (age one week), can fly continuously for 278 minutes. On one mg of glucose it can continue for 6.3 minutes (Wigglesworth, 1949). D.

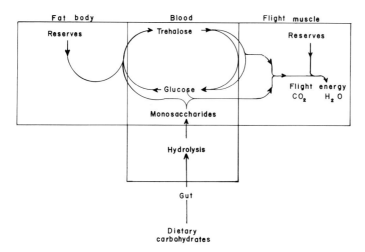

Fig. 6. Diagram of the transfer of carbohydrates among various tissues during flight (from Clegg and Evans, 1961).

repleta, a species weighing 3.5 mg, can continue for only 0.6 minutes on the same amount (Williams, Barness, and Sawyer, 1943). Flights of these durations would cover considerable distances. A female mosquito, *Culex pipiens* form *berbericus,* for example, could fly a straight line distance for 4,300 meters on 0.099 mg. of glucose (Clements, 1955).

In Diptera energy for flight is derived solely, if not exclusively, from aerobic oxidation of carbohydrates (Chadwick and Gilmour, 1940; Davis and Fraenkel, 1940; Williams, Barness, and Sawyer, 1943; Chadwick, 1947; Wigglesworth, 1949; Hocking, 1953; Clements, 1955; Hudson, 1958). It had always been assumed that glycogen stored in the flight muscles and fat body was the principal source of energy immediately mobilized for flight; and Wigglesworth had demonstrated that flying *Drosophila* depleted muscle glycogen while the crop remained full. On the other hand, flies flown to exhaustion could resume flight immediately upon having ingested sugar. Furthermore, Hudson's data revealed that in continuous flight the amount of carbohydrate used exceeded the amount of glycogen depleted. Clegg and Evans (1961) calculated that glycogen accounts for 20% of the time of continuous flight by *Phormia* and that the blood sugar, trehalose, accounts for 69%. On the basis of extensive analyses they proposed that carbohydrates were partitioned and exchanged among the compartments of the body according to the pathways diagrammed in Figure 6. Carbohydrates from a meal are hydrolized in the gut, absorbed in the blood mostly as monosaccharides, removed from the blood by the fat body, which stores them as glycogen and synthesizes trehalose, which is released into the blood on demand. Flight uses blood trehalose. The fat body can synthesize this sugar so rapidly from blood sugars that injection of sugar into the blood or a meal of sugar will enable an exhausted fly to resume flight almost immediately. Under normal circumstances, therefore, sugar stored in the crop is a most important reservoir of energy.

When the rate of utilization of carbohydrates exceeds the rate of mobilization from reserves, the fly stops, exhausted. After a short rest it can start again, but successive rest periods become longer and the flight attempts more feeble until all reserves are gone and complete and final exhaustion has ensued. If the fly is to continue its explorations, the carbohydrate must be replenished. Failure in this regard means death in two and one half days. However, there would appear to be a considerable margin of safety if one is to judge by the numbers of flies throughout the world. The fly has a very good chance of locating some food before collapsing. Either the method of searching is very efficient, or food is everywhere abundant, or both.

By and large, flies that have no immediate need of food or water and are not concerned with mating are going nowhere, doing nothing. By human standards they and most other animals are extraordinarily lazy. Nonsocial animals spend all their time in loafing, sleeping, searching for and consuming food or water, and mating. Grooming occupies a minor fraction of the time of some species, and nest construction is at best a seasonal occupation. Female houseflies spend 12.7% of their time walking or flying, 2.5% feeding, 29.7% regurgitating, 14.5% grooming, and 40.6% resting. For males the corresponding figures are: 24.3%, 4.1%, 23.0%, 20.7%, and 27.9% (Barber and Starnes, 1949). Most of the time is spent in pointless idleness and in simply existing. Only when some level of deprivation becomes critical does locomotor activity begin to increase.

There is probably no clear distinction between basal activity and activity associated with deprivation. The transition between the two states is gradual and continuous, and the appetitive phase can be identified only if there is knowledge of the fly's state of need or its goal. The stimuli guiding the final stages of the "search" are different when needs are different (water as opposed to food, for example), but the patterns of behavior may be indistinguishable. Furthermore, after a dominant need has been met, and locomotion temporarily suspended, a subordinate need may now take its place so that complex periodicities of activity ensue. In the life of the fly, however, the most common, predictable, and overriding state of deprivation, and the one most profoundly modifying its activity, is that which relates to food. Then, when food is nearby it provides stimulation that begins to order behavior. Thereafter, activity relates to a goal.

The setting of the biological clock determines that the fly's appetitive behavior will be conducted during daylight hours. General environmental factors, temperature, humidity, wind velocity, cloud cover, and conditions of light and shade channel the activities into certain hours, levels of intensity, particular localities, certain days, and particular seasons (e.g., by initiating or terminating diapause). More specific stimuli bring about changes in the direction and pattern of flying and the frequency with which flying, resting, walking, and grooming alternate.

Visual features of the environment are particularly important. It is probable that the intensity and wave length of light reflected by surfaces, the form of objects, and motion, more than any other stimuli are responsible for stimulating the periodic landing of the fly and directing where it lands. It is no wonder that the eyes occupy a larger proportion of the body than any other sensory system and that the neural tissue comprising the optic system is greater in volume than the entire brain. When Cajal first looked at the neural network in the eyes of flies he exclaimed in amazement at their enormous complexity (Cajal and Sanchez, 1915). These great eyes, containing more than 7000 retinal cells and rivaled in structure, complexity, and visual capacity only by the eyes of squids, octopi, and vertebrates, enable the fly to see all wavelengths from the ultraviolet to the red, to detect the plane of polarization of the light of the sky, and to resolve flickering movements alternating as rapidly as 250 times per second. Only with respect to form perception do they fall behind their vertebrate counterparts.

They are superlatively adapted to detect motion, as anyone can prove to himself by trying to catch a fly. The ability to resolve a moving pattern is as essential to normal flight as to escape. Particular neurons in the region between the medulla externa and the medulla interna are specialized as motion detectors (Bishop and Keehn, 1966; McCann and Dill, 1969). Some are directional; they increase their basal activity in response to motion in one direction and suppress it to the opposite motion. A unit that responds to right-to-left movement also responds to downward movement and to an open hand that approaches rapidly; a unit that responds to left-to-right movement responds to upwards and to near-to-far movements. Other interneurons are nondirectional (Bishop and Keehn, 1967). Still others respond maximally to vertical edges, others to horizontal edges, and others to form and motion. Altogether, McCann and Dill (1969) located and studied forty distinct classes of interneurons involved in intensity, form, and motion perception by *Calliphora phaenicia* and *Musca domestica*. These and more with still different characteristics (Mimura, 1971) are instrumental in enabling the fly to detect many form-motion and edge-motion relationships (Bishop, Keehn, and McCann, 1968). The results of electrophysiological studies of these interneurons were put into a behavioral context by correlating them with earlier studies of the responses of flies to moving stripes (optomotor responses). Tethered flies attached to a device that measured yaw torque registered their attempts to orient to various moving patterns and so revealed the capacities of the eye to detect different kinds of motion (Fermi and Reichardt, 1963; McCann and MacGinitie, 1965; McCann and Fender, 1964). These studies showed, among other things, that flies are particularly attracted by spots of light turned on and off. They are also strongly attracted to small black objects of limited size, the limited size being "fly-size" (Wiesmann, 1960, 1962a). This latter ability provides the basis for their "herd instinct."

The spontaneous preference of flies for particular objects and patterns suggests that there is an efficient neural mechanism for pattern discrimination. Detailed analyses of this mechanism have been undertaken with *Drosophila, Musca,* and *Calliphora* (Götz, 1971; Götz and Wenking, 1973; Jander and Schweder, 1971; Reichardt, 1970; Wehner, 1972).

Black and red are particularly attractive to flies. At any given moment there are more flies on black and red objects than on those of any other color (Freeborn and Berry, 1935; Waterhouse, 1948; Wiesmann, 1960, 1962b; Pospíšil, 1962; Hecht, 1963, 1970). Wiesmann (1962b) reported that red was more attractive than black, but later studies failed to confirm this observation.

The question of the fly's ability to perceive red and the means whereby this is accomplished is still a topic of lively discussion (Autrum and Burkhardt, 1961; Burkhardt, 1964; Goldsmith and Fernandez, 1966; Mazokhin-Porshnyakov, 1960a, 1960b, 1969; Langer and Thorell, 1966; Bishop, 1968; McCann and Arnett, 1972). *Calliphora* is unusual in that the spectral sensitivity curve of its eye, as measured electrophysiologically, exhibits three maxima: near ultraviolet, blue-green, and red (Autrum and Stumpf, 1953; Walther and Dodt, 1957, 1959). Whether there is a red-sensitive retinal cell or whether the peak in the action spectrum arises from neural interaction of other elements or from light filtering through red screening pigments is undecided. Whether there is a behavioral sensitivity to red is also an open question (cf. Schneider, 1956; Wiesmann, 1962b; Bishop, 1968).

There is no question, however, as to the attractiveness of black. Observations of caged houseflies showed that they orient to dark surfaces from an appreciable distance. If a black square was placed on the floor of the cage and a glass plate held over it, flies settled on that area of the glass directly over the square even when the distance separating glass from black square was as great as 35 cm (Arevad, 1965). Edges and corners were especially attractive; however, the length of edge relative to area was not important. Broken and subdivided forms did not attract more flies than solid areas with shorter boundary dimensions.

In three-dimensional situations vertical boundaries were found to be more attractive than horizontal ones. However, the choice of landing place is a more complicated matter than these laboratory studies suggest because there are many modifying conditions. Among them are age, state of excitation, time of day, lighting conditions, climatic conditions, reproductive state, nutritional state, etc. (Keiding, 1965). Generally speaking, the wavelength and the amount of light reflected from a surface are the most important powerful determinants under all conditions.

Landing itself is controlled entirely by visual stimulation of the compound eyes (Goodman, 1960; Braitenberg and Taddei Ferretti, 1966). As a fly approaches a potential landing site, the legs are lowered from their retracted flight position in response to visual stimulation. The forelegs are lifted to both sides of the head and the hind pair are stretched backwards. Blinded flies are able to make adequate but clumsy landings because when they are on the point of colliding with the surface the wing-tips brush first, and this contact stimulates immediate lowering of the legs. In sighted flies the timing during the approach is not based upon an estimate of closing distance but rather on a decrease in the intensity of light. The effective stimulus is based on a multiplication of the change of intensity at successive ommatidia when the fly approaches a landing surface, the number of ommatidia so stimulated, and the rate of their successive stimulation. A given value of this product is required to evoke a landing response (Goodman, 1960). Intensity considerations tend to direct landing to corners, shady surfaces, and dark surfaces; however, a fly can land on a plain white wall. The product referred to above is a measure of the expansion of the pattern on the compound eye as the fly approaches and the decrement of light flux. The greatest distance from the pattern at which the fly reacts is proportional to the speed of the expansion (Fernandez Perez de Talens and Taddei Ferretti, 1970).

Although color, reflectance, and form determine the attractiveness of a surface, texture determines its acceptability as a landing site. Rough surfaces are better than smooth; plywood, wire screen, and cloth are better than glass (Barber and Starnes, 1949). Not all rough surfaces are equally acceptable. Sisal and cotton cords are clearly superior to jute and wool (Fay and Lindquist, 1954). Thus, texture and light reflectance are two attributes of an object that influence where and for how long a fly will rest from flight. To understand how these stimuli operate it is not sufficient merely to count how many flies are present at any time. One must, as Arevad (1965) has done, record the number of landings per unit time and the duration of rest on each surface. These data show that more flies arrive on dark surfaces than on light and that they remain longer on rough than on smooth surfaces (Arevad, 1967, 1969).

Before considering further what a fly does upon landing, and the relation of these activities to food-finding, let us return to it in its airborne state. What besides visual stimuli affect its flying behavior? We are not concerned here with the mechanisms that keep it stabilized, correct yaw, pitch, and roll, or ensure that it maintains sufficient air speed to prevent stalling. We are concerned with the directional aspects of flight. Few quantitative studies bear on this topic. Casual observation suggests that flight is random insofar as direction is concerned. Many of the species of flies that spend a long time in the air have flight paths that are continuous successions of curlicues, dives, and ascents (Wyman, 1970). More precise studies on other nonsocial, nonmigrating insects support this conclusion in general [Schwinck (1954) with the silkworm moth, *Bombyx mori,* and Steiner (1953) with the dung beetle, *Geotrupes stercorarius*]. The dung beetle's "search" flight is roughly circular and figure-eight in pattern, the direction being independent of the direction of the wind but distorted by strong wind. Velocity and altitude are regulated by visual cues and are related to wind velocity. The crucial point is that the flight of food-deprived insects is basically nondirectional. This is a good strategy for searching because it is one way of ensuring that large areas will be explored.

The critical moment (apart from periods when the fly is under the influence of powerful visual stimuli) comes when an odor of food is borne upon the wind. Now comes a spectacular change in the pattern of flight. The dung beetle switches to a zigzag pattern oriented upwind. The physics of the situation being what it is, the beetle finds itself in an ever-increasing concentration of odor. If it overshoots the dung, it reverts to random flight, which tends to bring it once more under the influence of odor. When the odor encountered is of high concentration, the beetle folds its wings and falls to the ground. If no dung is there, the beetle resumes flying.

Some early simple experiments with *Drosophila* suggested that relations between the direction of flight, the direction of wind, and odor are the same for flies and beetles (Otto, 1951; Flügge, 1934). For ease of manipulation the wings of *Drosophila* were clipped. This operation converted the situation to a walking one rather than a flying one. Flies walking on a horizontal surface showed no change in direction when their paths carried them into a stream of air issuing from a tube (Fig. 7). When the odor of banana was added to the air, flies encountering the stream immediately oriented upwind and increased their rate of running. In the absence of wind and odor they averaged 1.1 cm/sec. In the presence of a stream of wind the rate was unchanged (0.9 cm/sec). With the addition of odor the rate increased to 2.3 cm/sec. Only when the wind bears the odor of flowers, fermentation, offal, or feces, or when a walking fly stumbles onto food does locomotory behavior as-

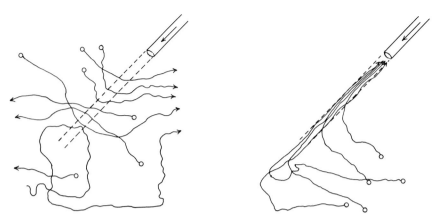

Fig. 7. Diagrams illustrating the pathways taken by *Drosophila melanogaster* walking through a stream of odorless air (left) and a stream of attractive odor (right) (from Flügge, 1934).

sume a directional aspect. It is extremely doubtful that concentration gradients of odor can be employed by flying insects for purposes of orientation, chiefly because of the great turbulence of air under even the calmest of conditions. Although odor emanating from a source is shaped into a plume by moving air, it is greatly contorted and broken. Anemotaxis (orientation to an air current) is of prime importance in enabling an insect to orient. Its mode of action has been demonstrated in flying *Drosophila* (Kellogg, Frizel, and Wright, 1962) and in the yellow fever mosquito *Aedes aegypti* (Kennedy, 1939; Wright, 1962) where the importance of being able to detect the moving pattern of the ground has been shown. Normally when the insect is flying upwind the pattern moves parallel to the line of flight from front to back. The direction of flight can be altered experimentally by moving patterns beneath the insect in different directions. Recently Farkas and Shorey (1972) in a study of the responses of female pink bollworm moths to sex pheromones were able to produce a momentarily stationary odor plume in still air and demonstrate the ability of the moths to orient to the source under these conditions. Moths encountering this plume oriented to the source by zigzag flying. Since the plume was a nonuniform mass of molecules, filamentous in nature, with the average molecular density higher at the longitudinal axis than at the middle, it is believed that the moths traversing the plume were stimulated to turn back into it every time they entered an area of lower concentration (klinokinesis). How they distinguished "upwind" from "downwind" is not known. Some of the conclusions that Farkas and Shorey drew from their experiments have been questioned by Grubb (1973) and in answer Farkas and Shorey conceded that more critical experiments were required to permit an unequivocal conclusion that anemotaxis is not required to provide directional cues to the odor trail (Farkas and Shorey, 1973).

More carefully designed experiments relating to optomotor anemotaxis in male moths (*Plodia interpunctella*) stimulated by wind-borne female sex pheromones have been conducted by Kennedy and Marsh (1974). The wind tunnel in which the pheromone plume was produced was equipped with a carpet of alternate black and orange stripes that could be moved forward or backward along the tunnel. When the stripes were stationary, a male released in the downwind end of the pheromone plume flew upwind in a series of diminishing irregular zigzags. When the stripes were moved downwind, thus giving the illusion to the moth of greater ground speed, the moth reduced its airspeed while still facing upwind and zigzagging. It is clear that the male was responding to the moving ground pattern, as the hypothesis of optomotor anemotaxis requires. The male could even be taken upwind of the source of the pheromone if the experimenter moved the stripes in an upwind direction. Additional experiments demonstrated that cessation of the chemical stimulus, occasioned by the moth's overshooting or flying out of the plume, resets the anemotactic angle. In other words, the moth's course deviates from the upwind direction, and the zigzagging track becomes perpendicular to the wind.

As Kennedy and Marsh have pointed out, the anemotacic hypothesis does not exclude some role for chemoklinokinetic and chemotactic responses; however, anemotaxis is still the most plausible guidance mechanism for orientation to sources of odor.

It is most likely that the fly is guided to odorous food initially by anemotaxis. A combination of klinotaxis and visual cues could provide the necessary stimulus situation for landing. As the concentration of odor became increasingly high, the flight path would assume ever tighter circles, the ground would approach closer in the visual field, and optical stimuli for landing would assume control of the situation. Having landed, the fly reverts to a terrestrial way of life.

Different sets of stimuli now become important. Wind and vision no longer play predominant roles. That is not to say that they cease entirely to be effective; a gust of wind causes a walking fly to grip more tightly; a steady wind also causes postural adjustments. Black and white and color variations in the substrate have little or no effect on the pattern of walking. When the fly is walking, the principal response to visual stimuli is to moving objects, and the response is an avoidance or escape reaction. The preeminently significant stimuli derive from the textural and chemical features of the substrate.

On a surface that is, from the fly's sensory point of view, chemically inert, locomotion consists of series of short runs the direction of which changes frequently and randomly. These runs are punctuated by periods of immobility, by grooming, and by extension of the proboscis. On dusty sticky surfaces or those on which locomotion may be difficult, the frequency of grooming increases. Extension of the probosics is more frequent on some surfaces than others, but the relation between this activity and surface texture has not been studied systematically.

If the surface is odorous, one of two basic patterns of behavior occurs. One is related to deterrence by unacceptable odors and the other is related to acceptance. The fly's immediate response to an unacceptable odor is to stop walking. This halt may be momentary and followed immediately by flight, by turning away, or by frantic grooming. In response to an acceptable odor the fly may respond in a number of different ways; it may stop walking, may walk in erratic circles, may extend its proboscis to the surface, may groom.

When a surface has upon it an acceptable nonvolatile chemical, the fly indulges in very interesting, precise, and highly predictable patterns of behavior. These patterns have been studied by employing water and sucrose as stimuli. The flies used in the experiments had had their wings clipped the day before to prevent their departing in the middle of a test. The results are the same as those obtained with intact flies. The first experiment consisted of painting, with an artist's brush, a ring of water about 10 cm in diameter on a horizontal sheet of paper (Fig. 8). Two and one half centimeters outside of this was painted a concentric ring of 0.1 M sucrose. A thirsty, hungry fly was then placed in the center of the rings. It began to walk in a straight line that soon brought it into contact with the water. As soon as a foot touched the water, the fly stopped, lowered its proboscis, and drank. Having drunk to repletion it resumed walking with no special regard to direction. It walked through the water as often as away from it. The significant point is that the water no longer had any effect upon locomotory behavior. As a consequence, the fly sooner or later walked through the water and encountered the ring of sugar. It stopped as soon as a foot touched the sugar, turned toward the stimulated foot, lowered its proboscis, and commenced eating. After it had fed to repletion, it ignored sugar, which thereupon had no further effect upon locomotory behavior.

As long as the flow of sugar into the proboscis was uninterrupted, the fly remained rooted to the spot until satiated. If the flow ceased, either because it had evaporated or been absorbed by the paper or completely imbibed, the proboscis was retracted, the fly took one or more steps, and the proboscis was lowered once more. The pattern of this activity can be demonstrated most clearly by painting a piece of paper lightly with a solution of sugar colored with methylene blue and

The Search: Appetitive Behavior

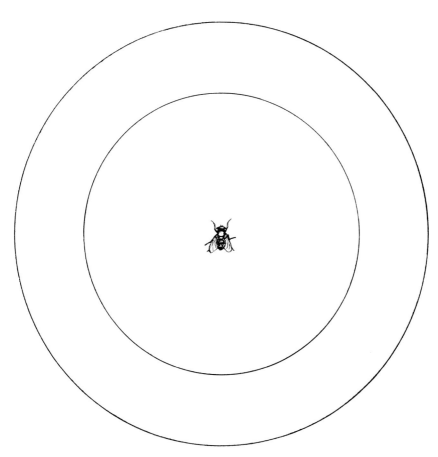

Fig. 8. Diagram illustrating the arrangement of rings of different fluids in tests of the responses of *Phormia* to substances encountered while walking.

releasing a hungry fly on the surface. Each time the fly has sucked one spot dry there is left a white imprint of its labellar lobes. These "lip" prints mark the trail of eating (Fig. 9). Compared with the pattern of walking on a nonchemical surface, this trail is erratic and convoluted.

The departure from linear progression in the presence of food ensures that when the proboscis is extended it will hit the target. Although a certain degree of lateral mobility by the proboscis is possible, the feeding process is clearly more efficient when the food is directly beneath the mouth. The behavior patterns that bring this about can be studied by painting sucrose solutions in various geometric configurations on nonabsorbant paper and releasing thereon a fly with clipped wings. Let us consider a few selected situations. In order to prevent the fly from becoming satiated in the course of an experiment it is desirable to use fairly dilute solutions (0.1 M) and thin lines. When a walking fly approaches a line of sugar, the initial encounter is

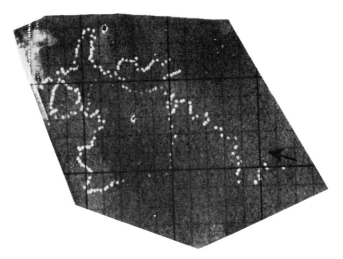

Fig. 9. "Lip" prints of *Phormia* as it walks and feeds upon a sugar solution mixed with methylene blue (from Dethier, Copyright 1957 by the American Association for the Advancement of Science).

with one of the forelegs. The fly immediately stops and pivots around the stimulated leg in a way that places the proboscis over the line. The proboscis is then extended, and drinking ensues. The three responses, stopping, pivoting, extension, follow each other so rapidly as to appear a single coordinated action. When the solution immediately beneath the proboscis is completely consumed, the proboscis is momentarily retracted, the fly steps directly forward, the proboscis is lowered once more. If the fly happens to be oriented in the line with the streak of solution and straddling it, the proboscis on its next extension again encounters sugar, and so on. Accordingly the fly can follow a trail of sugar with its proboscis. That the proboscis alone can keep the fly on the trail is demonstrated by amputating fore and middle legs, leaving only the hind pair for propulsion. The trail is followed accurately.

What happens when the fly arrives at the end of a trail is demonstrated by presenting interrupted trails, that is, trails of dots or dashes in which the lengths of the intervals are varied. So long as the interval does not exceed one fly-length, the trail is followed as though it were continuous. With a longer interval a change in locomotion occurs. When the extended proboscis fails to encounter sugar, the fly's "momentun" carries it forward one fly-length, the proboscis all the while being repeatedly extended. At this juncture forward progression stops, and the fly turns either to the right or the left. The turn is usually one hundred and eighty degrees and brings the fly back onto the lost trail. In this maneuver the forelegs again come into play.

The importance of the forelegs in trail-following is strikingly revealed by presenting the fly with Y-shaped trails in which the arms of the Y are greatly elongated and the angle between them varied from acute to obtuse. Consider the case where the angle is acute and the fly is released at the stem. It progresses along the stem in the usual fashion. Reaching the bifurcation, it proceeds along one arm, led on by its proboscis. Let us say it is following the left arm (Fig. 10). As the arms diverge, the fly soon reaches a point where its right foreleg steps in the right arm. It then abandons its pursuit of the left arm, turns toward the stimulated leg, and commences drinking from the right arm. This maneuver, however, soon causes the left leg to step in the left arm of the Y. Now the fly abandons the right for the left. As long as the two arms of the Y are within reach of the forelegs, the fly oscillates in its drinking from one to the other. Eventually this progression brings it to a point where the distance between the two arms of the Y exceeds the spread of the forelegs. From this point on the fly continues along whichever arm it happens to be following.

The pattern of behavior just described is similar but in reverse when the fly is initially released at the end of one arm of the Y (Fig. 11). It proceeds along this arm, for example, the right, gradually converging on the left, until one foreleg encounters the left. The fly turns to the left. From here on it oscillates from one arm to the other until the stem is reached, whereupon it proceeds uninterruptedly down the stem.

When an obtusely angled Y is substituted for the acutely angled one, the fly's procedure is the same (Fig. 12). The one important difference is that in going along the stem it opts for one arm sooner than before. Similarly, coming in the reverse direction (Fig. 13) along one arm it approaches closer to the junction before turning to the other arm. Measurements of the points at which the behavior changes in each case show, as might be expected, that the critical point is where the distance between the two arms exactly equals the span of the forelegs. The same relationship is demonstrated with the use of parallel lines of solution (Fig. 14). So long as the distance between the two does not exceed a foreleg-span, the fly oscillates between the two (unless of course they are so close together as to be out of range of the legs on the medial side). At greater distances one line is abandoned.

A final experiment to demonstrate the role of the legs in the finding of food involves a repetition of the foregoing variations, and employs flies from which the forelegs have been amputated. The span of the middle legs is greater. The decision point for these flies is the place where divergences are equal in distance to the span of the middle legs (Fig. 14).

The value of an arrangement involving the legs is immediately apparent when one presents the fly with a meandering trail or with

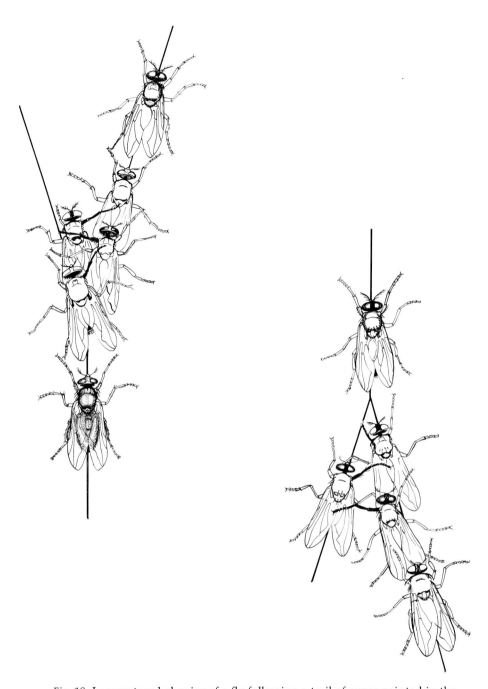

Fig. 10. Locomotory behavior of a fly following a trail of sugar painted in the form of an acutely angled Y. The fly encountered the stem of the Y first. The fly begins by aligning itself on the trail. The black leg indicates that while the fly is on one trail that leg encounters a new trail.

Fig. 11. The behavior of a fly encountering one arm of an acutely angled Y of painted sugar. The black leg indicates an encounter with a new trail.

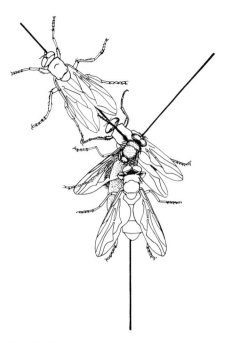

Fig. 12. The behavior of a fly encountering the stem of an obtusely angled Y of sugar solution.

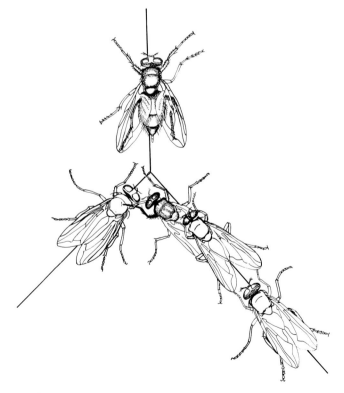

Fig. 13. The behavior of a fly encountering one arm of an obtusely angled Y.

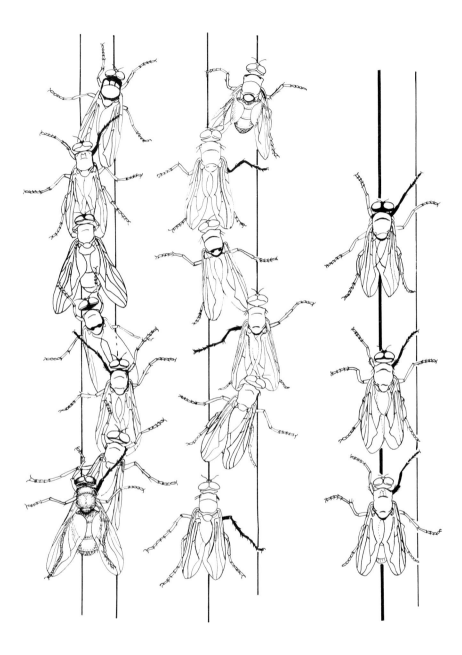

Fig. 14. The walking behavior of a fly in the presence of two parallel lines of solution. In the left and middle pairs the concentration of sugar is the same in both tracks. In the right pair the right track is more dilute than the left. The fly walking the middle paired tracks has had the prothoracic legs removed. Note the greater width between the lines. In each case the black leg indicates that the fly has encountered a line different from the one where it is eating at the time.

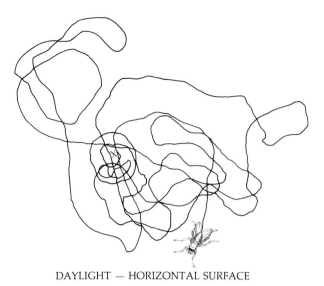

DAYLIGHT — HORIZONTAL SURFACE

Fig. 15. The "dance" performed by a fly that has briefly encountered and consumed a small drop of sugar (X) (from Dethier, Copyright 1957 by the American Association for the Advancement of Science).

series of dots of solution randomly spaced. By turning every time the proboscis or leg loses contact with the food and by pivoting around a stimulated leg, the fly increases the probability of locating food that is unevenly distributed on the substrate.

There is still another pattern of locomotion that increases the chances of finding food. This has been referred to as the "fly dance" (Dethier, 1957). Once again experiments were conducted with walking flies. When a hungry fly was presented with a drop of sugar too small to do more than whet its appetite, the fly resumed walking. Whereas the pattern of locomotion prior to encountering sugar consisted of a series of short straight or gently curved lines connected in random fashion insofar as general direction was concerned, the path now assumed the form of repeated irregular convolutions that resembled a crude, formless dance (Fig. 15). The fly seemed to be "looking for" the lost sugar. That the action is completely stereotyped rather than purposeful was demonstrated by holding a fly in the hands and stimulating it with sugar. Immediately upon being released on a horizontal surface, it began "searching" actions on the spot, which bore no relation to the special location of the previous stimulus. The action is purely automatic; nevertheless, it constitutes a very effective search response by enhancing the probability of again encountering the "lost" sugar or other drops in the vicinity (Dethier, 1957; Mourier, 1965).

The intensity and duration of this dance are affected by three conditions: the concentration of the stimulus; the threshold of the central

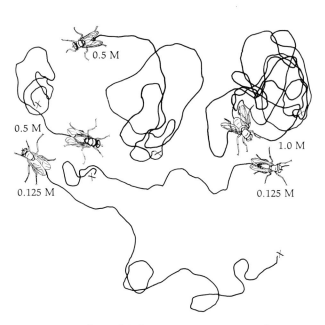

Fig. 16. The effect of different concentrations of sugar on the form of the "dance" (from Dethier, Copyright 1957 by the American Association for the Advancement of Science).

nervous system; and the time lapse between the withdrawal of stimulation and the onset of response.

For example, after stimulations with 0.1 M glucose, there were few turnings, of short duration, before the fly resumed its former random-like mode of running (Fig. 16). After stimulation with 0.5 M glucose the fly performed a more convoluted dance of longer duration. Stimulation with 1.0 M glucose provoked still greater convolutions and longer persistence of action. These patterns of locomotion do not differ in the angular acuteness of the turns but in the number of turns per unit time and the total duration of action. The concentration of the stimulus can easily be deduced from the pattern of the dance.

For any particular concentration of stimulating solution the intensity and duration of response is related to the threshold of the central nervous system. Any change in the physiological state of the fly that alters this threshold is reflected as a change in response. The most influential state is the nutritional one. A starved fly performs more active gyrations in response to any given concentration than a fly that has recently fed. Flying also affects the intensity of response, as might be expected from the fact that it unbalances the nutritional status. A fly that has flown continuously for one hour responds more vigorously than one that has flown only ten minutes. The importance of the third variable, time, with respect to the vigor of response, is related to the decay of intensity. In other words, the rate of turning gradually dimin-

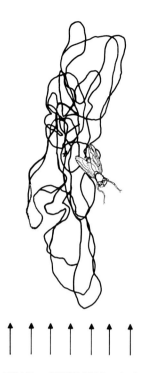 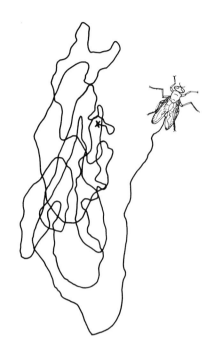

LIGHT BEAM – HORIZONTAL SURFACE   VERTICAL SURFACE – DARKNESS

Fig. 17. The effect of light on the form of a "dance" being performed on a horizontal surface in darkness. *Fig. 18.* The effect of gravity on the form of a "dance" being performed on a vertical surface in darkness (both from Dethier, Copyright 1957 by the American Association for the Advancement of Science).

ishes as the action proceeds until the fly resumes a straight path. Accordingly, any isolated segment of the dance characteristically represents the elapsed time between the end of stimulation and the beginning of that segment. Since the rate of turning diminishes with time, a diffuse segment of the pattern represents a long time lapse, and a tightly convoluted segment represents a short time lapse. Furthermore, the longer a fly is prevented from responding after stimulation (by being held in the hand, for example), the less intense is the response. Prevention by flying, however, is a different matter. A fly that is induced to fly immediately after momentary stimulation with sugar will, upon landing, perform the dance it would have done had it not flown.

Although the dance lacks a directional component, a bias can be imposed upon it by subjecting the fly to the continuing influence of some directional stimulus while it is dancing. In the presence of a beam of light the dance becomes elongated in a plane parallel to the beam (Fig. 17). Less spectacular but nonetheless real is the deformation imposed by gravity. The dance of a fly performing on a vertical surface in darkness is elongated in the direction of the vertical axis of the substrate (Fig. 18). Light on the vertical substrate destroys the directional component effected by gravity.

It is clear that there are striking parallelisms between the gyrations of flies and the dances of honeybees. It is tempting to imagine that this behavior of the fly is a primitive and commonly occurring response associated with the search for food and an evolutionary forerunner of the dances of honeybees.

A complex interplay of light, form, color, and motion perception together with anemotaxis, olfactory perception, discrimination of texture, and contact chemoreception have steered the fly—perhaps not very efficiently or directly—to a source of energy. Having found food the fly begins to eat. Eating is basically antagonistic to walking and flying. Appetitive behavior has for the time being come to an end.

Literature Cited

Arevad, K. 1965. On the orientation of houseflies to various surfaces. *Ent. Exp. Appl.* 8:175–188.

Arevad, K. 1967. Laboratory investigations on the behaviour of newly emerged houseflies. *Arsberetning Ann. Rpt.,* 1967, Gov't Pest Infestation Lab., Lyngby, Denmark, pp. 56–58.

Arevad, K. 1969. Behaviour of houseflies. *Arsberetning Ann. Rpt.,* 1969, Gov't Pest Infestation Lab., Lyngby, Denmark, pp. 54–55.

Autrum, H., and Burkhardt, D. 1961. Spectral sensitivity of single visual cells. *Nature* 190:639.

Autrum, H., and Stumpf, H. 1953. Elektrophysiologische Untersuchungen über das Farbensehen von Calliphora. *Zeit. vergl. Physiol.* 35:71–104.

Barber, G. W., and Starnes, E. B. 1949. The activities of houseflies. *J. N. Y. Ent. Soc.* 57:203–214.

Barton Browne, L., and Evans, D. R. 1960. Locomotor activity in the blowfly as a function of feeding and starvation. *J. Insect Physiol.* 4:27–37.

Bishop, L. G. 1968. Spectral response of single neurones recorded in the optic lobes of the housefly and blowfly. *Nature* 219:1372–1373.

Bishop, L. G., and Keehn, D. G. 1966. Two types of motion sensitive neurons in the optic lobe of the fly. *Nature* 212:1374–1376.

Bishop, L. G., Keehn, D. G. 1967. Neural correlates of the optomotor response in the fly. *Kybernetik* 3:288–295.

Bishop, L. G., Keehn, D. G., and McCann, G. D. 1968. Motion detection by interneurons of optic lobes and brain of the flies *Calliphora phaenicia* and *Musca domestica. J. Neurophysiol.* 31:509–525.

Brady, J. 1972. The visual responsiveness of the tsetse fly *Glossina morsitans* Westw. (Glossinidae) to moving objects: the effects of hunger, sex, host odour and stimulus characteristics. *Bull. Ent. Res.* 62:257–279.

Brady, J. 1974. The pattern of spontaneous activity in the tsetse fly *Glossina morsitans* Westw. (Diptera, Glossinidae) at low temperatures. *Bull. Ent. Res.* 63:441–444.

Braitenberg, V., and Taddei Ferretti, C. 1966. Landing reaction of *Musca domestica* induced by visual stimuli. *Naturwiss.* 53:155–156.

Burkhardt, D. 1964. Colour discrimination in insects. *Adv. Insect Physiol.* 2:131–173.

Cajal, S. R., and Sanchez, D. 1915. Contribucion al conocimiento de los centros nerviosos de los insectos. *Trabajas del Lab. de Investig. Biologicas Univ. Madrid* 13:1–167.

Chadwick, L. E. 1947. The respiratory quotient of *Drosophila* in flight. *Biol. Bull.* 93:229–240.

Chadwick, L. E., and Gilmour, D. 1940. Respiration during flight in Drosophila repleta Wollaston: the oxygen consumption considered in relation to the wing-rate. *Physiol. Zool.* 13:398–410.

Clegg, J. S., and Evans, D. R. 1961. The physiology of blood trehalose and its function during flight in the blowfly. *J. Exp. Biol.* 38:771–792.

Clements, A. N. 1955. The sources of energy for flight in mosquitoes. *J. Exp. Biol.* 32:547–554.

Danilevsky, A. S., Goryshin, N. I., and Tyshchenko, V. P. 1970. Biological rhythms in terrestrial arthropods. *Annu. Rev. Ent.* 15:201–244.

Davis, R. A., and Fraenkel, G. 1940. The oxygen consumption of flies during flight. *J. Exp. Biol.* 17:402–407.

Dethier, V. G. 1957. Communication by insects: physiology of dancing. *Science* 125:331–336.

Farkas, S. R., and Shorey, H. H. 1972. Chemical trail-following by flying insects: a mechanism for orientation to a distant odor source. *Science* 178:67–68.

Farkas, S. R., and Shorey, H. H. 1973. Odor-following and anemotaxis. *Science* 180:1302.

Fay, R. W., and Lindquist, D. A. 1954. Laboratory studies on factors influencing the efficiency of insecticide impregnated cords for house fly control. *J. Econ. Ent.* 47:975–980.

Fermi, G., and Reichardt, W. 1963. Optomotorische Reaktionen der Fliege *Musca domestica. Kybernetik* 2:15–28.

Fernandez Perez de Talens, A., and Taddei Ferretti, C. 1970. Landing reactions of *Musca domestica:* dependence on dimensions and position of stimulus. *J. Exp. Biol.* 52:233–256.

Flügge, C. 1934. Geruchliche Raumorientierung von *Drosophila melanogaster. Zeit. vergl. Physiol.* 20:463–500.

Freeborn, S. B., and Berry, L. B. 1935. Color preferences of the housefly *Musca domestica* L. *J. Econ. Ent.* 28:913–916.

Goldsmith, T. H., and Fernandez, H. R. 1966. *The Functional Organization of the Compound Eye.* Pergamon Press, London.

Goodman, L. J. 1960. The landing responses of insects. I. The landing response of the fly *Lucilia sericata* and other calliphorniae. *J. Exp. Biol.* 37:854–878.

Götz, K. G. 1971. Spontaneous preference of visual objects in *Drosophila. Drosophila Information Service* 46:62.

Götz, K. G., and Wenking, H. 1973. Visual control of locomotion in the walking fruitfly *Drosophila. J. Comp. Physiol.* 85:235–266.

Green, G. W. 1964a. The control of spontaneous locomotor activity in *Phormia regina.* I. Locomotor activity patterns of intact flies. *J. Insect Physiol.* 10:711–726.

Green, G. W. 1964b. The control of spontaneous locomotor activity in *Phormia regina*. II. Experiments to determine the mechanisms involved. *J. Insect Physiol.* 10:727–752.

Grubb, T. C. 1973. Odor-following and anemotaxis. *Science* 180:1302.

Hecht, O. 1963. On the visual orientation of house-flies in their search of resting sites. *Ent. Exp. Appl.* 6:107–113.

Hecht, O. 1970. *Ecologia y comportamiento de moscas domesticas.* Parte I. *Musca domestica* L. Depto. Publ. Inst. Politecnico Nacional, Mexico, D. F.

Hinton, H. E. 1951. A new chironomid from Africa, the larva of which can be dehydrated without injury. *Proc. Zool. Soc. London* 121:371–380.

Hocking, B. 1953. The intrinsic range and speed of flight of insects. *Trans. Roy. Ent. Soc. Lond.* 104:223–345.

Hudson, A. 1958. The effect of flight on the taste threshold and carbohydrate utilization of *Phormia regina* Meigen. *J. Insect Physiol.* 1:293–304.

Jander, R., and Schweder, M. 1971. Über das Formunterscheidungsvermögen der Schmeissfliege *Calliphora erythrocephala*. *Zeit. vergl. Physiol.* 72: 186–196.

Jones, M. D. R., Cubbin, C. M., and Marsh, D. 1972. Light-on effects and the question of bimodality in the circadian flight activity of the mosquito *Anopheles gambiae*. *J. Exp. Biol.* 57:347–357.

Keiding, J. 1965. Observations on the behaviour of the housefly in relation to its control. *Rivista Parassitologia* 26:45–60.

Kellogg, F. E., Frizel, D. E., and Wright, R. H. 1962. The olfactory guidance of flying insects. IV. Drosophila. *Canad. Ent.* 94:884–888.

Kennedy, J. S. 1939. The visual response of flying mosquitoes. *Proc. Zool. Soc. London*, Ser. A, 109:221–242.

Kennedy, J. S., and Marsh, D. 1974. Pheromone-regulated anemotaxis in flying moths. *Science* 184:999–1001.

Langer, H., and Thorell, B. 1966. *The Functional Organization of the Compound Eye.* Pergamon Press, London.

McCann, G. D., and Arnett, D. W. 1972. Spectral and polarization activity of the Dipteran visual system. *J. Gen. Physiol.* 59:534–558.

McCann, G. D., and Dill, J. C. 1969. Fundamental properties of intensity, form, and motion perception in the visual systems of *Calliphora phaenicia* and *Musca domestica*. *J. Gen. Physiol.* 53:385–413.

McCann, G. D., and Fender, D. H. 1964. Computor data processing and systems analysis applied to research on visual perception. In *Neural Theory and Modeling* R. F. Reiss, ed., pp. 232–252. Stanford University Press, Stanford, Calif.

McCann, G. D., and MacGinitie, G. F. 1965. Optomotor response studies of insect vision. *Proc. Roy. Soc. London* B163:369–401.

Mazokhin-Porshnyakov, G. A. 1960a. Colorimetric study of color vision in housefly. *Biofizika* 5:295–303.

Mazokhin-Porshnyakov, G. A. 1960b. Color vision in *Calliphora*. *Biofizika* 5:697–703.

Mazokhin-Porshnyakov, G. A. 1969. *Insect Vision*. Plenum Press, New York.

Mimura, K. 1971. Movement discrimination by the visual system of flies. *Zeit. vergl. Physiol.* 73:105–138.

Mourier, H. 1965. The behaviour of houseflies (*Musca domestica* L.) towards "new objects." *Vidensk. Medd. fra Dansk naturh. Foren.* 128:221–231.

Norris, K. R. 1966. Daily patterns of flight activity of blowflies in the Canberra district as indicated by trap catches. *Austral. J. Zool.* 14:835–854.

Nuorteva, P. 1966. The flying activity of *Phormia terraenovae* R.-D. (Dipt., Calliphoridae) in subarctic conditions. *Ann. Zool. Fennici* 3:73–81.

Otto, E. 1951. Untersuchungen zur Frage der geruchlichen Orientierung bei Insekten. *Zool. Jahrb.*, Abt. 3, *Allgem. Zool. Physiol.*, 62:65–92.

Pospíšil, J. 1958. Some problems of the smell of saprophilic flies. *Acta Soc. Entomol. Bohem.* (Čsl.) 55:316–334.

Pospíšil, J. 1962. On the visual orientation of the housefly (*Musca domestica*) to colours. *Acta Soc. Entomol. Bohem.* (Čsl.) 59:1–8.

Reichardt, W. E. 1970. The insect eye as a model for analysis of uptake, transduction, and processing of optical data in the nervous system. *The Neurosciences. Second Study Program*, F. O. Schmitt, ed., pp. 494–511. Rockefeller University Press, N.Y.

Reynierse, J. H., Manning, A., and Cafferty, D. 1972. The effects of hunger and thirst on body weight and activity in the cockroach (*Nauphoeta cinerea*). *Animal Behav.* 20:751–759.

Schneider, G. 1956. Zur spektralen Empfindlichkeit des Komplexauges von *Calliphora*. *Zeit. vergl. Physiol.* 39:1–20.

Schwinck, I. 1954. Experimentelle Untersuchungen über Geruchssinn und Strömungswahrnehmung in der Orientierung bei Nachtschmetterlingen. *Zeit. vergl. Physiol.* 37:19–56.

Steiner, G. 1953. Zur Duftorientierung fliegender Insekten. *Naturwiss.* 40:514–515.

Sytshevskaya, V. I. 1962. On the daily dynamics of the specific composition of synanthropic flies within a season. *Entomol. Obozrenie* 41:545–553 (in Russian).

Walther, J. B., and Dodt, E. 1957. Elektrophysiologische Untersuchungen über die Ultraviolettempfindlichkeit von Insektenaugen. *Experientia* 13:333.

Walther, J. B., and Dodt, E. 1959. Die Spektralsensitivität von Insekten-Komplexaugen im Ultraviolett bis 290 mμ. *Zeit. Naturf.* 14b:273–278.

Waterhouse, D. F. 1948. The effect of colour on the numbers of houseflies resting on painted surfaces. *Austral. J. Sci. Res., B, Biol. Sci.* 1:65–75.

Wehner, R. 1972. Spontaneous pattern preferences of *Drosophila melanogaster* to black areas in various parts of the visual field. *J. Insect Physiol.* 18:1531–1543.

Wiesmann, R. 1960. Zum Nahrungsproblem der freilebenden Stubenfliegen, *Musca domestica* L. *Zeit. angew. Zool.* 47:159–181.

Wiesmann, R. 1962a. Untersuchungen über den "fly factor" und den Herdentrieb bei der Stubenfliege, *Musca domestica* L. *Mitt. Schweiz Entomol. Ges.* 35:69–114.

Wiesmann, R. 1962b. Neue Erkenntnisse aus der Biologie von *Musca domestica* L. im Zusammenhang mit der Insektizidresistenz. *J. Hyg. Epid. Microbiol. Immunol. (Prague)* 6:302–321.

Wigglesworth, V. B. 1949. The utilization of reserve substances in *Drosophila* during flight. *J. Exp. Biol.*, 26:150–163.

Williams, C. M., Barness, L. A., and Sawyer, W. H. 1943. The utilization of glycogen by flies during flight and some aspects of the physiological aging in *Drosophila*. *Biol. Bull.* 84:263–272.

Wright, R. H. 1962. The attraction and repulsion of mosquitoes. *World Rev. Pest Control.* 1:20–30.

Wyman, R. J. 1970. Patterns of frequency variation in dipteran flight motor units. *Comp. Biochem. Physiol.* 35:1–16.

3
To Each His Taste: Discrimination and Preference

My only difficulty is to choose or reject.
Dryden, Preface to *Fables, Ancient and Modern*

The sources of potential energy are almost limitless. It might fairly be said that there is nothing that is not consumed by some organism. Even such bizarre items as horn, beeswax, cayenne pepper, cured tobacco, and poison ivy can be eaten. And though nutrition may not be the goal, there are, nonetheless, insects that chew lead pipes and cables, rubber insulation, plastic containers, and other synthetic materials. Larvae of the fly *Psilopsa petrolei* breed in crude petroleum pools (but probably eat other organisms trapped in the oil). Larvae of the beetle *Trichogenius* breed in argol, the crude tarter consisting of potassium tartrate that forms as sediment in wine casks (they probably eat the sedimented yeasts). Some caterpillars live in the base of horns of African antelopes whose keratin they ingest. The diet of human beings is no less fantastic. One need only peruse the menus of gourmet restaurants, study the eating habits of different cultures, or consult medical records of patients with unusual addictions to appreciate the variety.

Potential energy is everywhere if only an organism possesses the key to unlock the particular form. Not everyone can extract energy from wood. Not everyone can ravage the green plant profitably. Furthermore, the dwarfs cannot consume the giant unless they are as multitudinous as malarial protozoa in the blood of man or the fungal spores in the phloem of elm trees infected with Dutch elm disease. Nor can the giant consume the dwarfs unless the dwarfs are as common as the grains of sand and the giant has some ingenious device to collect and concentrate them. Many marine and fresh water animals sieve out of the water the most minute of organisms. The baleen whale, a giant among ocean dwellers, manages to accumulate a sufficient quantity of tiny crustacea to fuel its great bulk by straining them through its baleen fringe. Man, a moderately large terrestrial species, eats some of the smallest of grains, which he harvests and concentrates through his mechanical ingenuity.

No two animals have precisely the same diet. Each species possesses some of the qualities of a gourmet. The common categorization as herbivore, carnivore, and omnivore depicts only the outline of the story. The details of diet reveal the enormous complexity of the feeding processes and diverse and precise specifications to which the physiological and biochemical machinery must be designed. Some animals are extraordinarily restricted in their diets. Caterpillars of the monarch butterfly eat only milkweeds. Caterpillars of the wax moth eat only wax of honeycombs. And lest one be inclined to think that only invertebrates are finicky, there is the three-toed sloth of the New World tropics that subsists solely on the leaves of the *Cecropia* tree and the koala bear of Australia that restricts its diet to eucalyptus. There are also those animals whose diets change with the seasons, outstanding examples being aphids that feed on herbaceous plants in summertime and woody plants during the fall.

It is commonly believed that the omnivores are gourmands rather than gourmets; however, they are more selective than their designation implies. Plague locusts, notorious for denuding a landscape, for leaving no green blade standing, are, given the opportunity, creatures of discrimination and preference. Vertebrate herbivores are equally selective. Woods ponies of Britain, turned loose in a lush pasture, will eat first the prickly plants, then various woody materials, and lastly the lush grass. Rabbits eat first dandelions and plaintains, and only after these are consumed is other foliage eaten. Man, the ultimate omnivore, obviously exhibits all degrees of finickiness and food preference, as any parent is all too painfully aware.

Geography and habit determine the options available to an animal. Although nutrition is the basic reason for all feeding, it alone has not been the sole evolutionary force molding diverse feeding habits. Except for aberrations, each habit has evolved in harmony with a particular environment and in response to the unique selective and competitive pressures to which each organism is subjected. An animal does not eat a certain food *because* it is nutritionally best for him; it may be nutritionally best for him because he has evolved the most efficient way of utilizing it. Different feeding habits, different latitudes of diet (e.g., monophagy vs. polyphagy), and different kinds of diets are all different ways of achieving the same end, given the circumstances under which they must operate.

Feeding systems, by which is meant here the digestive system proper, the structures for seizing, sucking, chewing, and the behavioral attributes associated in any way with food procurement, have been built in a manner that ensures that the specific energy requirements of the animal are met. Whatever the system, it must ensure successful locating and identification of food, discrimination as to kind and quality, a particular temporal pattern of feeding whether it be continuous, hourly, daily, or monthly, control of the quantity ingested at any one time, and adjustment to transient needs. Transient needs develop after heavy demands upon stored energy, attendant upon violent exercise, at times of such specialized activities as reproduction, and in anticipation of future energy demands occasioned by migration, hibernation, or aestivation. Many passerine birds prior to migration consume unusual quantities of food, which is stored as fat. The woodchuck is a popular example of a hibernator that eats excessively in the fall and converts this excess to a special kind of fat, brown fat, that sustains it during the winter months. Locusts are also known to lay in large stores of fat preparatory to migratory flights. Unlike flies and honeybees that derive energy for flying solely from carbohydrates, locusts use fat as fuel.

Selection of a food obviously involves the ability to distinguish all foods from non-food. Nearly every animal has this capacity. Those that do not, ingest whatever comes their way and rely upon mechanical or biochemical selective mechanisms to separate the nutrients. Indiscriminate feeding of this character is particularly widespread among marine invertebrates. Oysters by means of the cilia of the mantle create currents that sweep all particulate matter into the oral region. Large particles and indigestible particles are rejected after they have entered. Many so-called continuous feeders, like oysters, discriminate only after the fact. Other organisms, among them worms and some fishes, ingest mud and employ biochemical means to leach out the nutrients. In one respect this is true of all animals, ourselves included, in that only some components of ingested food are utilized. When we eat spinach, the contents of the leaf cells are extracted by enzymatic action while the cell walls and other cellular components pass through the digestive tract unaltered. In contrast to nonselective feeders, however, a selection has been exercised before the food is subjected to enzymatic action. It is the mechanism of this selection with which we are now concerned.

A logical procedure for studying an animal's feeding behavior would be to observe it in the field, ascertain by laboratory experiments its nutritional requirements, and finally search for underlying physiological mechanisms. This is what has been attempted with *Phormia*.

Curiously enough, rather little information is available concerning the habits of wild black blowflies, considering their ubiquity (cf., Norris, 1965). It is known that many lay eggs in sores of sheep and cattle where the maggots cause serious infection. It is suspected, though not proven, that they are carriers of disease insofar as man is concerned. Until the 1950s sterile blowfly maggots were still used to debride stubborn wounds in human patients. But what do black blowflies do in nature? They are rarely captured except around dead or decaying matter where females oviposit. Their principal foods are obviously substances rich in carbohydrate, yet these flies are less common visitors to flowers than are related flies. Horning (1966) has reported capturing flies on the flowers of red osier dogwood (*Cornus sericea* form *stolonifera* Michaux), buckwheat (*Eriogonum ovalifolium* Nuttall), and cow parsnip (*Heracleum lanatum* Michaux). They were also found congregating on the foliage of *Populus trichocarpa* Torrey and Gray, where they apparently were feeding on the honeydew secreted by aphids. In fall days they come to fermenting and decaying apples in orchards. The related species *Phormia terraenovae* has been observed feeding on *Saxifraga oppositifolia* L. and *Arnica alpina* (Kevan, personal communication). The sugar concentration in the nectar of the purple saxifrage averaged 62.5%.

On the assumption, subsequently verified, that the basic food requirement was carbohydrate, Hassett, Dethier, and Gans (1950) undertook a comparative study of the nutritive values of various carbohydrates and the flies' taste thresholds for these compounds. Table 1 summarizes the nutritive values in terms of longevity on 0.1 M concentrations. Obviously, flies do not encounter pure and simple solutions in nature; nevertheless, it is informative to investigate the fly's preferences when more than one sugar is available at a time.

A statement about words is advisable at this point. The words "preference," "selection," "choice," and "discrimination" are used almost interchangeably in the literature on this subject. In contemplating their proper use it is very easy to become enmeshed in semantic shrubbery. On the other hand, a thoughtful analysis by Irwin (1958) has shown that more than one concept is involved in the types of comparative experiments usually conducted and that a realization of this and the careful use of the appropriate term can lead to the design of more definitive and informative experiments. Irwin's ideas can be summarized most succinctly by reference to his example of the behavior of rats in a jumping apparatus. There are two situations, S_1 and S_2, in each of which there are two panels indicated by a circle and a triangle respectively. In S_1 the triangle is on the left and the circle on the right. S_2 is the reverse. Behind one panel in each situation is food. If a rat in situation one (S_1) jumps to the left, he obtains food; if to the right, no food. The situation is reversed in S_2. If the rat responds with significantly different relative frequencies in S_1 and S_2, it has exhibited differential response. If in addition the response is contingent upon some difference in outcome (e.g., receiving or not receiving food), *dis-*

Table 1. Nutritive Values of Carbohydrates and Related Compounds Compared with the Acceptance Thresholds for *Phormia*.

Compound	Comparative nutritional value[a]	Molar concentration at acceptance threshold[b]
D-maltose	++	0.0043
D-fructose	++	0.0058
Sucrose	++	0.0098
Melezitose	++	0.064
α-methylglucoside	++	0.069
L-fucose	−	0.087
D-glucose	++	0.132
L-sorbose	−	0.140
D-arabinose	−	0.144
Inositol	+	0.194
Raffinose	++	0.200
L-xylose	+	0.337
D-xylose	+	0.440
D-galactose	++	0.500
L-arabinose	+	0.536
Cellobiose	+	5.01
D-mannose	++	7.59
D-ribose	+	8.99
D-lyxose	+	42.27
Melibiose	++	*
Sorbitol	++	*
Glycerol	+	*
Dulcitol	+	*
Meso-erythritol	+	*
Penta-erythritol	−	*

From Hassett, Dethier, and Gans, 1950.
[a] A minus sign indicates no nutritive value.
[b] An asterisk indicates no acceptance at any concentration.

Table 2. Hypothetical Results for a Single Rat in Lashley's Jumping Apparatus.

	Situation (S)[a]			
Direction of jump	S_1 △○	S_2 ○△	S_1' △○	S_2' ○△
Percentage to left	96 (+)	22 (−)	10 (−)	95 (+)
Percentage to right	4 (−)	78 (+)	90 (+)	5 (−)
Total	100	100	100	100

From Irwin, 1958.
[a] △ and ○ = the positions of the panels in the given situation.
+, obtains food; −, does not obtain food.

crimination has been demonstrated. It is demonstrated by reversing the situations in order to disassociate situations and outcomes (Table 2). There may in addition be a bias in behavior. In Table 2, the rat jumps more often to the correct figure, but the ratio of correct to incorrect jumps when the food is behind the triangle, differs from that than when the food is behind the circle. There is a bias toward the triangle. If the bias is such that one outcome occurs more often than the other, the rat is said to have shown a *preference.*

The implication of these distinctions can best be appreciated after the feeding experiments with the fly have been discussed in detail. Until then, in the descriptions that follow, the words "preference" and "selection" are employed in the context of drinking more of one solution than of another when more than one is available. Nothing more is implied.

If an animal is to select, it must be able to discriminate; therefore in preference studies we are also concerned with discrimination. The most direct way to investigate the ability and propensity of an animal to discriminate and select is to present it with two or more varieties of food simultaneously, observe its behavior toward them, and measure the quantity of food eaten. With some animals it is also possible to reveal discriminative ability by employing conditioning techniques whereby the animal is trained to associate a particular food with reward or punishment. This technique has proved very effective with honeybees. Simple observations can tell something about preference by indicating whether an animal spends more or less time with a given food, visits it more or less often, or shows some other sign of "interest"; however, more significant information can be derived only by employing some quantitative measurement of feeding. One can measure either the amount of food ingested or the amount of feces produced. The latter is, to say the least, an indirect method. Though commonly employed, especially, by entomologists, it is not without its sources of error (e.g., Waldbauer, 1964).

The art of measuring how much food an animal eats is not so simple as it appears. As Pfaffmann (1963) has emphasized, all ingestive behavior and preferences are controlled by three events: the sensory properties of substances; (2) the physical properties of the post-ingestive load; (3) the nutritive condition of the animal. Furthermore, there are many environmental determinants of preferences—for example, position—and they may be very subtle in their actions.

Each animal must be approached as a special case, and very small animals are particularly difficult subjects. The classical prototype is the two-bottle technique so fruitfully employed with rats by Richter and Campbell (1940a, 1940b), Richter (1943), Young (1941, 1948), Soulairac (1947), Lepkovsky (1948), and innumerable other research workers

Fig. 19. Apparatus employed in the first preference-aversion tests with groups of flies (from Dethier and Rhoades, 1954).

since then. This technique was adapted for blowflies (Dethier and Rhoades, 1954) and subsequently improved upon by Dethier (1961a), Strangways-Dixon (1961a), and Belzer (1970).

The construction of apparatus for ascertaining the fluid intake of blowflies presented two problems: (1) design of a "feeder" or potometer that would allow only the proboscis to be exposed to the liquid during feeding; (2) development of a technique for measuring accurately the volume of fluid removed. The first problem was solved by bending 5-ml volumetric pipettes halfway between the bulb and the capillary opening to form a U with an approximate radius of 2.5 cm

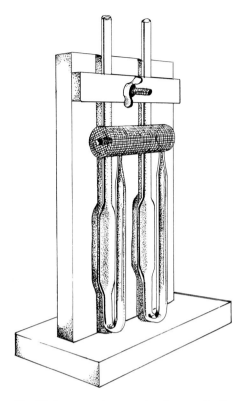

Fig. 20. Improved apparatus for conducting preference-aversion tests with individual flies.

(Fig. 19). The capillary tip was brought upward until it was level with the top of the bulb. With this arrangement, the hydrostatic pressures of the two columns allowed an effective maximum reserve of 2.5 ml of fluid in the pipette. There was presented at the capillary opening a sphere of fluid large enough for the fly to feed upon but small enough to prevent excessive loss by evaporation and by spilling. The drop remained accessible until 2.5 ml had been removed from the reservoir.

Measurement of the volume of fluid removed was obtained by inserting the needle of a calibrated syringe filled with water into the capillary opening of the pipette and forcing fluid out of the syringe until the fluid in the pipette was returned to its original level. The volume was then read directly from the syringe.

Pipettes were filled by immersing them completely in fluid-filled cylinders. Upon removal, each was given a sharp tap to dislodge bubbles and relieve any pressures in the stem. After the outside was washed in distilled water and dried, the level of the fluid was marked with a grease pencil.

Equal numbers of anaesthetized male and female flies were placed in a quart mason jar fitted with a screen top in which two holes had been made to receive and support the pipettes. The metal screw lids were split in three places to convert the lid to a snap-on type. Next the two pipettes were stood upright in the jar. The cap with its screen was slipped over the stems of the pipettes, lowered into the jar, and snapped tightly into place. Measurements were made every 24 hours. The loss due to evaporation was measured in identical pipettes, containing the same solutions, set up in empty jars. With careful attention to details this relatively crude technique yielded remarkably reproducible results. Subsequently, the technique was modified when studies of protein ingestion were undertaken. The jar was replaced by a cylindrical nylon mesh cage 2.5 by 5 cm. The two pipettes were supported in a holder, and they in turn supported the nylon cage, which rested on the capillary ends projecting 1 mm through openings in the floor of the cage. One fly occupied each cage (Fig. 20).

Strangways-Dixon (1961a) developed a vastly improved technique, which had the great merit of permitting continuous reading but provided comparative rather than absolute measures. Each fly was housed in a glass tube 7.5 cm long and 2.5 cm in diameter, closed at one end with nylon mesh and at the other with a cork through which passed two thick-walled capillary tubes, the ends of which were nearly flush with the inside end of the cork. These tubes were filled with the fluids to be tested and a mark made to indicate the level of the fluid. The amount consumed by the fly was measured as centimeters of fluid. One hundred of these tubes were supported in a Perspex frame slanted so that they were maintained at a slight incline. This arrangement provided gravity feed of fluid as the fly imbibed. Light, temperature, and evaporation were controlled in the usual manner. In the hands of Belzer (1970) a modification of this yielded very precise data (Fig. 21). The apparatus consisted of precision-bore capillary tubes held firmly in place by rubber tubes on a Acrylite platform 1.9 cm in thickness. The dimensions of the tubes were: length, 30.5 cm; inside diameter, 0.9 mm $\pm$ 0.0076 mm; outside diameter, 7–8 mm. Twenty-five pairs were mounted on each platform. Cages for individual flies were constructed of cellulose nitrate centrifuge tubes, 3 cm in diameter and cut off to a length of 7.5 cm. The cut end of each was covered with nylon marquisette mesh. Mesh was also glued to the inside to provide footing for the fly. The open end of the cage was closed with a rubber stopper through which were drilled two holes to receive the capillaries. These protruded into the cage. The platform, which was hinged to an Acrylite base, was tilted by inserting a wedge in order to provide an incline, which permitted fluid to flow toward the tip as the fly imbibed. Surface tension prevented overflowing.

The capillaries were filled by means of a hypodermic needle and

Fig. 21. Refined apparatus employed by Belzer (1970) to study the responses of flies in a two-choice situation.

syringe and thoroughly dried on the outside. A "zero" reading was made just before the cage was fitted into place. This was not done until one to two hours after filling because evaporation was most variable in this early period. At least three feeding units were used as controls for evaporation. Readings were made as desired during the ensuing 24 to 32 hours.

The entire apparatus was housed in an Acrylite chamber in which lighting was maintained at 305 or 350 foot-candles by means of overhead fluorescent lamps; relative humidity was maintained at 60–65% by means of salt baths in the chamber; and temperature was kept at $23 \pm 1°C$ by thermostatically controlling the temperature of the whole room.

Because the *rate* of evaporation was influenced by the incline of the tubes, the level of fluid, and the frequency with which the chambers were opened (actual variation among evaporation controls was less affected by these factors), the standard deviation of the average rate of evaporation that occurred among the various experiments did not accurately represent the variation. The *average range* among evaporation controls provided a more accurate characterization of the actual variation. This value was derived by arbitrarily selecting 100 time intervals from five different experiments and calculating the range between the least and the greatest amount of evaporation for each time interval. These ranges were then converted to microliters per hour, averaged, and the standard deviation computed. Actual measurements from many evaporation tests fell within the calculated limits (0.19 ± 0.09 microliters per hour for 0.1 M sucrose and 0.09 ± 0.06 for 10% yeast extract).

Before these refinements were introduced, a great deal of information was obtained from the original crude experiments. All two-bottle experiments are successful only if the two pipettes are identical except with respect to the solutions within. In experiments employing mason jars, inactive flies congregate on the cover; active flies wander around the sides and bottom of the jar. Visits to pipettes are made either by direct hops or short flights from the walls or by crawling upward from the floor. When all physical factors and visual cues inside and outside of the jar are symmetrical with respect to the two pipettes, flies make approximately the same number of visits to each pipette if the contents are identical. Also, the volumes of fluid removed from each are approximately equal.

Other variables had also to be evaluated before definitive experiments could be undertaken. How many flies represented the optimum number per jar? In what ways did they influence one another? The amount of fluid drunk per fly varied significantly with the number of flies in the jar. Maximum consumption per fly occurred in a population of 20. When there were 50 flies, consumption decreased because there was too much competition for drinking places and too many interruptions caused by the general hustle and bustle. Consumption was also low when there were only 10 flies in a jar. In this case, less activity occurred and fewer exploratory trips were made than when there were more flies to jostle and excite one another.

Flies exert other subtle influences on each other that could confound the experiments. "Fly-factor" (Barnhard and Chadwick, 1953) (Fig. 22) and "ganging-up" (Dethier and Rhoades, 1954; Dethier, 1955b) or "herd instinct" (Wiesmann, 1962) could bias the visits to pipettes. Of the two, "fly-factor" was potentially the more serious in the close confines of a quart jar and also in the improved apparatus involving single flies, because a pipette could be made more attractive by repeated and prolonged visitations as well as by fouling with saliva and feces. A simple experiment proved, however, that this concern was groundless. A test was made in which two pipettes were used, one clean and one that had been exposed to flies for a week; each pipette contained aliquots of the same solution. A comparison showed no differences in the number of visits made or in the volume drunk.

One final obstacle had to be cleared, namely, possible effects of variable light. Routine experiments were to be conducted under natural lighting conditions. Three conditions were compared: continuous darkness, continuous light, and normal diurnal fluctuations. In continuous darkness and in normal lighting the consumption per fly was essentially the same. This is not so odd as it might at first appear. We

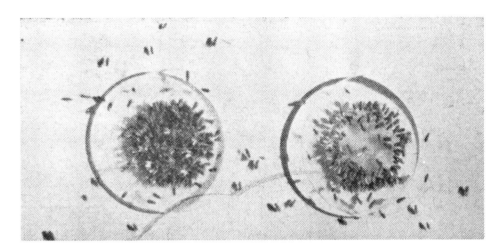

Fig. 22. The effect of previous feeding on the attractiveness of proteose peptone to flies. The dish on the left had previously been exposed to flies for twenty minutes; the dish on the right, for two minutes (from Barnhard and Chadwick, Copyright 1953 by the American Association for the Advancement of Science).

have seen that even in continuous darkness a spontaneous circadial rhythm of activity persists. It will also be recalled that in continuous light the normal periods of inactivity are absent. It is not surprising, therefore, that the unit comsumption of sugar increases markedly in constant light. Of further interest is the fact that flies exhibit normal feeding behavior in constant darkness. This shows that discrimination and preference can operate in the absence of visual cues.

Having assessed the variables, we could now proceed with the business of investigating the ability of the fly to discriminate among kinds of solutions. Is the fly's selection influenced by the number of options available? To what extent does concentration affect choice? How accurately can the fly discriminate? Of what importance is the nutritional value in a situation involving choice? As the experiments progressed, a wealth of ancillary information emerged that prompted further questions and provided further insight into the characteristics of feeding behavior. The number of comparisons that can be made is almost limitless. Only the more fundamental will be discussed here. These include: one-choice vs. two-choice situations, sucrose vs. sucrose at equimolar concentrations, sucrose preference-aversion (vs. water), sucrose vs. sucrose at unequal concentrations, glucose preference-aversion, glucose vs. glucose at unequal concentrations, and sucrose vs. glucose.

In the first comparison in which both pipettes contained equimolar (0.01 M) concentrations of sucrose, the fly drank equal amounts from the two, thus confirming the symmetry of the situation. The volume consumed was 2.75 times that drunk when only one pipette was available. In general, when two sources of sugar were available, irrespective of the kind, concentration, or combination, more was drunk than when only one source was available. When one situation contained a pipette of sucrose paired with an empty pipette and another had a pipette of sugar paired with one of water, the consumption of sugar was the same in both situations. The small amount of water drunk had no effect on sugar intake.

So-called preference-aversion experiments derive their name from the fact that when some compounds over a wide range of concentrations are paired with water, less water may be taken than the test compound at one concentration but more water than test compound at another concentration. Not every compound has a rejection range nor does every compound have an acceptance range. When each of several concentrations of sucrose was paired with water, it was clear that sugar was always preferred within the range investigated (Fig. 23). The curves represent the average 24-hour intake at each concentration over a 4-day period. Several interesting results emerge from these comparisons. The most obvious is that the volume of sucrose taken increases as the concentration increases, up to a point. Beyond this point (0.01 to 0.1 M) there is a progressive decline in the volume. The second result of note is that the amount of water drunk is fairly constant throughout; that is, there is not an enormous reduction in water intake as more sugar is imbibed. The drop below zero in the curve for water results in part from regurgitation. In short, as the concentration of sugar increases, up to a point the fly takes in more fluid in total than it does of water alone. This suggests among other things that the limitation on water intake and on the intake of dilute sugar is not controlled by volume. A third point is that the fly is able to distinguish water from sucrose when the sucrose is as dilute as 10^{-8}–10^{-7} M. A comparable experiment in which the sugar was glucose gave similar results but at different concentration levels (Fig. 24).

At first glance it would appear that the decline in the volume of sugar taken at higher concentrations compensates for the increase in concentration. If, however, the intake is recalculated in units of weight instead of volume, the weight of sugar ingested steadily increases. Thus, even though smaller volumes are drunk at higher concentrations, the concentration is, in a manner of speaking, increasing faster than the reduction in intake. In another insect, the honeybee, there is some tendency for the gram intake to decrease at 1.54 M sucrose and 2.82 M glucose (Wykes, 1952). Neither the honeybee nor the fly reduces intake at high concentrations sufficiently to ensure a constant weight intake. The rat, on the other hand, adjusts its ingestion of

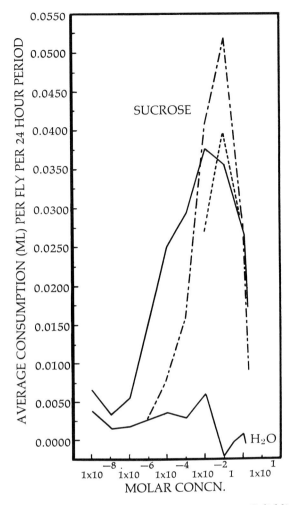

Fig. 23. Sucrose preference-aversion curves. Solid lines, sucrose and water intake in a two-choice situation. Dashed line (---), intake of sucrose when sucrose is paired with glucose; data from Table 3, column 3. Broken line (---), intake of preferred concentration when sucrose is paired with itself; data from Table 3, column 3 (from Dethier and Rhoades, 1954).

sugar so that the gram intake is remarkably constant over a wide concentration range (Richter and Campbell, 1940a, 1940b).

When a fly is presented with two concentrations of the same sugar the higher of the two is always drunk in greater quantity (Table 3). Thus, once again, on a short-term basis, that is, a 24-hour period, there is no apparent compensation for the amount of sugar by weight. Furthermore, the quantity of preferred solution may not always be identical with the amount that would be imbibed were that solution paired with water. Clearly, therefore, the degree of preference, if indeed

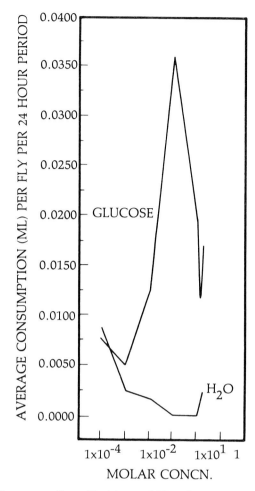

Fig. 24. Glucose preference-aversion curve (from Dethier and Rhoades, 1954).

quantity ingested is an accurate criterion, depends very much upon what options are available. On the other hand, the absolute preference does not vary significantly. If a comparison is made of the volumes taken of the more concentrated member of each pair, a preference-aversion curve is obtained that is similar to the one already described with sugar-water pairs. Compare the solid and broken lines in Figure 23. As before, the results obtained with solutions of glucose are similar to those just described.

When sucrose is paired with glucose in all possible combinations, additional information about discrimination and preference is obtained. Usually the higher concentration is preferred irrespective of the value (Table 4). At equimolar concentrations sucrose is the preferred sugar. On a weight basis (Fig. 25) more sucrose is ingested at all concentrations. The point of great interest, however, is that sucrose

Table 3. Amount of Solution Consumed when Two Concentrations of Sucrose are Paired.

Molar concentrations paired		Quantity consumed (ml/fly/24 hr.)	
I	II	I	II
3.0 M	1.0 M	0.0116	0.0073
3.0	0.1	0.0121	0.0066
3.0	0.01	0.0126	0.0028
3.0	0.001	0.0137	0.0015
1.0	0.1	0.0250	0.0014
1.0	0.01	0.0241	0.0017
1.0	0.001	0.0226	0.0006
0.10	0.01	0.0398	0.0006
0.10	0.001	0.0404	0.0000
0.01	0.001	0.0234	0.0039

From Dethier and Rhoades, 1954.

and glucose can be made equally acceptable at any point in the concentration range by choosing the correct pair of concentrations. Presumably members of these pairs are indistinguishable by the fly. With data obtained from these matching experiments a curve can be derived that is analagous to the "isosweetness" curves derived from experiments with human beings (Fig. 26). As will be explained shortly, poor intensity discrimination at certain points in the concentration range limits the precision of this operation. It is clear, nevertheless, that glucose is less acceptable than sucrose. The data also show that acceptability alters with changing concentration and that the changing relationship of concentration for equally acceptable solutions of sucrose and glucose is logarithmic. A similar relationship was found by Dahlberg and Penczek (1941) and confirmed by Cameron (1947) for the sweetness to the human taste of sucrose, glucose, and several other sugars. The studies on man were based on determinations over the range 0.15–0.75 M whereas the range for the fly extended from 1×10^{-5} to 1 M.

The concept of "isosweetness," however, must be viewed with caution because more subtle experimental techniques may reveal qualitative differences among sugars. This caveat is prompted by the finding of Pfaffmann (1970) that the squirrel monkey can distinguish sucrose from fructose by quality as well as by intensity. No concentration of fructose was ever found that was preferred to 0.3 M sucrose. Sucrose was the more effective sugar in behavioral tests, but the less effective electrophysiologically.

The ability of the fly to discriminate among different concentrations of the same compound can also be assessed by the "two-bottle"

Table 4. Sucrose and Glucose Paired in All Concentration Combinations.

Concentrations paired		Quantity consumed (ml/fly/24 hr.)		Significance[a]
Sucrose	Glucose	Sucrose	Glucose	
3.0 M	1.0 M	0.0090	0.0097	—
3.0	0.1	0.0073	0.0065	+
3.0	0.01	0.0115	0.0055	+
1.0	1.0	0.0252	0.0039	+
1.0	0.1	0.0301	0.0010	+
1.0	0.01	0.0289	0.0004	+
1.0	0.001	0.0302	0.0023	+
0.1	1.0	0.0065	0.0257	+
0.1	0.1	0.0516	0.0012	+
0.1	0.01	0.0529	0.0001	+
0.01	1.0	0.0002	0.0237	+
0.01	0.1	0.0083	0.0696	+
0.01	0.01	0.0442	0.0084	+
0.01	0.001	0.0380	0.0022	+
0.001	1.0	0.0023	0.0315	+
0.001	0.1	0.0031	0.0698	+
0.001	0.01	0.0186	0.0156	—
0.001	0.001	0.0135	0.0063	+
0.0001	0.01	0.0066	0.0108	—
0.0001	0.001	0.0087	0.0080	+
0.0001	0.0001	0.0122	0.0046	+
0.00001	0.001	0.0045	0.0033	—
0.00001	0.0001	0.0019	0.0025	—
0.00001	0.00001	0.0033	0.0034	—

From Dethier and Rhoades, 1954.

[a] A plus sign indicates that the difference in the amounts consumed is statistically significant; a dash, not significant.

method. Discrimination was measured at selected concentrations in the range 1×10^{-5} to 1 M by presenting the flies with pairs of concentrations and in successive experiments gradually closing the gap until there was no significant difference in consumption. For each concentration selected a determination was made of the least increase in concentration that could be detected and also the least decrease. The tabulations in Table 5 show that the amount by which the concentration must be changed in order to be detected ($\Delta I/I$) is greatest at low concentrations (I), decreases to a low value when I is 0.1 M, and begins to increase as I approaches 1 M. In other words, discrimination is most acute at the middle range of concentration. It is optimal at the concentration for which there is maximum preference.

The lowest value of $\Delta I/I$ obtained was 0.12. This is lower than most values heretofore reported for insects. Previous determinations had

To Each His Taste: Discrimination and Preference

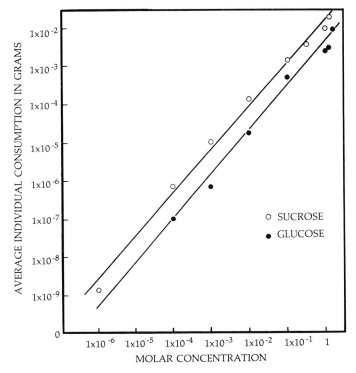

Fig. 25. Weight of sugar consumed as a function of concentration (from Dethier and Rhoades, 1954).

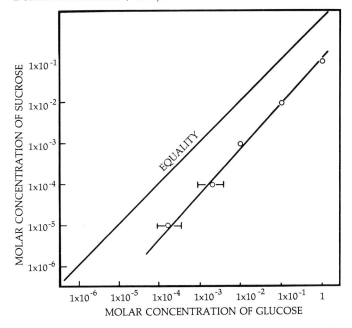

Fig. 26. Concentrations of sucrose and glucose that are equally acceptable. This curve is analogous to those that relate the sweetness to man of glucose and sucrose (from Dethier and Rhoades, 1954).

Table 5. Intensity Discrimination among Different Concentrations of Sucrose.

Intensity	Nearest concentration that can be distinguished	$\Delta I/I$	$\Delta I/I$ (geometric mean)
0.001	0.000007	0.99	2.4
0.001	0.007	6.0	
0.01	0.007	0.3	0.387
0.01	0.015	0.5	
0.1	0.088	0.12	0.12
0.1	0.112	0.12	
1.0	0.70	0.3	0.346
1.0	1.40	0.4	
2.0	1.0	0.5	0.418
2.0	2.7	0.35	

From Dethier and Rhoades, 1954.

been made for insects only at concentrations representing median acceptance thresholds or limited ranges on either side of these, but, as has been shown, this is the area of optimum discrimination. For honeybees von Frisch (1934) found a value of 0.25. Frings (1946) tested responses of the American cockroach (*Periplaneta americana* L.) to salt solutions in a series of steps, each successive concentration being 1.25 times greater than the previous one. On the assumption that the cockroaches could distinguish between these in the range of threshold, $\Delta I/I$ is 0.25. Beck (1965b) obtained a minimum value of 0.05 for glucose and sucrose and 0.08 for fructose with larvae of the European corn borer. The average values for man are 0.3 (Moncrieff, 1944) although lower values have been reported from time to time. Lemberger (1908) recorded 0.15, and Dahlberg and Penczek (1941) gave comparably low values for sucrose. The last-named authors found that the value was lowest over the range 0.3–0.6 M and increased beyond these limits in either direction. For *Phormia* the lowest values lay in the range 0.001–1.00 M and increased beyond these limits to 2.4 at 1×10^{-3} M and 0.418 at 2 M. These figures represent the discrimination factor over nearly the entire spectrum of sensitivity to sucrose.

A word of caution must be injected here, however. The U-shape of this curve may be partly spurious because, as will be discussed later, the ascending limb of preference-aversion curves probably represents a set of conditions different from that for the descending curve. Furthermore, there is little doubt that viscosity is asserting itself at the highest concentrations.

To this point it does not appear that the nutritive value of foods is a compelling determinant of selection; however, more incisive tests are required. These have been provided in part by an extention of the

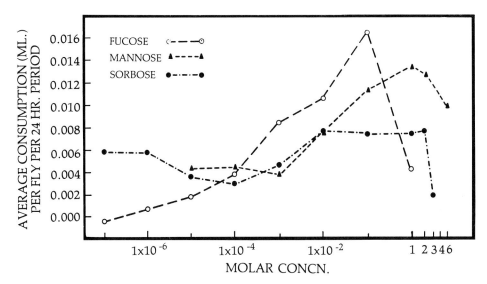

Fig. 27. Preference-aversion curves for fucose, mannose, and sorbose (from Dethier, Evans, and Rhoades, 1956).

preference studies to include sugars of markedly different nutritive value or no value at all (Dethier, Evans and Rhoades, 1956). The sugars chosen for study were fucose, sorbose, and mannose. Fucose is a methyl pentose that is rather effectively stimulating for the fly's gustatory organs. The median acceptance threshold with stimulation of the tarsal chemoreceptors is 0.087 M. It has no nutritional value (Hassett, Dethier, and Gans, 1950). Sorbose, a heptose, also stimulates the taste receptors (tarsal threshold = 0.14 M) and also is not utilized. Mannose, another heptose, is an extremely poor source of stimulation (tarsal threshold = 7.59 M), but is nutritionally highly effective.

For all three sugars there is a concentration below which they are not distinguished from water. As the concentration is increased, greater volumes are ingested until a point is reached beyond which a decline in intake occurs (Fig. 27). Although the curve for sorbose is rather flat, the trends are the same as previously found for sucrose and glucose. This is especially true of the highly stimulating nonnutritive sugar fucose. A cursory examination of the curves reveals no apparent relation between the volume intake and either the nutritional value or the relative stimulating effectiveness (for the tarsi). Of the three sugars, the maximum intake is greatest for fucose and least for sorbose. None is consumed in so great quantities as glucose or sucrose.

An unexpected characteristic of these curves is an inversion at very low concentrations, at which water may be taken in preference to sugar. With fucose, sorbose, and mannose the inversion occurs respectively at 1×10^{-7} M, 1×10^{-7} M, and 1×10^{-3} to 1×10^{-4} M. Bimodal

a.

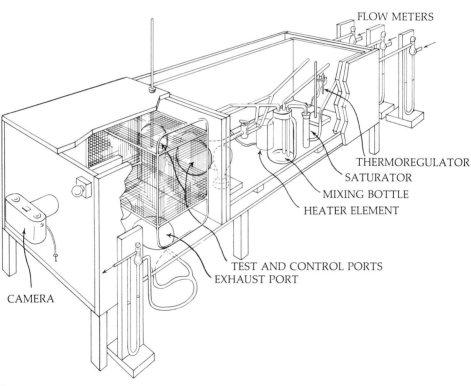

b.

Fig. 28. The olfactometer employed in studying the behavioral responses of *Phormia* to compounds in the gas phase. (a) Photograph. (b) Schematic diagram (from Dethier and Yost, 1952).

preference-aversion relationships of sugars were first noted by Beck (1956a) in his studies of the larvae of the European corn borer (*Pyrausta nubilalis* Hbn.). A reexamination of the raw data of Dethier and Rhoades (1954) revealed similar bimodal relationships. The meaning of rejection at low concentrations is still not understood.

All of the experiments just described involved nonodorous compounds. The role of odor must be considered as well. The quantitative behavioral study of olfaction has always been beset with formidable difficulties, not the least of which has been perfecting a means of delivering known concentrations of vapors to test organisms. Over the years numerous olfactometers have been designed, each specifically for use with a particular species of insect. The instrument employed successfully with *Phormia* is illustrated in Figure 28 (Dethier and Yost, 1952). Basically it consisted of a box with two compartments, one containing a cage of flies and the other containing delivery tubes, mixing jars, and funnels involved in the presentation of odors.

The principle of operation was relatively simple. A source of air was split into three streams, each of which was metered. These entered a reaction chamber through two ports. One stream passed directly to the port that was to serve as a control. Streams two and three went to the experimental or test port. Of these two, one was passed through saturators containing the compound to be tested; the other was used to dilute the first to the desired concentration. A system of three-way stopcocks permitted directing the control stream, and similarly the test stream, to either right or left port. Odor concentration was regulated by varying the ratio of the rates of flow of streams two and three and also the rate through the saturator. In all experiments the rate of the air flow through the control port was made equal to the total flow (streams two and three combined) through the test port.

Insects to be tested were confined in a cage within the reaction chamber. The cage was placed in such a position that the two air streams, control and test, passed through it with little or no mixing. Beneath the floor of the cage were two exhaust ports. Since this compartment was not airtight, it was possible to exhaust gases at a rate greater than that at which they entered. In this manner turbulence and dead air areas were reduced to a minimum. Lights shining through the ports attracted flies to that side of the cage. In a control experiment, when pure air passed through both ports, the flies distributed themselves on the wall of the cage adjacent to the ports in such a way that equal numbers appeared before each port. When odor replaced the air at one port, the distribution of flies shifted so that a greater number congregated at the test port if the odor was attractive and a lesser number if it was repellent. In the former case the concentration that caused the number of flies at the test port to be increased by 50%

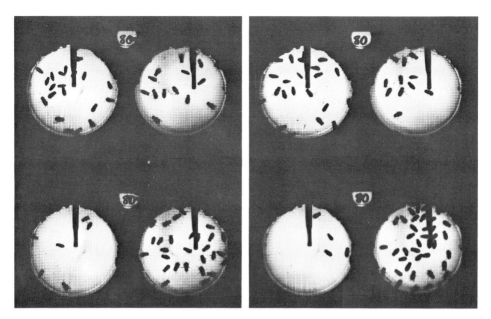

Fig. 29. Two typical records illustrating olfactory responses. Upper pairs of photographs show the distribution of flies when pure air is entering both ports. The lower photographs show the unbalanced distribution of flies when supraliminal concentrations of pentanol vapor enter each left port. The flies are repelled (from Dethier and Yost, 1952).

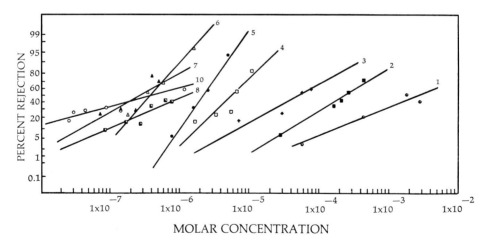

Fig. 30. Distribution of olfactory thresholds (rejection) of *Phormia* for different aliphatic alcohols. The curve for each alcohol is designated by a number representing the number of carbon atoms (from Dethier and Yost, 1952).

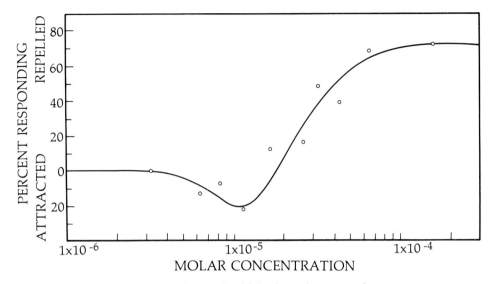

Fig. 31. The stimulating effect of iso-valeraldehyde in the vapor phase on *Musca domestica*. The compound is slightly attractive at low concentrations and repellent at high (from Dethier, 1954).

was termed the median acceptance threshold. Conversely, the least concentration that caused the number of flies at the test port to be decreased by 50% was termed the median rejection threshold. Records of the distribution were made photographically, ten consecutive exposures being made for each experiment (Fig. 29). The criterion of response used in this investigation was the median rejection threshold. Additional details concerning construction and operation of the apparatus, sources of variability, and methods of statistical analysis are given in the original report of this work (Dethier and Yost, 1952).

The distribution of thresholds within a population of flies is normal with respect to the logarithm of concentration (Fig. 30). By employing the median threshold of rejection or acceptance it is possible to compare the stimulating effectiveness of a variety of compounds. When, for example, the rejection thresholds for a homologous series of normal alcohols are plotted against their respective chain lengths on logarithmic coordinates, the relation is linear, the alcohols being rejected at increasingly lower concentrations as the molecules become larger.

A comparable study of the stimulating effect of normal aldehydes revealed similar relationships. These aldehydes are attractive at low concentrations and repellent at high (Fig. 31). Both acceptance and rejection thresholds decrease as the chain length of the molecule increases (Dethier, 1954).

The foregoing experiments, however, are concerned with the effects of odor on orientation. Its actual effect on ingestion is another matter.

The most extensive investigations of odors acting strictly as feeding stimuli (as opposed to cues for orientation) have been carried out with plant-feeding insects (e.g., Dethier, 1941, 1953), but the nature of the process is still in doubt (Thorsteinson, 1960; Dethier, 1973). Insofar as the blowfly is concerned there is no doubt that water vapor and some odors can elicit proboscis extension, the first act in the feeding sequence of those insects possessing retractable mouthparts. Minnich (1924) had observed that the cabbage butterfly (*Pieris rapae* L.) extended its proboscis to the odor of apple, a source of carbohydrate upon which the butterfly would feed in the laboratory. The same response occurred to warm, humid air (Verlaine, 1927). *Phormia* extends its proboscis in response to vapors from malt extract, isovaleraldehyde, and water. It responds in a similar manner to compounds that are not natural to its feeding environment. Among these are a number of essential oils, especially oil of caraway (Saxena, 1958). Evans (1961) confirmed those observations of Saxena but pointed out that the character of the proboscis extension resembled cleaning behavior, that the vapors were toxic, and that ingestion of sugar in the presence of these vapors was not increased. These results are not consonant with the idea that these odors elicit feeding. They reaffirm the idea that proboscis extension may be concerned with several different aspects of behavior and cannot be interpreted solely in terms of feeding. It had earlier been demonstrated with *Pieris* that irritants and interference with the antennae by coats of foreign material (e.g., Vaseline) cause proboscis extension (Minnich, 1924).

The converse action, reduction in ingestion caused by repellent odors, has not been generally observed. The acceptance of sugar by honeybees, for example, is unaffected by the presence or absence of essential oils (von Frisch, 1934). An exception to the general rule is *n*-butanol. Vapor in concentrations 10^3 times that necessary for repellency (Dethier and Yost, 1952) does reduce the intake of sucrose solutions by *Phormia*. The effect is abolished when antennal olfactory receptors are extirpated.

None of these observations throws much light on the question of whether or not odors occurring naturally in a feeding context play any positive role in ingestion. In an attempt to settle this question an extensive series of experiments to test the effect of ethanol on ingestion were undertaken (Dethier, 1961b). Ethanol was selected because it is a ubiquitous constituent of fermenting substances upon which *Phormia* feeds and because it is nutritionally adequate. Furthermore, its choice permitted a direct comparison with this aspect of feeding behavior of the rat.

Four kinds of experiments were conducted: (1) measurements were made of the volume consumed in a single drink; (2) the amount consumed ad libitum over a 24-hour period was measured; (3) daily intake was measured in two-choice situations; (4) the number of visits made to each pipette in the choice situations was counted. All experiments were carried out on normal flies and on flies rendered anosmic by ablation of the antennae and labellum. Measurement of the volume of fluid taken at single drinks was made by removing and weighing the crop after ingestion or, when plans had been made to employ the fly in further experimentation, by timing the duration of drinking. The latter measurement provides a usable measure of intake because a relatively constant relation exists between the duration of feeding and the volume ingested (Evans and Dethier, 1957). For measurements of long-term drinking or in choice situations, individual flies were placed in small nylon mesh cages as previously described. Except in choice experiments the two pipettes contained identical solutions.

One major difficulty that attended these experiments was the inability of flies to survive for long periods on some of the solutions tested. Flies live as long as 60 days on 0.1 M ethanol but survive only 2.5 days on water. In order to extend the period available for study, each experiment (in which water was always the control solution and ethanol in water the test solution) was duplicated with 0.1 M sucrose added as a nutrient in both solutions. It was reasoned that if the odor of ethanol contributed to feeding behavior, its effect was not likely to be swamped by the addition of sucrose. The parallel results obtained in the two series of experiments verified this assumption.

A subsidiary study made at the same time consisted of drawing on a sheet of paper two parallel lines of liquid, one of water and one of 0.1 M ethanol, or of 0.1 M sucrose and one of 0.1 M ethanol in sucrose. These lines were drawn 2 mm apart, and a fly was placed facing down the parallel tracks. Locomotory and drinking behavior were then observed.

The number of factors that theoretically might influence the ingestion of alcohol is great because it stimulates both olfactory and contact chemoreceptors, is a source of energy, and acts as a general stimulant and depressant. The difficulty of separating these effects is considerable. Insofar as ingestion is concerned, all of the studies revealed that ethanol is accepted at certain concentrations and rejected at others (Fig. 32). The reasons for rejection seem fairly clear. Observations that flies in an olfactometer were repelled by a concentration of 2.4×10^{-4} M ethanol vapor (Dethier and Yost, 1952), that normal flies made fewer visits to one pipette in a pair when it contained a concentrated ethanol solution, and that normal flies avoided a pipette with concentrated ethanol although anosmic flies did not, all agree in showing that high vapor concentrations are repellent (i.e., flies turn away from the source). The action results from stimulation of the olfactory receptors.

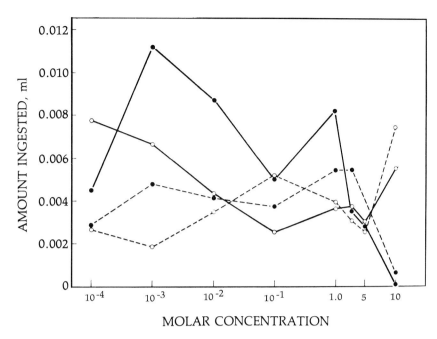

Fig. 32. Average volume of ethanol solution drunk per fly per 24-hour period during a 6-day test in a choice-situation in which both ethanol and water were present: ●—●, ethanol drunk by normal flies; ○—○, water drunk by normal flies; ●---●, ethanol drunk by anosmic flies; ○---○, water drunk by anosmic flies (from Dethier, 1961b).

Thus it may be concluded that strong odors of ethanol may prevent a fly from ever coming to the point of ingestion.

If a fly is not prevented by odor from coming to a solution and sampling it, the taste of the solution may prevent or limit ingestion. Measurements showing that even anosmic flies ingest only small quantities of concentrated solutions indicate that rejection of a solution may be mediated by contact chemoreceptors in the legs and mouthparts. The concentration at which ethanol is rejected in these experiments (1.5–2.5 M) is in the same range as the tarsal rejection thresholds (1.3–2.4 M) reported by Dethier and Chadwick (1947) and Dethier (1955a). That there is any ingestion of these solutions at all may be attributed to a need for water and food. When two unacceptable concentrations are presented simultaneously, the lesser is always taken in preference to the greater.

Behavior with respect to concentrations in the acceptable range is somewhat more obscure. During a 24-hour period or a period of days or weeks (Table 6) a normal fly ingests more of an ethanol solution than water or more of a mixture of ethanol and sucrose than of sucrose alone. This is true whether the solutions are presented simulta-

Table 6. Amounts of Various Concentrations of Ethanol Ingested by the Blowfly at a Single Feeding.

Compound	Molar concentration	No. of animals	Amount ingested (mg/fly)	Duration of feeding (sec)
Ethanol in H_2O	5.0	10	0	—
	2.5	10	0	—
	1.0	20	5.7	—
	0.1	20	4.0	—
	0.01	20	5.0	—
	0.001	20	4.8	—
H_2O	—	20	3.8	—
Ethanol in 0.1 M sucrose	5.0	10	3.7	21.30
	2.5	10	5.4	36.26
	1.0	20	8.0	51.45
	0.1	20	10.0	51.50
	0.01	20	10.1	62.46
	0.001	10	10.0	60.12
Sucrose	0.1	20	9.8	56.91

From Dethier, 1961.
Dash: no data.

neously or separately. When the solutions are paired, the fly makes more visits to the pipette containing ethanol. Clearly, then, the fly orients to the ethanol by odor. The behavior of a walking fly confronted with parallel lines of solution, one of which contains ethanol, confirms this. The fly tends to turn more often, if not exclusively, to the ethanol solution.

Although Dethier and Yost (1952) did not demonstrate attraction by ethanol in olfactometric experiments (cf. Fig. 31), it is not unlikely that ethanol is indeed a weak attractant. Wieting and Hoskins (1939) demonstrated that it is a weak attractant for house flies. It has long been known to be a strong attractant for *Drosophila* (Reed, 1938).

The normal blowfly not only visits a pipette containing ethanol more often than one without when the two are paired, it also drinks more often when an ethanol solution is the only one available. The simplest explanation is that the odor of ethanol stimulates feeding. The failure of anosmic flies to consume as much ethanol as normal is in agreement with this idea. There is no evidence that removal of the antennae interferes with the ingestion of nonodorous substances. It is certainly true that under certain circumstances the odor of ethanol may cause proboscis extension, but this in itself does not necessarily relate to ingestion.

Two series of experiments, however, are at variance with the foregoing interpretation. First, it is only in choice situations that anosmic flies fail to consume more ethanol than control solution (Fig. 28). Second, a normal fly does not drink more ethanol than water at a single drink, nor, having drunk water to repletion, does it then drink ethanol. Similarly, a fly will not imbibe at one drink more of a mixture of sucrose and ethanol than of ethanol alone.

An alternative explanation is that alcohol after it is consumed shortens the intervals between drinks by some unknown mechanism. If this is true, anosmic flies should behave like normal flies when ethanol is the only solution available; that is, they should take more of the ethanol than they would of a control solution over a prolonged period. Measurements over a period of 20 days show that they actually do. Their failure to behave like normal flies in a two-choice situation may be explained by supposing that they take less ethanol than normal flies because of their inability to orient preferentially to ethanol. Since they take less, the post-ingestive effect is less and they drink less often. Thus, for anosmic flies the total fluid intake is greatest when ethanol is the only fluid available and least when no ethanol is available.

Although at all concentrations in the acceptable range more ethanol than water is taken, the fluid intake decreases as the ethanol concentration increases. The same is true of solutions made up in sucrose. The decrease cannot be attributed to sensory rejection because the volume of ethanol ingested always remains in excess of water. Moreover, the effect is not observed when measurements are based upon single drinks. Nor is it likely that the effect is based on caloric value. Even though the fly takes less of concentrated solutions, the caloric intake is actually greater. Looking at the situation in reverse, one might argue that the fly takes the maximum volume of concentrated ethanol, thereby ensuring optimum caloric intake, and then increases its volume intake at lower concentrations as a device that keeps up the number of calories. Yet the fact that flies live longer on 0.1 M than on 2.5 M ethanol suggests that the caloric value of the former is adequate (or that 2.5 M is toxic). Furthermore, it was found in earlier studies of carbohydrate intake that even though flies ingested less of a concentrated sugar they actually acquired a greater weight of sugar, and this relationship held whether the sugar was metabolizable or not (Dethier and Rhoades, 1954; Dethier, Evans, and Rhoades, 1956). Since there was no evidence here of caloric regulation, it does not seem reasonable to postulate it in the case of ethanol.

A remaining possibility that might explain some of the results is that ethanol exerts some systemic effect on the fly, comparable to intoxication, narcosis, and so on, and that as a consequence the entire behavior is altered. There is ample anecdotal evidence pointing to the capacity of ethanol for intoxicating insects. It has been known by collectors of insects for many generations that insects attracted to fermenting baits become helplessly intoxicated. Observations of *Phormia* reveal that this fly is no exception. On a continuous diet of 0.1 M ethanol, flies become lethargic; on a diet of 1 M their coordination disintegrates and they may die within 3 to 6 days.

The factors affecting ethanol ingestion by the blowfly appear to be more complex than those operating in the rat, although the end results for the two animals are nearly identical. The rat shows an ethanol preference at low concentrations and a rejection at high (Kahn and Stellar, 1960). For the rat, however, the preference appears to be based primarily on olfactory stimulation whereas olfaction can account for only part of the preference displayed by the blowfly. The preference diminished in rats from which the olfactory bulbs had been removed. Rejection in rats, as in flies, is partly olfactory and partly gustatory since anosmic rats reject at higher concentrations than do normal rats.

Olfaction assuredly plays an important part in enabling a fly to discriminate among potential foods (while in flight or walking). At this point of ingestion, however, the contact chemical senses are the primary, if not the sole, sources of information.

Whether or not a food will be ingested or which of several will be preferred depends upon the stimulating effectiveness of that food and the nutritional state of the fly at the time. The nutritional value of the food is not a causal factor.

As stated at the beginning of this chapter, the words "preference," "discrimination," and so forth have been employed in a fashion that was not intended to refer to any special psychological concepts associated with them. It is interesting, however, to consider what the outcome might be of an experiment designed after the paradigm of Irwin's hypothetical jumping-rat experiment. The experiment that was eventually conducted was designed in five steps. First, two identical unpainted pipettes (designated as "white") containing identical sucrose solutions (0.1 M) were presented to flies, and the volume of sugar imbibed from each was measured. Second, two identical pipettes (white) were paired, but one contained 0.1 M sucrose and the other 1.0 M. Third, a black and a white pipette, each containing 0.1 M sucrose, were paired. Fourth, a black pipette containing 0.1 M sucrose was paired with a white pipette containing 0.1 M sucrose. Fifth, a black pipette containing 0.1 M sucrose was paired with a white one containing 1.0 M sucrose. There were ten replications of each set of conditions, and each experiment continued for 4 days. Consumption was measured every day. When the pipettes and the solutions in them (i.e., 0.1 M sucrose) were identical, the volumes taken from each were the same. When the pipettes were identical but the solutions of different concentrations, the concentrated solution was 18 times more effective than the weak one. When the comparison was between a black and a white pipette each containing the same concentration, black was approximately 2.4 times more effective. When the concentrated solution was in the black pipette, that combination was 37 times more effective. When the concentrated solution was in the white pipette, that combination was 16 times more effective. These results suggest that, in Irwin's terminology, the flies have a *bias* toward black but a *preference* for the more concentrated solution. In this context their behavior is very similar to that of rats in a comparable situation.

Literature Cited

Barnhard, C. S., and Chadwick, L. E. 1953. A "fly factor" in attractant studies. *Science* 117:104–105.

Beck, S. D. 1965a. A bimodal response to dietary sugars by an insect. *Biol. Bull.* 110:219–228.

Beck, S. D. 1956b. Nutrition of the European corn borer, *Pyrausta nubilalis* (Hbn.). IV. Feeding reactions of first instar larvae. *Ann. Ent. Soc. Amer.* 49:399–405.

Belzer, W. R. 1970. The control of protein ingestion in the black blowfly, *Phormia regina* (Meigen). Ph.D. dissertation, University of Pennsylvania, Philadelphia.

Cameron, A. T. 1947. The taste sense and the relative sweetness of sugars and other sweet substances. Sugar Res. Foundation, New York, *Sci. Rpt. Ser.* 9:1–72.

Dahlberg, A. C., and Penczek, E. S. 1941. The relative sweetness of sugars as affected by concentration. *N.Y. State Agr. Exp. Sta. Tech. Bull.* 258:3–12.

Dethier, V. G. 1941. Chemical factors determining the choice of food plants by *Papilio* larvae. *Amer. Naturalist* 75:61–73.

Dethier, V. G. 1953. Host plant perception in phytophagous insects. *Trans. 9th Int. Congr. Ent.,* Amsterdam, 1952, Vol. 2, pp. 81–88.

Dethier, V. G. 1954. The physiology of olfaction in insects. *Ann. N.Y. Acad. Sci.* 58:139–157.

Dethier, V. G. 1955a. The physiology and histology of the contact chemoreceptors of the blowfly. *Quart. Rev. Biol.* 30:348–371.

Dethier, V. G. 1955b. Mode of action of sugar-baited fly traps. *J. Econ. Ent.* 48:235–239.

Dethier, V. G. 1961a. Behavioral aspects of protein ingestion by the blowfly *Phormia regina* Meigen. *Biol. Bull.* 121:456–470.

Dethier, V. G. 1961b. The role of olfaction in alcohol ingestion by the blowfly. *J. Insect Physiol.* 6:222–230.

Dethier, V. G. 1973. Electrophysiological studies of gustation in lepidopterous larvae. II. Taste spectra in relation to food-plant discrimination. *J. Comp. Physiol.* 82:103–134.

Dethier, V. G., and Chadwick, L. E. 1947. Rejection thresholds of the blowfly for a series of aliphatic alcohols. *J. Gen. Physiol.* 30:247–253.

Dethier, V. G., Evans, D. R., and Rhoades, M. V. 1956. Some factors controlling the ingestion of carbohydrates by the blowfly. *Biol. Bull.* 111:204–222.

Dethier, V. G., and Rhoades, M. V. 1954. Sugar preference-aversion functions for the blowfly. *J. Exp. Zool.* 126:177–204.

Dethier, V. G., and Yost, M. T. 1952. Olfactory stimulation of blowflies by homologous alcohols. *J. Gen. Physiol.* 35:823–839.

Evans, D. R. 1961. The effects of odours on ingestion by the blowfly. *J. Insect Physiol.* 7:299–304.

Evans, D. R., and Dethier, V. G. 1957. The regulation of taste thresholds for sugars in the blowfly. *J. Insect Physiol.* 1:3–17.

Frings, H. 1946. Gustatory thresholds for sucrose and electrolytes for the cockroach, *Periplaneta americana* (Linn.). *J. Exp. Zool.* 102:23–50.

Frisch, K. von 1934. Über den Geschmackssinn der Biene. *Z. vergl. Physiol.* 21:1–156.

Hassett, C. C., Dethier, V. G., and Gans, J. 1950. A comparison of nutritive values and taste thresholds of carbohydrates for the blowfly. *Biol. Bull.* 99:446–453.

Horning, D. S. 1966. Insects of Craters of the Moon National Monument, Idaho. Master's thesis, University of Idaho.

Irwin, F. W. 1958. An analysis of the concepts of *discrimination* and *preference*. *Amer. J. Psychol.* 71:152–163.

Kahn, M., and Stellar, E. 1960. Alcohol preference in normal and anosmic rats. *J. Comp. Physiol. Psychol.* 53:571–575.

Lemberger, F. 1908. Psychologische Untersuchungen über den Geschmack von Zucker und Saccharin. *Pflügers Arch.* 123:293–311.

Lepkovsky, S. 1948. The physiological basis of voluntary food intake (appetite?). *Adv. Food Res.* 1:115–148.

Minnich, D. E. 1924. The olfactory sense of the cabbage butterfly, *Pieris rapae* Linn., an experimental study. *J. Exp. Zool.* 39:339–356.

Moncrieff, R. W. 1944. *The Chemical Senses.* Leonard Hill, London.

Norris, K. R. 1965. The bionomics of blow flies. *Annu. Rev. Ent.* 10:47–68.

Pfaffmann, C. 1963. Taste stimulation and preference behavior. In *Olfaction and Taste,* Y. Zotterman, ed, Vol. I, pp. 257–273. Pergamon Press, Oxford.

Pfaffmann, C. 1970. Physiological and behavioral processes of the sense of taste. In *Taste and Smell in Vertebrates,* G. E. W. Wolstenhome and J. Knight, eds, pp. 3–30. J. A. Churchill, London.

Reed, M. R. 1938. The olfactory responses of *Drosophila melanogaster* Meigen to the products of fermenting banana. *Physiol. Zool.* 11:317–325.

Richter, C. P. 1943. Total self regulatory functions in animals and human beings. *The Harvey Lectures,* Series 38:63–103.

Richter, C. P., and Campbell, K. H. 1940a. Sucrose taste thresholds of rats and humans. *Amer. J. Physiol.* 128:291–297.

Richter, C. P., and Campbell, K. H. 1940b. Taste thresholds and taste preferences of rats for five common sugars. *J. Nutrition,* 20:31–46.

Saxena, K. N. 1958. Location of the olfactory receptors of the blowfly. *Proc. Nat. Inst. Sci. India* B24:125–132.

Soulairac, A. 1947. La physiologie d'un comportement: L'appétit glucidique et sa régulation neuro-endocriniénne chez les rongeurs. *Bull. Biol. France et Belgique* 81:273–432.

Strangways-Dixon, J. 1961a. The relationship between nutrition, hormones and reproduction in the blowfly *Calliphora erythrocephala* (Meigen). I. *J. Exp. Biol.* 38:225–235.

Thorsteinson, A. J. 1960. Host selection in phytophagous insects. *Annu. Rev. Ent.* 5:193–218.

Verlaine, L. 1927. Le déterminisme du déroulement de la trompe et la physiologie du gout chez lepidoptéres. *Bull. Ann. Soc. Ent. Belge.* 67:147–182.

Waldbauer, G. P. 1964. Quantitative relationships between the number of fecal pellets, fecal weights and the weight of food eaten by tobacco hornworms, *Protoparce sexta* (Johan.) (Lepidoptera: Sphingidae). *Ent. Exp. Appl.* 7:310–314.

Wiesmann, R. 1962. Untersuchungen über den "fly factor" und den Herdentrieb bei der Stubenfliege, *Musca domestica* L. *Mitt. Schweiz Ent. Ges.* 35:69–114.

Wieting, J. O. B., and Hoskins, W. M. 1939. The olfactory responses of flies in a new type of insect olfactometer. *J. Econ. Ent.* 32:24–29.

Wykes, G. R. 1952. The preference of honeybees for solutions of various sugars which occur in nectar. *J. Exp. Biol.* 29:511–519.

Young, P. T. 1941. The experimental analysis of appetite. *Psychol. Bull.* 38:129–164.

Young, P. T. 1948. Appetite, palatability and feeding habit: A critical review. *Psychol. Bull.* 45:289–320.

4
The Etiquette of Eating: Mechanisms and Control of Ingestion

Where the bee sucks, there suck I.

Shakespeare, *The Tempest*

The process of transporting food to that place or places in the body where enzymatic action can commence is as complicated a phase of feeding as any. Natural man eats by bringing food to his mouth with his hands. He has learned by experience that certain visual and olfactory cues indicate acceptable food. His organs of taste pass final judgment after the food is placed in his mouth. Hydra captures food with its tentacles and stuffs it into its oral cavity. Crabs use their pincers to bring food to the mouth. Mantids employ their forelegs, as do also mice, squirrels, racoons, monkeys, and apes. Elephants use their trunks. Some of these animals rely upon visual cues to select food; some learn by experience what to eat and what to avoid. Animals lacking appendages to convey food to the mouth convey the mouth (in the case of starfish, the stomach) to food. The process is facilitated in some species by possession of a mouth that can be extended for feeding and retracted when not needed. Into this category falls the fly.

Many actions, simple in themselves, must be coordinated to ensure successful ingestion. Food in the hand, so to speak, or a high concentration of odor suggesting the immediate proximity of food arrests locomotion. These stimuli, food touching the feet or odor wafted onto the antennae, evoke extension of the proboscis. Spontaneous extension, even when the fly is in the throes of extreme starvation, is not known to occur. When the proboscis touches the substrate, the oral lobes (Fig. 33) are spread apart. The oral surface is pressed to the surface, and sucking commences. Sucking is accomplished by a muscular pump associated with the pharynx. The main feature of this cibarial pump is a set of short stout muscles inserted on the upper wall of the pharynx. When these muscles contract, the diameter of the pharynx is increased, thus creating a partial vacuum in that portion leading to the

Fig. 33. A photomicrograph made with the scanning electron microscope to show the oral surface of the labellum and the fringe of aboral chemosensory hairs.

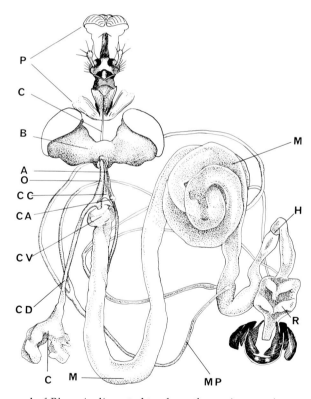

Fig. 34. The alimentary canal of *Phormia* dissected to show the various regions. P, proboscis; O, oesophagus; B, brain; A, aorta; CC, corpora cardiaca; CA, corpus allatum; CD, crop duct; C, crop; CV, cardiac valve; M, midgut, MP, malpighian tubules; H, hindgut; R, rectum.

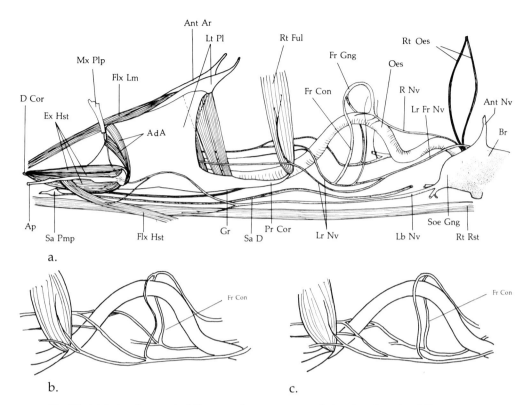

Fig. 35. (a) Lateral aspect of the muscles, nerves, and structural parts of the rostrum of *Phormia*. Ad A, adductors of apodeme; Ant Ar, anterior arch of fulcrum; Ant Nv, antennal nerve; Ap, apodeme; Br, brain; D Cor, distal cornu of fulcrum; Ex Hst, extensors of haustellum; Flx Lm, flexor of labrum; Flx Hst, flexor of haustellum; Fr Con, frontal ganglion connective; Fr Gng, frontal ganglion; Gr, gracilis muscle; Lb Nv, labial nerve; Lr Fr Nv, labrofrontal nerve; Lr Nv, labral nerve; Lt Pl, lateral plate of fulcrum; Mx Plp, maxillary palpus; Oes, oesophagus; Pr Cor, proximal cornu of fulcrum; R Nv, recurrent nerve; Rt Ful, retractor of fulcrum; Rt Oes, retractors of oesophagus; Rt Rst, retractor of rostrum; Sa D, salivary duct; Sa Pmp, salivary pump; Soe Gng, suboesophageal ganglion. (b) and (c), Alternative distribution of nerves in the region of the frontal connectives (Fr Con) of some individuals (from Dethier, 1959).

orifice between the labellar lobes. As fluid rushes up this tube it is pumped from the pharynx into the oesophagus, where peristalsis takes over the work of driving the fluid still deeper into the alimentary canal (Fig. 34).

The proboscis, including the portion with the cibarial pump, is a versatile and complex structure. It consists of three segments. From proximal to distal they are: rostrum, haustellum, labellum (Figs. 35–38). The proboscis can be extended fully and rapidly, as when the stimulus is concentrated sugar, or slowly and partially, as when the

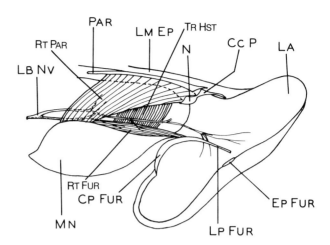

Fig. 36. Lateral aspect of the muscles and nerves of the oral disc. Cc P, cochleariform process; Cp Fur, central process of furca; Ep Fur, epifurca; La, labellum; Lb Nv, labial nerve, Lm Ep, labrum-epipharynx; Lp Fur, lateral process of furca; Mn, mentum; N, nodulus; Par, paraphysis; Rt Fur, retractor of furca; Rt Par, retractor of paraphysis; Tr Hst, transverse muscle of haustellum (from Dethier, 1959).

stimulus is weak. Partial extension can assume many forms. The rostrum may be extended while the other three segments remain flexed, or the labellum may extend while the other three segments remain flexed. The proboscis is also capable of limited lateral movements. Ten pairs of direct muscles activate the multiple movements (Figs. 37 and 38). The rostrum is extended by air pumped into extensive air sacs by unidentified muscles elsewhere in the body. Pricking the rostral membrane with a pin ruptures the pneumatic system, allowing air to escape and preventing extension. Transfer of blood can also cause extension. This hydraulic feature can be neatly demonstrated by squeezing the fly's thorax between thumb and forefinger, whereupon the probocis extends to a degree determined by the pressure applied. Other segments of the proboscis are extended by direct action of the extensors of the haustellum, the adductors of the apodemes, the retractors of the paraphyses, the transverse muscles of the haustellum, and the retractors of the furca. In all, five pairs of direct muscles are involved in complete extension. Retraction also may be slow and partial or rapid and complete. Slow retraction is accomplished in part by elastic recoil. Rapid contraction relies upon five other pairs of muscles: the retractors and accessory retractors of the rostrum, the retractor of the fulcrum, the flexor of the labrum, and the flexor of the haustellum.

All of these complex actions must be initiated by inputs from a galaxy of sense organs (chemoreceptors, mechanoreceptors, proprioceptors), constantly monitored by them, and finally terminated by them.

The Etiquette of Eating: Mechanisms and Control of Ingestion

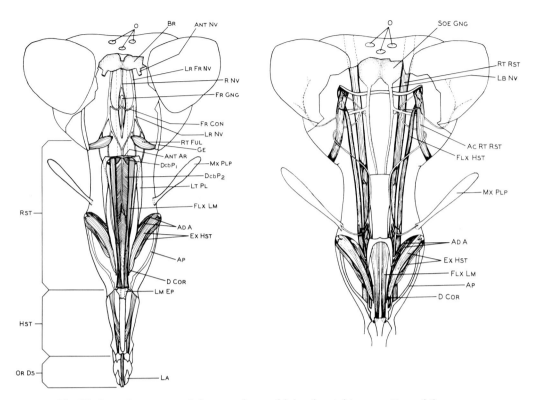

Fig. 37. Anterior aspect of the muscles and labrofrontal innervation of the proboscis of *Phormia*. Rst, rostrum; Hst, haustellum; Or Ds, oral disc; O, ocelli; Br, brain; Ant Nv, antennal nerve; Lr Fr Nv, labrofrontal nerve; R Nv, recurrent nerve; Fr Gng, frontal ganglion; Fr Con, frontal ganglion connective; Lr Nv, labral nerve; Rt Ful, retractor of fulcrum; Ge, gena; Ant Ar, anterior arch of fulcrum; Dcb P_1, DCb P_2, dilators of cibarial pump; Mx Plp, maxillary palpus; Lt Pl, lateral plate of fulcrum; Flx Lm, flexor of labrum; Ad A, adductors of apodeme; Ex Hst, extensors of haustellum; Ap, apodeme; D Cor, distal cornu of fulcrum; LM Ep, labrum-epipharynx; La, labellum (from Dethier, 1959).

Fig. 38. Anterior aspect of the muscles and labial innervation of the proboscis of *Phormia*. O, ocelli; Soe Gng, suboesophageal ganglion; Rt Rst, retractor of rostrum; Lb Nv, labial nerve; Ac Rt Rst, accessory retractor of rostrum; Flx Hst, flexor of haustellum; Mx Plp, maxillary palpus; Ad A, adductors of apodeme; Ex Hst, extensors of haustellum; Flx Lm, flexor of labrum; Ap, apodeme; D Cor, distal cornu of fulcrum (from Dethier, 1959).

Before we can turn our attention to events that initiate and coordinate these actions, it is necessary to describe the sensory systems involved. Here the contact chemoreceptors of the legs and labellum play a major role. The story of their discovery and identification is a fascinating one.

As far back as the turn of the eighteenth century, biologists were debating the question of the existence of olfactory and gustatory senses in insects (for complete references see Dethier and Chadwick, 1948; Dethier, 1955). The idea that insects possessed distinct senses of smell and taste and that the most important site of olfactory receptors was the antenna gradually carried the day. The actual identity of the receptors remained a mystery, although there was a great deal of speculation based upon the kind of structure that chemoreceptors were presumed to have. In 1946, when the study of feeding behavior of the blowfly was begun in our laboratory, only one unequivocal report existed of the identification of a chemoreceptor in insects. This was a taste receptor. In this paper, Minnich (1926b) had stated that "stimulation of the tip of a single hair (on the aboral surface of the labellum of the flies *Phormia* and *Calliphora*) may be sufficient for the response." The response referred to was extension of the proboscis. More than a decade elapsed before anybody even looked at these hairs. Then Tinbergen (1939) made a histological study of the homologous hairs in the fly *Calliphora erythrocephala*. He discovered that the hairs possessed two lumina. Other hairs have only one. His illustrations depict the dendrites of three neurons entering the thin-walled lumen (in fact they enter the thick-walled cavity). The end of the dendrite of another neuron is attached to the base of the hair where it swings in a socket (Fig. 39). Tinbergen correctly surmised that this neuron registers movement of the hair, whereas those extending up the lumen are concerned with chemoreception.

In the same era Eltringham (1933) had examined the legs of butterflies by means of standard histological techniques and had discovered and described delicate, thin-walled, recurved hairs projecting from among the scales on the tarsi of the butterfly *Pyrameis*. Many years later I was able to examine his original slides at Oxford. The hairs were indeed those that were subsequently proven to be chemosensitive.

A turning point in the study of chemoreception and feeding behavior in insects was Minnich's earlier (1921) discovery, followed by extensive behavioral studies (1926a, 1929), that the legs of butterflies and flies bore chemoreceptors on their tarsal segments. Barrows (1907) had earlier arrived at this conclusion with respect to *Drosophila* and had guessed that two transparent hairs beneath the claws were the sense organs involved (these correspond to the chemosensory D hairs in *Phormia*). By observing with intuition what many others had undoubtedly seen before him, namely, that butterflies standing on flowers and flies standing in food extend their mouthparts and feed, Minnich concluded that the feet are chemosensitive. His observations and experiments did not pass unchallenged (cf. Verlaine, 1927) and it was only after a series of rigorously controlled experiments (Minnich, 1929)

The Etiquette of Eating: Mechanisms and Control of Ingestion

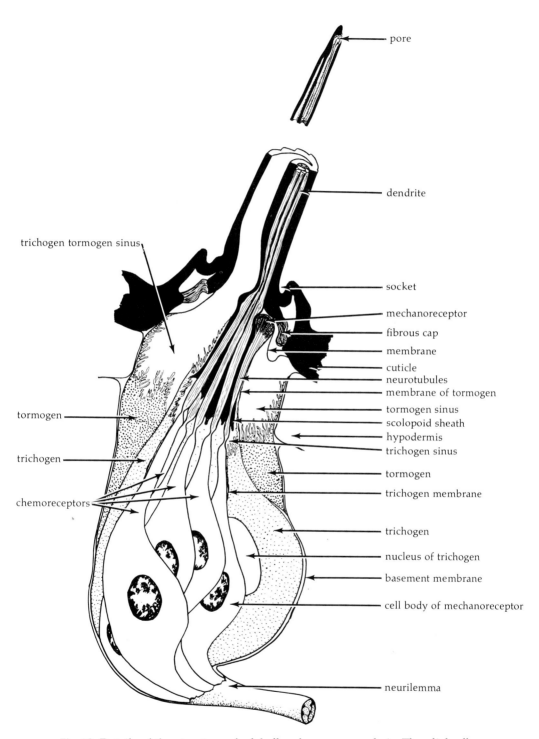

Fig. 39. Details of the structure of a labellar chemosensory hair. The glial cells, tracheoles, and pigmented syncitium are omitted.

that the existence of tarsal chemoreceptors was accepted. During the ensuing period many other investigators were attracted to the study of tarsal chemosensitivity, among them Abbott (1928, 1932), Krijgsman (1930), Hertwick (1931), Crow (1932), Haslinger (1935), Deonier (1938a, 1938b, 1939a, 1939b), Deonier and Richardson (1935). Nearly twenty-five years elapsed, however, before the search for the tarsal receptors was resumed. Thin-walled, multiply innervated, socketed hairs, were seen on the tarsi of *Musca domestica* by Hayes and Liu (1947) who suggested that they might be the chemoreceptors; however, no experimental confirmation was obtained. In a single year, however, Grabowski and Dethier (1954) and Lewis (1954) confirmed the identity of tarsal chemoreceptors. Employing extension of the proboscis as a criterion of sensitivity to sugars, they demonstrated that the taste receptors on the legs and labellum of flies are blunt-tipped hairs characterized by the presence of two lumina. These are the hairs that Tinbergen had described.

The two lumina result from the fact that the scolopoid sheath that encloses the dendrites, and in shorter sensilla (e.g., pegs) either terminates at the base of the peg or extends centrally to the tip, is here eccentrically situated and fused to one inside wall of the hair throughout its length. As will be seen later, this sheath serves as a very effective insulator between two conductors, the dendrite and the fluids of the lumen of the hair, and thus prevents what would otherwise be great cable losses in a dendrite as long as those present in these hairs. The eccentric position of the sheath (Figs. 40, 41) is probably a mechanical necessity because a centrally located sheath in a hair as long at 300 μ would lack support. Fusion with the wall of the hair provides this.

The taste hairs on the legs of *Phormia* number 308, 208, and 147 on each of the first, second, and third pair of legs, respectively. They are concentrated on the ventral and ventrolateral surfaces of the tarsi. A few also occur on the dorsal surface of the tibia. Not only does their number decrease progressively from front legs to back but also from distal to proximal on each leg (Table 7). They range in length from 28 to 264 microns. On the basis of location, size, and morphology, Grabowski and Dethier (1954) distinguished four types designated as A, B, C, and D hairs (Fig. 40). The first detailed electrophysiological studies indicated that there are functional distinctions as well (McCutchan, 1969). As additional electrophysiological analyses were made, more subtle differences were revealed. Shiraishi and Tanabe (1974) have reclassified the hairs into six types: A, B_1, B_2, and B_3, C, and D. The dorsal B-type of Grabowski and Dethier has been designated C, the other B-type hairs have been subdivided into types B, B_2, and B_3. The original C and D types have been combined as one type (D). The designation of type A is unchanged.

The Etiquette of Eating: Mechanisms and Control of Ingestion

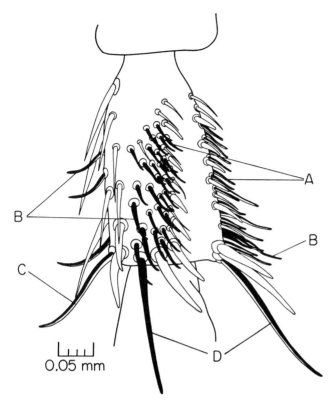

Fig. 40. One tarsal segment showing the distribution of chemosensory hairs (shaded). A, B, C, and D refer to different anatomical types (from Grabowski and Dethier, 1954).

Fig. 41. Photomicrographs of type A tarsal hairs made with the scanning electronmicroscope. (a) ×1,500; (b) ×8,300.

Table 7. Distribution of Two-toned Hairs on the Legs of *Phormia regina*.

Segment	Prothoracic Leg No. 1				Prothoracic Leg No. 2				Prothoracic Leg No. 3			
	A	B	C	D	A	B	C	D	A	B	C	D
5th tarsomere												
ventral	12		2		16		2		15		2	
ventro-lateral		9				10				12		
lateral		3		2		5		2		2		2
dorsal		2	2			2	2			2	2	
Total	12	14	4	2	16	17	4	2	15	16	4	2
4th tarsomere												
ventral	35				36				41			
ventro-lateral		16		2		14		2		11		2
lateral		4	2			3	2			4	2	
dorsal		2				2				2		
Total	35	20	2	2	36	19	2	2	41	17	2	2
3rd tarsomere												
ventral	34				42				43			
ventro-lateral		17		2		16		2		12		2
lateral		2	2			2	2			4	2	
dorsal		4				4				2		
Total	34	23	2	2	42	22	2	2	43	18	2	2
2nd tarsomere												
ventral	38				51				47			
ventro-lateral		21		2		16		2		18		2
lateral		2	2			3	2			2	2	
dorsal		4				4				4		
Total	38	27	2	2	51	23	2	2	47	24	2	2
1st tarsomere												
ventral												
ventro-lateral		34	1	5		27		5		30		5
lateral		1	1	1		3	2			1	2	
dorsal		6				6				6		
Total	0	41	2	6	0	36	2	5	0	37	2	5
Tibia		31	2			20				26		
Femur												
Totals	119	156	14	14	145	137	12	13	146	138	12	13
Grand totals		303				307				309		

From Grabowski and Dethier, 1954.

On the aboral surfaces of lobes of the labellum there are on the average 245 hairs in the male and 257 in the female. They range in length from 41 to 429 μ and from 50 to 500 μ in the related fly *Calliphora* (Peters and Richter, 1965). Their distribution and neuroanatomical connections (Fig. 42) have been described in detail by Wilczek (1967). On *Phormia* they are arranged in a definite pattern and can be divided topographically into the following groups: prickles, bristles,

Table 7. (continued)

Segment	Prothoracic Leg No. 4				Mesothoracic Leg No. 1				Metathoracic Leg No. 1			
	A	B	C	D	A	B	C	D	A	B	C	D
5th tarsomere												
ventral	14		2		10		2		8		2	
ventro-lateral		11				6				6		
lateral		1		2		2		2				2
dorsal		1	2				2			1	2	
Total	14	13	4	2	10	8	4	2	8	7	4	2
4th tarsomere												
ventral	31				22				16			
ventro-lateral		17		2		12		2		9		2
lateral		2	2				2				2	
dorsal		2				2				2		
Total	31	21	2	2	22	14	2	2	16	11	2	2
3rd tarsomere												
ventral	38				24					26		
ventro-lateral		14		2		13		2				2
lateral		5	2				2				2	
dorsal		4				3				4		
Total	38	23	2	2	24	16	2	2	0	30	2	2
2nd tarsomere												
ventral	39				21					21		
ventro-lateral		17		2		15		2				2
lateral		3	2				2				2	
dorsal		4				4				4		
Total	39	24	2	2	21	19	2	2	0	25	2	2
1st tarsomere												
ventral										18		2
ventro-lateral		31		5		31		2				2
lateral		3	2				2				2	
dorsal		6				6						
Total	0	40	2	5	0	37	2	2	0	18	2	4
Tibia		18				15				8		
Femur												
Totals	122	139	12	13	77	109	12	10	24	99	12	12
Grand totals		286				208				147		

intermediate hairs, large hairs, largest hairs, and marginal hairs (Figs. 43, 44).

Elucidation of the innervation of the hairs proved very difficult. Early studies relied upon special histological techniques employing methylene blue staining or silver impregnation. Originally it was believed that each hair was innervated by three neurons (Grabowski and Dethier, 1954; Dethier, 1955). Later the number was variously

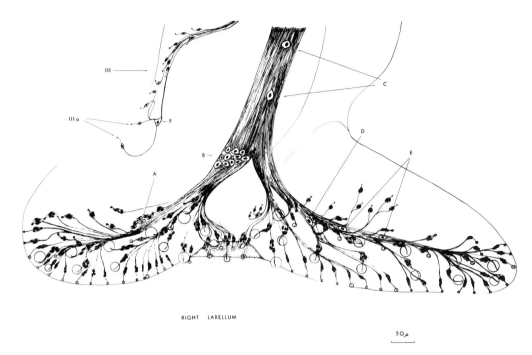

Fig. 42. Neuroanatomy of the aboral region of the labellum (from Wilczek, 1967).

reported as three to six (Peters, 1961, 1963; Stürckow, 1962; Larsen, 1962a; Larsen and Dethier, 1965). The decision that there were six neurons associated with some hairs was based upon cross-sections of the shaft that appeared to show five dendrites (plus the mechanoreceptor at the base) (Larsen, 1962a). Frequently, however, there are supernumerary structures that may be mistaken for dendrites. Longitudinal sections show that these are herniations of the dendrites, possibly fixation artifacts, in which the dendritic membrane bulges out in a variety of shapes. No neurotubules enter these extrusions (Fig. 45).

The consensus now, supported by studies with the electronmicroscope, is that the normal complement of each healthy hair at the prime of life is five (Peters and Richter, 1965; Stürckow, 1965; Dethier, 1971). Abnormalities, both structural and functional, do occur. It is also possible that some hairs may occasionally be found that have fewer neurons because of developmental anomalies. Since all of the neurons are derived from the same parent cell, the number of daughter cells could vary, depending upon the timing of mitosis with reference to the completion of adult development.

Studies with the electronmicroscope have now provided a fairly complete picture of the fine structure of the labellar and tarsal hairs. It is possible to correct some of the inaccuracies that arose as a consequence of limitations of light microscopy, but not all mystery has

The Etiquette of Eating: Mechanisms and Control of Ingestion

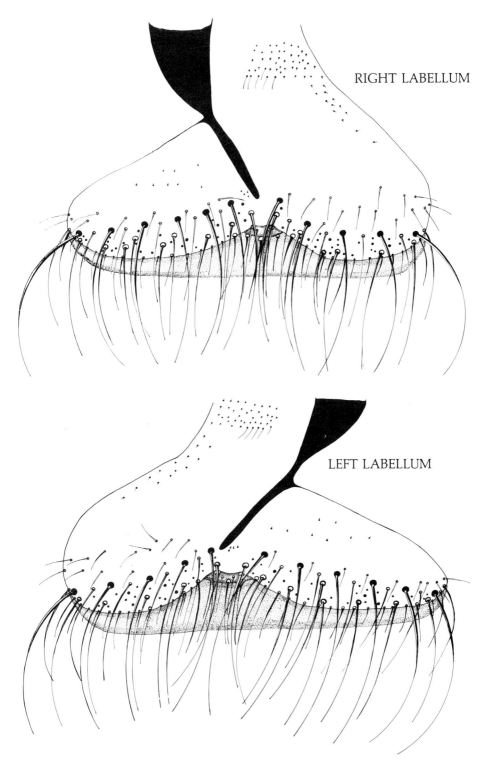

Fig. 43. The labellar chemoreceptor hairs of *Phormia* (from Wilczek, 1967).

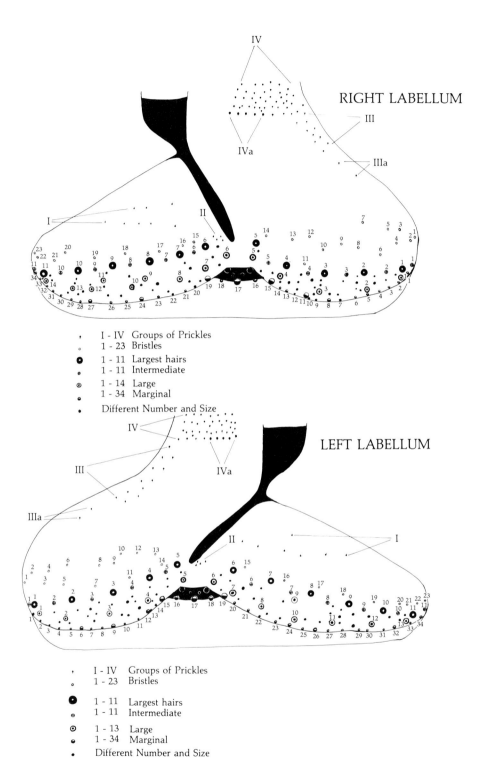

Fig. 44. Maps illustrating the distribution of the different types of labellar hairs (from Wilczek, 1967).

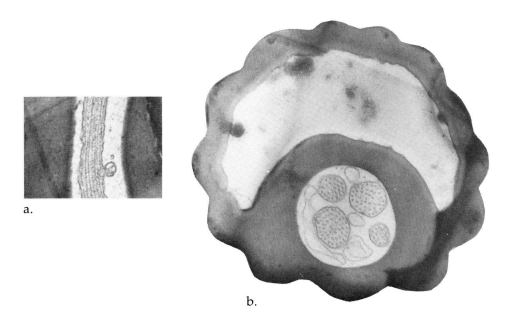
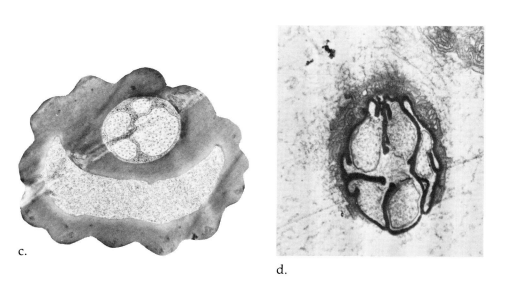

Fig. 45. Photographs of sections of labellar hairs viewed with the electron-microscope. (a) Longitudinal section showing herniations of a dendrite; (b) and (c) cross sections through the shaft of a hair; (d) cross section through the region of the scolopoid sheath.

been swept away by any means. The picture shown in Figure 39 is a composite based upon the studies of Grabowski and Dethier (1954), Adams (1961), Adams, Holbert, and Forgash (1965), Stürckow (1962, 1970, 1971), Larsen (1962a), Stürckow, Holbert, and Adams (1967), Peters and Richter (1965), Rees (1968), Stürckow *et al.*, (1973), and more recent work by Dethier.

The five neurons together with the trichogen and tormogen, which in the course of development secreted the hair and socket respectively, constitute a conspicuous subhypodermal sac hanging below the hair. All of the sacs in the labellum are enveloped by a yellowish brown syncitium, the pigment of which is contained in spherical globules. As far as is known, this pigmented layer is not directly involved in the primary processes of reception. Each sac is separated from the haemocoele by a membrane that is continuous with the basement membrane of the hypodermis and with the neurilemma which surrounds the bundle of five axons leaving the sac.

The arrangement of cells within the sac is very complicated. It can be appreciated best in cross sections. The neurons lie in a group within the trichogen, which in turn lies within the tormogen. As the dendrites proceed distally from the cell bodies each is separated from the trichogen by a membrane. The identity of these membranes is not clear. They could be extensions of a neurilemma cell inserted among the dendrites. They could also be finger-like extensions of the trichogen, such as occur in some sensilla of the larvae of mosquitoes (Zacharuk, Yin, and Blue, 1971). Continuing distally, the dendrites pass through a sinus of the trichogen cell. No membranes other than their own plasma membranes envelop them at this point. From here they continue into the scolopoid sheath, a cuticular tube that is coextensive with the thick-walled lumen of the hair. The scolopoid sheath at its distal end is scalloped and compartmentalized so that each entering dendrite is partially isolated from its companions. Soon four of the five come together in a common lumen. At this point the whole scolopoid structure is surrounded by the plasma membrane of the trichogen. In other words, as the dendrites proceed distally from their cell bodies they are first individually surrounded by membranes of the trichogen or neurilemma, then lie free in a sinus of the trichogen, then are encased in the scolopoid sheath.

At the base of the hair, in the region of the socket, there is still another thin membrane that extends from the rim of the socket in all directions and forms an incomplete diaphragm; one region is intimately associated with the termination of the mechanoreceptor. This membrane might be the methylene-blue-staining socket described by Grabowski and Dethier (1954). The mechanoreceptor, which lies within its own scolopoid sheath at this point, is characterized by a terminal structure consisting of parallel tubules in an electron-dense matrix. This "tubular body" is a common feature of mechanoreceptors (Thurm, 1965). In the socket there is also a fibrous cap extending from the root of the hair to the adjacent body cuticle.

Before the chemoreceptive dendrites enter the scolopoid sheath they constrict. In all other chemoreceptors studied this constriction marks the location of the junction body, that region of transition between the ciliary structure at the base of the dendrite and the distal microtubular region (Zacharuk, Lin, and Blue, 1971). Ciliary structures in the contact chemoreceptors of flies were first reported by Tominga, Kabuta, and Kuwabara (1969). They found them in the labellar hairs and interpseudotracheal papillae of the fleshfly *Boettcherisca peregrina*. We have also found ciliary structures in the labellar sensilla of *Phormia*.

Since the scolopoid sheath gradually merges with the thick-walled lumen in the shaft of the hair, the dendrites eventually come to lie within that lumen. In this manner they extend to the tip of the hair.

The extreme difficulty of obtaining good fixation of tissue in narrow, enclosed, impervious hairs is an obstacle to any accurate interpretation of structure within the shaft. Much of what is seen may be artifactual. With this caveat in mind one can describe the dendritic relationships of the labellar hairs of *Phormia* as follows. There is extreme variation in the cross-sectional shape and diameter of a dendrite from one level to the next. At some points the dendrites almost completely fill the lumen; at others they comprise only a small fraction of the volume (see Fig. 45). In longitudinal sections they are seen to vary in diameter, to twist and contort, and to have herniae. These resemble the lateral vesicles in the dendrites of the sensory cones of the antennae of mosquito larvae (Zacharuk, Lin, and Blue, 1971). Characteristic of the dendrites throughout their length distad of the junction bodies are conspicuous neurotubules. There are fifty on the average.

No desmosomes or tight junctions have yet been observed in the shaft or elsewhere between dendrites, an important point because of the potential for electrical "cross talk" that such junctions might afford. In some other insects, septate desmosomes between dendrites and trichogens have been seen in the region of the trichogen sinus, and unions answering the description of tight junctions have been observed between dendrites (Zacharuk, Lin, and Blue, 1971). In *Phormia* there are desmosomes between the trichogen and tormogen.

The structure of the tip of the hair and the relations of the dendrites to the pore at the tip are even more poorly understood than the rest of the hair's structure because of the enormous difficulty of obtaining ultrathin longitudinal sections in the plane desired for electronmicroscopial examination. Adams (1961), Adams and Holbert (1963), and Stürckow, Holbert, and Adams (1967) have had the most success, and our knowledge is based primarily upon their study of the stable fly (*Stomoxys calcitrans* L.). Here the pore is 0.25 μ in diameter and enlarges to a chamber in which the dendrites terminate. They lie about 3 μ from the surface. No specialization of their structure can be seen.

It is generally believed that the dendrites are exposed to the air (except for a fluid covering); however, in some cross sections of labellar hairs an extremely thin membrane-like structure can be seen between the dendrites and the wall of the lumen. It may be an artifact, but Zacharuk and Blue (1971) described a similar structure in peg sensilla on the antennae of larval mosquitoes, *Aedes aegypti* (L). In those sensilla the membrane actually extends over the otherwise exposed tips of the dendrites.

The fluid that surrounds the dendrites in the lumen of the hair may occasionally extrude (Morita and Takeda, 1957; Stürckow, 1967), sometimes in copious amounts (Fig. 46). Moulins (1967) has described in hypopharyngeal sensilla of the cockroach *Blaberus* a canal surrounding the distal prologations of the dendrites, which is an extracellular space lined with the plasma membrane of the trichogen cell, not in communication with its vacuole but nonetheless probably the exit for secretory material destined for the tip of the sensillum. Slifer (1970) has pointed out that there is a fluid-filled space between the tormogen and trichogen that may be continuous with the fluid in the thick-walled and thin-walled cavities. The so-called vacuole in the tormogen cell (Larsen, 1962a) is actually this space and the "intracellular fibrillae" of Larsen are actually extracellular microvilli of the tormogen and the trichogen extending into this space (Slifer, 1970).

The five neurons are daughters of the same cell that itself was a hypodermal cell. A most important characteristic of these cells is their being primary neurons, that is to say, they extend from the periphery well into the central nervous system before synapsing. Travelling from the bases of the hairs the axons come together into ever larger bundles, eventually forming, together with motor fibers, the labial nerve in the labellum and the leg nerve in the tarsi. The best evidence currently available supports the idea that each axon retains its individuality until it reaches the central nervous system. Neurons whose somata lie more or less between the periphery and the brain have been found in the labial nerve (Peters, 1961, 1962; Wilczek, 1967; Stürckow, Adams, and Wilcox, 1967). Peters suggested that these were second-order neurons with which the primary chemosensory neurons synapsed; however, two later studies fail to support this contention. Stürckow, Adams, and Wilcox (1967) counted the neurons seen in cross sections of the labial nerve and found that the number agreed with the number of neural cell bodies in the periphery. In a similar study we found that there were 2254 neurons in the labial nerve (Fig.

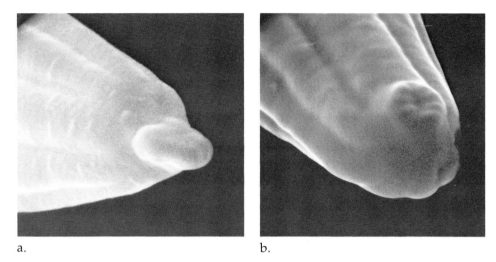

Fig. 46. (a) One chemosensory hair with a viscous exudate at the tip; (b) one without (×16,500).

47). By assigning five neurons to each chemosensitive labellar hair, four to each oral papilla, one to each mechanoreceptor, and one to each of the mystery cells in the nerve trunk, we arrived at a total of 2040, which compares favorably with the number in the nerve. This agreement argues against the idea that the cells of Peters are second-order neurons, because if they were the interneurons collecting primary axons from the periphery, there would be a much lower number of axons in the nerve proximad of these cells. Additionally, electron-microscopy has thus far failed to reveal synapses in this region. A more likely role for these cells is that of stretch receptors, monitoring the various positions that the very mobile labellum can assume. Similar cells have been found in the legs and in the nerves from the oral papillae (Fig. 48).

Labellar and tarsal hairs are not the only contact chemoreceptors associated with feeding by the blowfly. When the lobes of the labellum fold back, exposing the oral surface, they reveal 132 minute papillae (Figs. 49 and 50) (Wilczek, 1967). Their existence had been known for a century, and they were suspected of being organs of taste (Künckel d'Herculais, 1879). The first actual demonstration of a taste function came from behavioral experiments (Dethier, 1955). It was shown that when all labellar receptors except the papillae were covered with paraffin the fly was still capable of discriminating among water, sugar, and salts at first contact before the solutions were ingested.

The papillae are homologous with the labellar hairs and generally similar in basic structure. They are small, hollow, cuticular pegs, 10 μ long, innervated by four bipolar neurons (Dethier, 1955; Larsen 1962a,

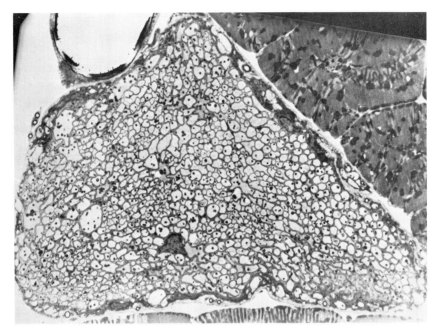

Fig. 47. Cross section of the labial nerve in the labellum as viewed with the electronmicroscope.

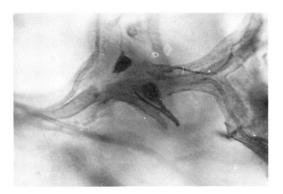

Fig. 48. Three of the presumed stretch receptors in a branch of the labial nerve from the oral papillae; stained with methylene blue.

The Etiquette of Eating: Mechanisms and Control of Ingestion

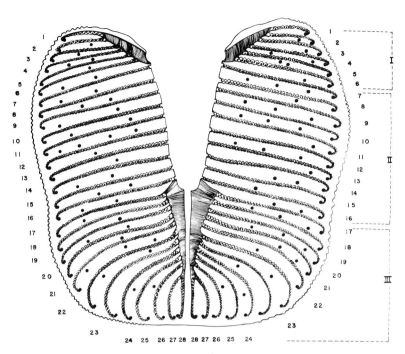

Fig. 49. Map showing the location of the interpseudotracheal papillae (from Wilczek, 1967).

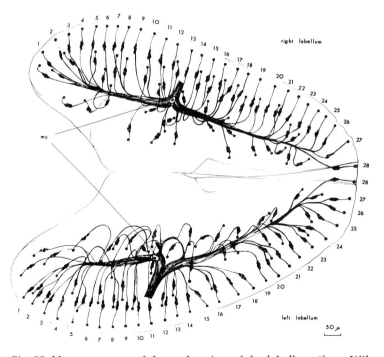

Fig. 50. Neuroanatomy of the oral region of the labellum (from Wilczek, 1967).

b; Larsen and Dethier, 1965; Peters and Richter, 1965; Tominga, Kabuta, and Kuwabara, 1969) (Figs. 51 and 52).

There may be still more gustatory receptors farther back in the pharyngeal region. Arab (1957) found that the fly was still capable of some descrimination even after all labellar receptors had been removed. A search with the scanning electonmicroscope has so far failed to reveal any likely candidates, but no final conclusion can be drawn. Other hairs sensitive to stimulation by sodium chloride have been found on the costal margin of the wings (Wolbarsht and Dethier, 1958) and on the ovipositor (Wolbarsht and Dethier, 1958; Wallis, 1962). They are not involved in feeding behavior.

Finally, mention should be made of the olfactory receptors even though they play a subordinate role in feeling behavior. The principal site of olfactory receptors in flies, as in the majority of insects is the antennae. In the absence or impairment of the antennae, olfactory acuity is drastically reduced. Experiments involving ablation and threshold measurements have proved that the antennae of *Phormia* mediate responses to normal aliphatic alcohols and aldehydes and a variety of natural products (Dethier, 1952b, 1954, 1961; Dethier, Hackley, and Wagner-Jauregg, 1952). Action potentials elicited by 0.01 M aqueous solutions of NH_4Cl and $NaCl$ have been recorded from some of the antennal sensilla (Wolbarsht and Dethier, 1958). Slow potentials representing summed receptor potentials have been recorded from an antennal pit of *Lucilia sericata* in response to stimulation with homologous alkyl thiols from C_2 to C_6, hexane, cyclohexane, cyclo-octane, methanol, geraniol, ethyl acetate, and *d*-camphor (Kay, Eichner, and Gelvin, 1967). More precise experiments utilizing more relevant stimuli have been reported briefly by Boeckh, Kaissling, and Schneider (1965). These workers found two types of pegs (sensilla basiconica), odor specialists, on the surface of the antennae of male and female *Calliphora erythrocephala*. They are odor specialists in that each responds only to a few special odors. One responded with a low threshold to meat, carrion, and cheese and to some alcohols, aldehydes, and mercaptans. It was inhibited by propionic butyric, and valeric acids and styrolyl acetate. The other was excited by styrolyl acetate and inhibited by carrion.

There are also odor specialists in pits on the antennae (Kaib, 1974). Large pits are populated with longitudinally grooved sensilla that are definitely olfactory. Small pits contain sensilla basiconica that are presumed to be olfactory. Receptors that respond to meaty odors are insensitive to flowery odors, and vice versa. The two kinds also differ in their responses to series of pure compounds; that is, they have different reaction spectra. Furthermore, on the basis of their reaction spectra, the receptors sensitive to meaty odors may be subdivided into

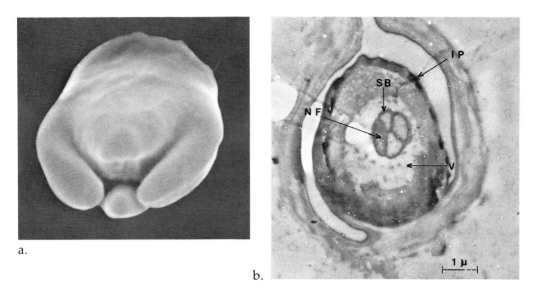

Fig. 51. (a) Interpseudotracheal papilla (×18,000). (b) Cross section through an interpseudotracheal papilla in the region of the scolopoid sheath (from Larsen, 1962b. Copyright by the American Association for the Advancement of Science).

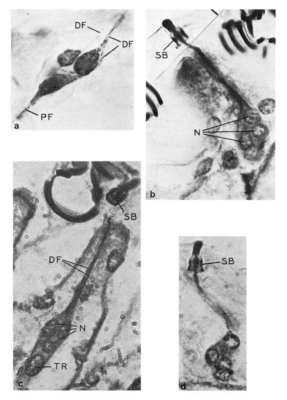

Fig. 52. Structure of the oral papillae. (a) Methylene blue preparation showing three of the four receptor cells; (b)–(d) silver preparation. N, nuclei of neurons; SB, sensillum basiconicum; DF, dendrites; PF, axons; TR, trichogen.

six different types and the receptors sensitive to flowery odors into three different types.

Although some of the morphological types of olfactory receptors have been identified, the identity of all the olfactory receptors from among the thousands present is not known. In addition to the sensilla basiconica found by Boeckh and Kaissling (1965) and Kaib (1974) there are many candidates for the role, as the histological studies of Lieberman (1926) and more recent electronmicroscopic examinations show.

The antennal sensilla of *Phormia* are massed on the large, fleshy, apical segment (flagellum), which also bears a conspicuous plumose arista, a sort of antennal panache. On the dorsal, ventral, and inner lateral areas of the distal half of the segment are groups of pits, 9 to 11 in males, 11 to 16 in females. Some are simple pocket-like cavities; others are subdivided into chambers. The entrances to pits are protected by rings of short, stout spines. These serve, no doubt, to prevent the entrance of debris. In all there are six morphologically distinct types of sensilla, three in the pits, three on the general antennal surface (Dethier, Larsen, and Adams, 1963). The pit sensilla, of which there is only one type per pit, are: thin-walled pegs, thick-walled pegs, and coronal pegs. The surface sensilla are: thin-walled pegs, thick-walled pegs, and stellate pegs (Fig. 53).

Coronal pegs, so called because in cross-section they appear as nearly perfect crowns with projections usually numbering twelve, are beautifully ornate structures. Longitudinal sections show that the projections are extensions of the peg wall. The projections do not, as originally believed, proceed outward at right angles to the long axis of the peg and then bend down to hang at right angles to the axis. They are in fact ridges between deep longitudinal grooves (Slifer and Sekhon, 1964). Coronal pegs should therefore be renamed "grooved" pegs. The stellate pegs, star-shaped in cross section, are also longitudinally fluted structures. Like the coronal pegs, they are few in number and the least likely candidates for olfaction of all the sensilla present. The thin-walled pegs, characteristic of the more proximal pits, are delicate structures with rounded tips averaging 13 μ in length and 1.5 to 1.8 μ in diameter. The wall at its thinnest point averages 0.08 μ. They will serve to illustrate the major features of olfactory sensilla because the thick-walled pit pegs, the thin-walled surface pegs, and the thick-walled surface pegs differ only in number, size, and minor details. Each peg is innervated by one, or sometimes two (four in the case of thick-walled surface pegs) bipolar neurons, the cell bodies of which lie 15 μ or more below the base of the peg. As the dendrites leave the soma, each one has the ciliary structure typical of sense cells. As Slifer and Sekhon (1964) have shown in their study of the related fly *Sarcophaga argyrostoma* R. -D., the cilium arises from a basal body that has rootlets

The Etiquette of Eating: Mechanisms and Control of Ingestion

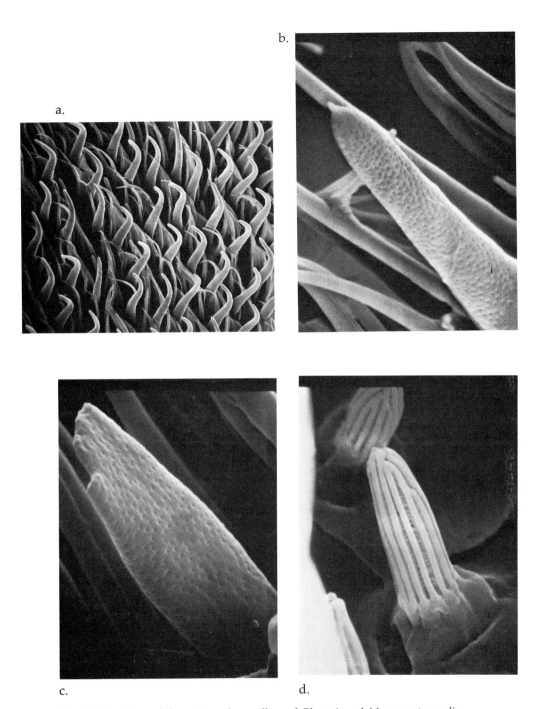

Fig. 53. Structure of the antennal sensilla. a-d *Phormia*, e-l *Musca autumnalis* De Geer (the face fly). (a) General surface of antenna showing mixed population of sensilla. (b) Surface thin-walled sensillum basiconicum showing pits in surface (×8,200). (c) Leaf-like sensillum basiconicum on surface (×12,600). (d) Coronal sensillum in pit (×12,600). (e) Small pit with sensilla basiconica

(Continued on next page.)

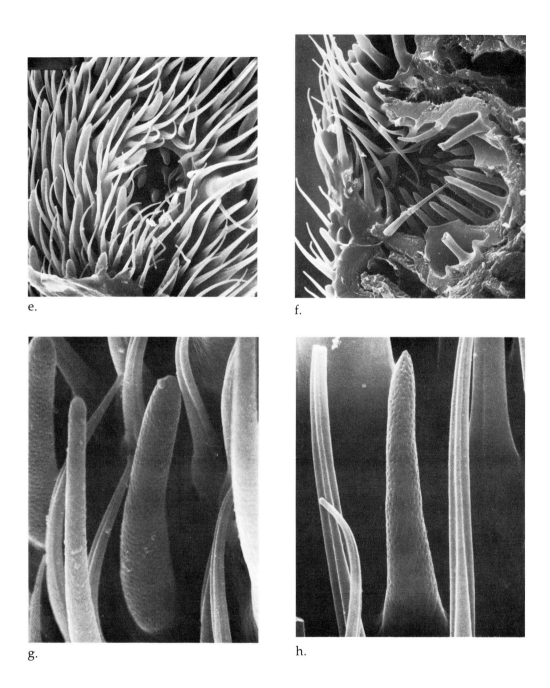

(*Fig. 53 continued*)
(×1,850). (f) Section through a large pit within thin-walled sensilla basiconica (×1,640). (g) Surface sensillum basiconicum, type 1, in center and one sensillum trichodeum to the left (×6,650). (h) Surface thin-walled sensillum basiconicum showing pits in surface (×6,850). (i) Surface thin-walled

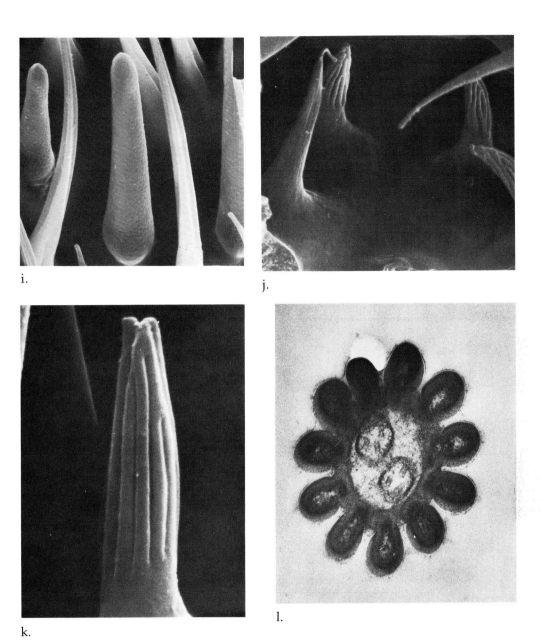

sensillum basiconicum, type 1 center, type 2 left (×5,760). (j) Three stellate (similar to coronal sensilla of *Phormia*) sensilla and one sensillum with a filamentous tip in a pit adjacent to the arista on the dorsolateral surface (×2,320). (k) Surface stellate sensillum (×18,280). (l) Cross section of stellate sensillum (equals coronal sensillum of *Phormia*) containing two unbranched dendrites (×37,420) (courtesy of D. E. Bay and C. W. Pitts).

with a periodic structure. The basal body in cross-section is seen to possess nine sets of triplet tubules peripherally situated. Two tubules from each set continue distally into the cilium to become nine sets of doublets. There is no central pair. Proceeding distally the ciliary structure gives way to a typical dendrite, which then enters a scolopoid sheath, which appears to fuse with the walls of the peg as it extends into it. When there is more than one dendrite, the sheath is compartmentalized to accommodate each separately. Within the peg the dendrites branch into many fine units, filling the lumen completely. The fine details of the interface between the dendrites have been carefully investigated by Slifer and Sekhon (1964, 1969a, 1969b) and Ernst (1969) in a variety of other insects. Each thin-walled pit peg has from 180 to 360 pits in its surface. These constitute from 7 to 14% of its surface. The orifice, 0.03 μ in diameter, opens into a larger basal chamber (0.088 μ). Here there are a number of filaments, each of which is less than 200 Å in diameter. Although these were originally believed to be the ultimate extensions of the dendrites, Ernst (1969) has shown that they are in fact cuticular canals. There exact relation to the dendrites remains obscure.

These, then are the chemosensory systems known to be concerned with feeding. Unfortunately, so little is known of the contribution of mechano- and proprioceptors that we shall have to omit them from the discussion until it is time to describe in detail the working of the cibarial pump.

Once knowledge of the location of gustatory receptors was acquired, a more meaningful description of ingestive behavior became possible. The degree of control that each receptor field exerts over the different components of the feeding pattern could also be assessed. The normal sequence of events as already described is: arresting of locomotion, turning toward the source of stimulation, extension of the proboscis, opening of the labellar lobes, and finally sucking. Stimulation of the tarsal receptors can arrest locomotion and effect turning and proboscis extension. It does not seem to be able by itself to initiate sucking. Stimulation of the aboral hairs of the labellum can stop locomotion, elicit proboscis extension, cause orientation of the labellum itself toward a stimulus applied from the side, and cause the lobes to open. Stimulation of the papillae initiates and drives sucking.

One of the most remarkable features of the gustatory receptors of the blowfly is that stimulation of a single sensory neuron in a labellar hair can elicit a complete and complex pattern of behavior. This means that nerve impulses ascend one axon of the labial nerve to the subesophageal ganglion, where they stimulate interneurons that eventually distribute messages to a minimum of five different sets of ipsilateral motor fibers and also cross over to five different sets of contralateral fibers (Figs. 37 and 38). Similarly, stimulation of a single sensory neuron in any of the legs causes impulses to ascend to the thoracic ganglion, and thence to the subesophageal ganglion where the same motor fibers as before are stimulated.

The different kinds of partial extension that weak stimuli can evoke indicate that different sets of muscles can be activated and to different degrees. In other words, the quantitative recruitment of motor fibers and the intensity of motor response are controlled in part by the intensity of sensory input; however, information from other sources must recruit the association neurons responsible for selecting those sets of muscles that are to be placed in operation in any given situation.

Muscles involved in sucking are serviced by fibers in the labral nerve. Thus, although it is possible that all fiber tracts utilized for proboscis extension lie entirely within the subesophageal ganglion, the act of sucking requires that afferent impulses from the interpseudotracheal papillae pas via the circumesophageal connectives (compare Fig. 131) to the brain. It would be interesting to separate the supra- and subesophageal ganglia surgically and observe whether indeed the supraesophageal ganglion is superfluous for muscular movements involving extension of the proboscis.

Since sucking is processed in the brain proper (supraesophageal ganglion), one might speculate that there is tighter control over sucking than that exercised over extension. From an adaptive point of view tighter control at this stage of feeding is obviously desirable. Certainly there is more than one sensory level at which solutions can be monitored. Unacceptable compounds that escape detection by the labellar hairs can be assessed again by the papillae. The sugar L-arabinose, for example, is acceptable insofar as labellar hairs are concerned but unacceptable from the point of view of the papillae. Finally there is involvement of hypothetical receptors farther along the alimentary canal. The experiments of Arab (1957) indicate that sucking is monitored when fluids reach the labral-epipharyngeal region. Sensilla here would route their fibers directly to the brain; consequently, at least spatially, very fine control should be possible. In addition, the cibarial pump is equipped with mechanoreceptors that modulate sucking. In view of Rice's (1970a) discovery of chemoreceptors in this region in the tsetse fly *Glossina austeni*, further examination in *Phormia* would be rewarding.

The act of ingestion is under tight sensory control, and there is no evidence that it can be initiated in the absence of chemosensory or mechanosensory input from peripheral receptors. Nor is it maintained for long when sensory input declines below some critical level or ceases completely. When this occurs, sucking stops, muscles holding the proboscis in the extended position relax, and the proboscis passively partially relaxes. By elastic recoil the labellar lobes close, the haustellum relaxes against the rostrum, and the rostrum becomes flaccid as the pneumatic pressure in the air sacs falls. Failure to maintain adequate sensory input may result from adaptation of the chemoreceptors, central adaptation, or inhibition from the stomatogastric nervous system.

Sucking ceases not only in the absence of sensory input but also as a consequence of stimulation of the papillae, labellar hairs, or tarsal hairs with compounds (e.g., salts, acids, etc.) that stimulate rejection receptors. Impulses from these receptors must inhibit centrally the initiation of potentials in the efferent fibers to the cibarial pump.

When powerful adverse stimuli are presented to the tarsal or labellar receptors, the proboscis can be retracted actively with great speed. Other responses to noxious stimuli will be considered later. This emergency type of response involves impulses routed to five sets of retractors and flexors. Present evidence seems to suggest that active retraction is all-or-none in nature, future work may show that in this case also there is a recruitment of muscles correlated with the intensity of stimulation.

All of the intricate behavior patterns described here suggest that a study of the relation between sensory input and motor output would be a rewarding endeavor. Rarely has electrophysiological activity in specific sensory cells been monitored simultaneously with resultant behavioral responses (Dethier, 1968). A start in this direction has been made by Getting (1971), Getting and Steinhardt (1972), Falk (personal communication) and Pollack (personal communication). Getting selected for initial study the sensory control of the extension of the haustellum because only two pairs of muscles are involved, the extensors of the haustellum and the adductors of the apodeme. Stimulation of a single sugar receptor and recording of its sensory output was accomplished by techniques described in the next chapter. Motor activity was recorded with an electrode placed under the proximal end of the apodeme and in contact with the extensor muscle.

Stimulation of a single labellar sugar receptor of a water-satiated fly starved for 72 hours results in motor activity in the extensor-adductor complex. A typical record (Fig. 54) depicts impulses arising from the sugar and the water receptor (C and W) and motor spikes from two units (A and B.). Whether A and B represent activity in two units of the extensor or one unit each from the extensor and adductor is not known. The record contains a wealth of information and by comparing it with similar records taken when the concentration of stimulating sugar was varied, it became possible to establish relationships between latency, interspike interval in the sensory nerve, and motor response. A single spike from the sugar receptor was never able to ini-

The Etiquette of Eating: Mechanisms and Control of Ingestion

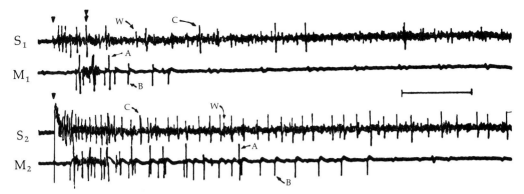

Fig. 54. Electrophysiological records of sucrose stimulation of a single labellar hair and the resultant motor activity in the extensor-adductor muscle complex of a water-satiated fly starved 72 hours. S_1 and M_1, receptor and motor activity, respectively, in response to 100 mM sucrose; S_2 and M_2, to 400 mM sucrose. C = sugar receptor; W = water receptor; A and B = two sizes of motor spike. ▼ = instant of stimulation; ↓ = last sugar spike for which the sensory interspike interval is less than 20 msec. Note cessation of activity in M_2 despite a constant ISI of < 20 msec. Time mark, 100 msec. (from Getting, 1971).

tiate a motor response; furthermore, two spikes could accomplish it only if the interval was less than 20 msec. The critical interval varied from one fly to the next, but an interval less that 15 msec. was almost invariably effective.

The delay or latency between the second sensory spike and the first motor spike was always the same irrespective of the interval between the two sensory spikes and of the number of sensory spikes preceding the motor response.

In flies starved for 62 to 66 hours, i.e., less hungry than before, the interval between the first two spikes had to be less than 5 msec. in order to evoke a motor response. An interval of more than 5 msec. (5 to 10) between the first two spikes was effective if a third spike followed. These results are in agreement with earlier behavioral studies, which showed that the threshold of response to sugar decreased more and more, the longer the fly was starved. They also agree with the finding that activity from the water receptor is ineffective in eliciting a response from a water-satiated fly (see Chapter 10).

Although Getting's experiments dealt only with extension of the middle third of the proboscis, they supply additional evidence that proboscis extension, the first act of ingestion, relies upon external sensory activity for its initiation. They further indicated that the duration and number of motor spikes, once started, are affected by the frequency of sensory firing (Fig. 55). There is a linear relation between motor and sensory activity until the motor response becomes maximal.

The duration of sensory input also affects the magnitude of

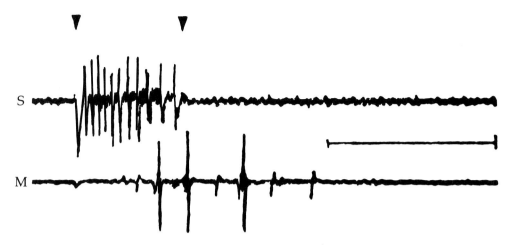

Fig. 55. Short 200 mM sucrose stimulation (S) of a single labellar hair and the resultant motor response (M) from a fly starved 65 hours. Arrows mark beginning and end of sensory stimulation. All ISI's are less than 20 msec. Motor response continues for 84 msec. after last sugar spike. Time mark, 100 msec. (from Getting, 1971).

response. Sugar receptors cease firing a few milliseconds after the sucrose solution is removed (Tateda and Morita, 1959), yet for a fly starved for 60 to 72 hours continuous stimulation by 0.2 M sucrose for only 50 msec. can trigger a motor response lasting 84 msec. (Fig. 56). Increased duration of sensory input can increase the duration of motor response, to a point. Eventually the duration of motor response reaches a maximum beyond which continued sensory input is ineffective. Since the interspike interval of the sensory receptor quickly attains a value greater than 20 msec. because of sensory adaptation, continued input cannot effect motor response. The critically short interspike interval can be maintained for a longer period of time by stimulating with more concentrated sugar; however, the motor response still terminates before the interval becomes too long. Termination in this instance must be due to adaptation farther along the line. In short, sensory input triggers motor response, and affects its magnitude and duration, but is neither a necessary nor a sufficient condition for its maintenance.

This sensory-motor preparation has been very useful for reinvestigating some of the phenomena, notably adaptation and summation, that were revealed and characterized by behavioral studies. It has confirmed many of the earlier conclusions and has been able to reach farther into the neural processes involved. For example, behavioral studies had indicated at least two levels of adaptation, one peripheral, the other central. The existence of peripheral adaptation was well

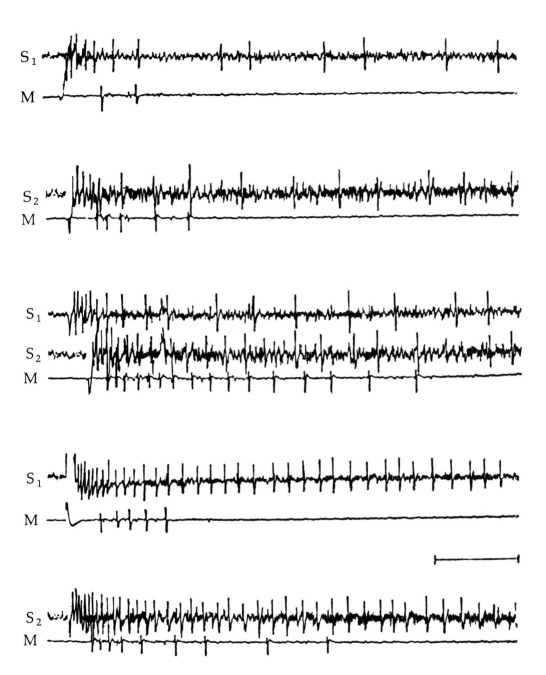

Fig. 56. Individual and simultaneous sucrose stimulation of two adjacent hairs and the resultant motor activity in a fly starved 73 hours. Top two traces (S_1 and M), 100 mM sucrose on hair 1. S_2 and M, 100 mM sucrose on hair 2; S_1, and S_2, simultaneous stimulation of both hairs. Stimulation of either hair 1 or hair 2 alone, to give a frequency equal to or greater than the frequencies during simultaneous stimulation yields a smaller motor response (bottom two sets of records). Time mark, 100 msec. (from Getting, 1971).

documented by many of the first electrophysiological studies (see Chapter 5). It occurs in two phases within 1 to 35 msec. Getting found, however, that the motor response still declined when stimulations were spaced one minute apart, giving time for the sugar receptors to undergo considerable disadaptation. When this happened, normal response could be reinstituted by stimulating a different hair. Electrophysiological monitoring ruled out sensory adaptation and motor fatigue as contributory causes in this case. Getting concluded, therefore, that habituation must have occurred at an interneuronal synapse in the central nervous system located at a point before the convergence of the axons from individual sensilla. Recovery of responsiveness at this point required a period of twenty minutes without stimulation.

The existence of spatial summation, that is, summation between receptor units, was also confirmed by the observation that the electrophysiological activity of motor units in response to stimulation of two sensilla exceeded that in response to either alone. The observation provided additional information: first, that dividing the sensory input between two channels was more effective than routing it through one; second, that summation was nonlinear, that is, the motor response to, say 0.1 M sucrose, applied to both sensilla simultaneously was greater than the sum of motor responses to 0.1 M sucrose applied to each sensillum individually (Fig. 55). Getting ruled out the possibility that the increased responsiveness of the motor units to stimulation of two sensilla might have been due partly to a central excitatory state. Prior stimulation by sucrose does not induce a change in central exicitory state insofar as the extensor-adductor muscles of the haustellum are concerned. In additional studies employing this preparation, Fredman and Steinhardt (1972) were unable to observe any change in motor output when salt was applied to one hair and sugar to another. They concluded that there was not central interaction, either summation or inhibition, between input from sugar receptors and that from salt receptors. Fredman (1975) repeated and extended these experiments and found that the original general conclusions were incorrect. Central interaction, both summation and inhibition, between input from water and sugar receptors and that from salt receptors was observed. The existence of central excitatory state was also confirmed, as was the inhibition of central excitatory state by input from salt receptors.

Another segment of the feeding response that is beginning to receive some attention is spreading of the labellar lobes. Recordings have been made from the retractors of the furca as well as from the motor nerves to these muscles (Pollack, personal communication). When only one chemosensory hair is stimulated by sucrose the concentration must exceed 2 M to be effective. When two hairs are stimulated, a 0.5 M solution is adequate. A 0.008 M solution is effective if applied to large numbers of hairs. Pollack has found that the increased effectiveness produced by stimulating more than one hair is not due simply to an increased number of sensory impulses that reach the central nervous system. For example, in one experiment simultaneous stimulation of largest hairs numbers 2, 3, and 4 with 0.125 M sucrose failed to effect spreading, whereas stimulation of numbers 2, 3, 4, and 5 with that concentration was effective. Stimulation of hair number 5 alone with 2 M sucrose was ineffective. Stimulation of hairs 2, 3, and 4 with 2 M sucrose elicited a total of 113 spikes in the first 500 msec.; stimulation of hairs 2, 3, 4, and 5 with 0.125 M sucrose elicited only 71 spikes in the same period of time. The latency of lobe spreading was less than 500 msec. Thus, the motor response is determined not solely by the number of spikes delivered to the central nervous system by the sensory system, but also by spatial summation. As judged by muscle potentials, increasing the concentration of the stimulus in one hair to a value above threshold leads to a more vigorous response.

Generally speaking, responses are ipsilateral. Only rarely does stimulation of hairs on one lobe elicit response from both lobes. Furthermore, observations of responses to electrical stimulation of nerves to the muscles indicate that summation of input to a muscle must occur before significant movement of the lobes is effected. Pollack has suggested that summation plus ipsilateral restrictions cooperate in producing fine control of lobe spreading and ensuring accurate positioning of the lobes with respect to a drop of food.

Much of what has been said about the regulation of proboscis extension may apply equally well to sucking. Less is known about the relation between activity in the chemosensory nerves and activity in the cibarial motor nerves. On the other hand, there is some knowledge of proprioceptive feed-back, a feature of which we are in total ignorance insofar as the labellum is concerned.

The details of sucking have been worked out experimentally by Rice (1970b) and Rice and Finlayson (1972) for the tsetse fly and the blowfly *Calliphora*. The system is basically two pistons operating out of phase. A ventral and a dorsal dilator muscle, short muscles that are attached to the roof of the cibarium and enlarge the lumen by contracting (Fig. 57), operate reciprocally. The labropharyngeal nerve, labelled the labral nerve by Dethier (1959), innervates these muscles and also two bilateral sets of multiterminal afferent fibers. The afferent neurons have numerous symmetrically arranged dendrites, desheathed at the tips, surrounded by connective tissue and set in the wall of the epithelium of the anterior pharyngeal wall. They are stretch receptors. Pumping, simulated or actual, evokes bursts of action potentials in these

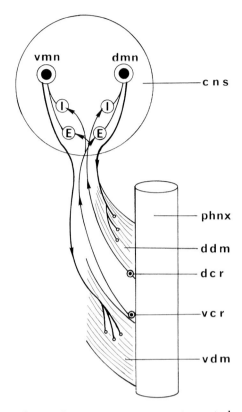

Fig. 57. Diagram of the control of cibarial pumping. vmn, motor nerve to ventral muscle; dmn, motor nerve to dorsal muscle; I, inhibitory interneuron; E, excitatory interneuron; vdm, ventral dilator muscle; ddm, dorsal dilator muscle; vcr, ventral cibarial stretch receptor; cns, central nervous system; phnx, pharynx (after Rice, 1970b).

neurons. The frequency of discharge of each is proportional to the extent of indentation of the wall and the point of stimulation. Stimulation of different areas elicits different patterns of firing.

The model (Fig. 57) illustrates how the cibarium works (Rice, 1970b). Sucking begins when the motor neuron (VMN) to the ventral muscle (vdm) is centrally activated. Its discharge causes dilator muscles vdm to contract. This contraction sucks food into the dilated ventral pharynx. Meanwhile the increased rate of firing of stretch receptor VCR inhibits the motor discharge (VMN) and excites the dorsal muscle neuron DMN so that as vdm relaxes ddm contracts, transferring food from the ventral to the dilated dorsal pharynx. Relaxation of ddm is ensured by the inhibitory influence of the dorsal stretch receptor (DCR), transferring food to the oesophagus. During this phase of the cycle food is prevented from flowing back down the pharynx because the dorsal plate of the fulcrum swings across the opening of the oesophagus and, simultaneously, the oesophagus is

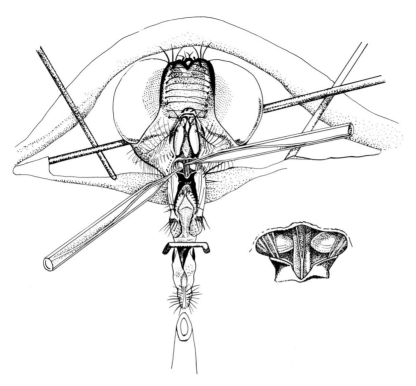

Fig. 58. Placement of electrodes for recording from the cibarial motor nerve. Inset, detail of nerve.

constricted. The excitatory influence of DCR on VMN starts the cycle again. For regurgitation DMN would be the first unit primed. Although this model does not completely explain pumping and its control, it could provide an automatic means of adjusting rate of pumping to the volume and viscosity of food and might also provide a means of monitoring the volume of food ingested. In *Rhodnius* the pump can regulate the size of a meal (Bennett-Clark, 1963). During the major part of a meal the pumping rate does not vary more than 20%; however, during the last 10 seconds it falls to less than one half of its maximum rate. The pump is stopped at a critical abdominal pressure (2.5 cm Hg) which prevents initiation of the emptying stroke (elastic recoil). The rate of feeding is limited by the rate at which the pump can fill, and the size of a meal depends upon the ease with which the abdominal cuticle can be stretched.

This description may adequately explain the mechanism in *Phormia*, but many details must still be elaborated. Some additional information has been provided by the experiments of Falk (personal communication), in which motor output has been monitored by recording from the cibarial motor nerve and extracellularly from single fibers of the cibarial muscle (Fig. 58). With each contraction of the cibarial

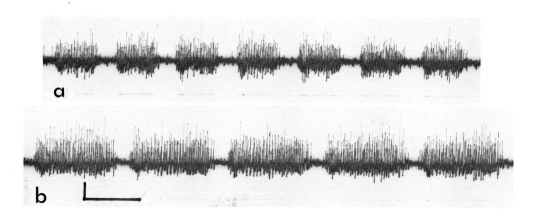

Fig. 59. Recordings from motor neurons supplying the cibarial dilators in an intact female fly. The pattern of bursts corresponds to the observed pattern of pumping. (a) A slightly viscous (approx. 1 centipoise) solution (0.1 M sucrose). (b) A more viscous (35 centipoises) solution (0.1 M sucrose in 3% methyl cellulose). 5 mm vertical = 100 μV. Horizontal bar = 100 msec. (courtesy of D. Falk).

muscle during pumping there is a burst of electrical activity lasting about 100 msec. The intervals between bursts last 25 msec. (Fig. 59). When the viscosity of the fluid imbibed is increased from 1 to 200 centipoises the duration of a burst doubles but the volume of fluid sucked decreases. As a consequence of these relations there is an eighteen-fold increase in the pressure exerted by the pump. The duration of bursts and cycles is also affected by the intensity of the chemical stimulus initiating the response, but the effect is much smaller than that exerted by viscosity.

When the cibarial nerve is cut so that there is no feedback from cibarial stretch receptors, and the labellar hairs are stimulated with sugar, the motor bursts recorded from the proximal end of the nerve are grossly similar to those obtained in an intact fly. They tend to be irregular in duration, and, of course, they do not adjust to differences in viscosity (Fig. 59). These experiments suggest that sensory input from labellar chemoreceptors is necessary to trigger and maintain sucking, that the motor activity effecting sucking is a centrally generated rhythmic pattern, and that the mechanosensory feedback is necessary for normal feeding. The mechanical stimulation in the cibarium is necessary to maintain continuous feeding, maintain the rhythmicity of bursts, change the frequency of bursting as a function of load, change the volume imbibed per burst as a function of load, change the pressure exerted as a function of load, and vary the spike frequency per burst as a function of load. Experience plays no role in cibarial behavior. Flies that had never fed behaved no differently from those that had.

One cannot help wondering why there are so many gustatory receptors if a single receptor can provide enough information to control segments of the feeding sequence. It is obviously wise to have spare sense organs in the event that some get destroyed or frayed beyond use in the business of living and aging. And there is indeed evidence that wear and tear exact their price and that receptors fail with age (Rees, 1970; Stoffolano, 1973). However, there are other advantages, not to mention necessities, in having receptor fields rather than solitary or few receptors. Multiple receptors allow for interactions that enrich the flow of information to the central nervous system. Summation, adaptation, and inhibition are among the interactions of importance. Much of what we know about them was initially learned from behavioral studies (Arab, 1959; Dethier, 1950, 1952a, 1952b, 1953, 1955) and subsequently augmented and refined by electrophysiology.

Although stimulation of one labellar hair will trigger extension of the proboscis, a very high concentration of sugar is required. A much lower strength of stimulus suffices if more than one hair is stimulated simultaneously. An accurate measurement of the exact concentrations of sucrose required to elicit proboscis extension showed that the median threshold with one largest labellar hair stimulated was 0.419 M; with two hairs, 0.0622 M; with three hairs, 0.0322M; with four hairs, 0.0127 M; with the entire labellum, 0.0164 M (Arab, 1959). The difference between one and two was statistically significant, as was the difference between two and four. The difference between four and all hairs was not significant. A probability analysis of the difference between one and two hairs when the two are adjacent revealed that the difference was greater than could be accounted for on a probability basis alone; therefore, it is likely that true neural summation rather than supplementation does occur with neighboring hairs. The same result was obtained when the comparison was between one hair and that hair plus another one located three hairs away. Summation between hairs more distantly separated or between hairs on opposite sides of the labellum was not investigated. Comparisons of this nature would be exceptionally interesting. It is unlikely that summation is the same throughout the entire labellum because the fly is able to localize accurately a point source in the labellum. Stimulation of hairs on one lobe results in an asymmetrical extension that rotates the proboscis toward the side stimulated. If hairs on the anterior surface are stimulated, the labellum is canted forward; if stimulation is posterior, the cant is backward. Also, Wilczek's (1967) neuroanatomical map showing that the axons from the hairs are collected together in special combinations suggests that there may be some meaningful order in central connections. This bears investigation.

Also bearing on the question of localization within a receptor field

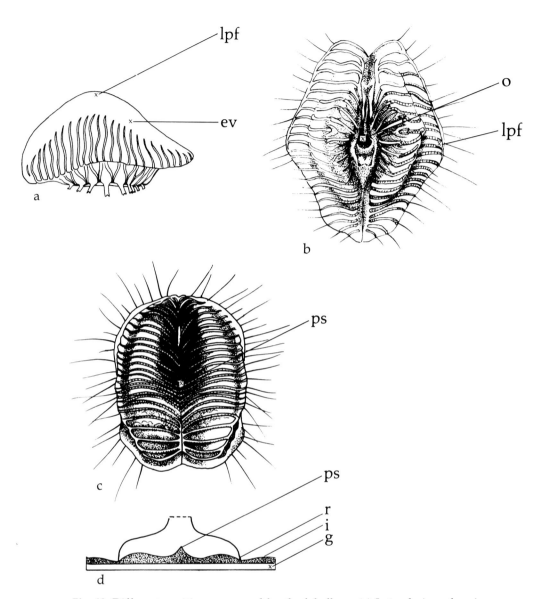

Fig. 60. Different positions assumed by the labellum: (a) Lateral view showing scraping position; (b) distal view showing direct-feeding position; (c) oral surface in cupping position; (d) diagrammatic transverse section of oral disc in the cupping position. ev, everted labellum; lpf, distal extremity of lateral process of the furca; O, oral aperture; ps, prestomal sulcus; r, rim of labellum; i, layer of India ink and syrup; g, glass plate (after Graham-Smith, 1930).

is the work of Graham-Smith (1930), who described in *Calliphora* the various attitudes that the labellum can take with respect to the substrate. Remarkably effective adjustments can be made to accommodate to the physical and textural features of a surface (Fig. 60). Flies allowed to feed on thin drying solutions of sugar containing finely ground

India ink arrange the labellum so that particles of ink can be filtered through a set of "teeth" located around the oral orifice. This is the filtering position. If the fluid is deeper, a cupping position is adopted. When presented with solid food, the fly more or less completely everts the labellum so that the teeth are brought to bear on the surface. After scraping the surface, the fly moistens the scrapings with saliva and then assumes the filtering position. With semisolid materials (e.g., feces) a direct feeding position is employed. In this position the labellum is so extremely everted that even the teeth are turned outward. The orifice is completely exposed and applied directly to the food. Even with this adjustment, liquids of high viscosity (7.74 to 48.5 centipoises) impose an impediment to feeding, resulting in a reduction in the volume of fluid imbibed per unit time. All of these fine adjustments to the physical features of surfaces indicate that the mechanoreceptors in the labellar hairs, as well as the various proprioceptors within the labellum, transmit a wealth of information to the central nervous system. The coordinated receptor field consisting of many units is much more versatile than a limited collection of receptors.

The intriguing feature of field activity is that it is not characterized by uniform redundancy, yet can permit summation. There can also be interactions between entire fields that have the same effect as summation. This can be shown by comparing the fly's threshold of response when one leg is stimulated, with that following stimulation of two (Dethier, 1953). The fly's acceptance threshold is lower when both forelegs are stimulated (Fig. 61). This result is similar to those obtained when thresholds for monocular and binocular vision in man were compared (Pirenne, 1943; Bárány, 1946a, 1946b). As was pointed out in those studies, the experimental procedure by its very nature ensures that two eyes see more clearly than one. The same source of bias is inherent in the procedure as applied to the determination of thresholds in other sense modalities (Smith and Licklider, 1949). Is there true neural summation or can the lowering of threshold be accounted for on the basis of bias introduced by the experimental procedure?

Let us assume that x number of sugar receptors on the legs of the fly must be activated in order to ensure extension of the proboscis. This is assuming that there is intraleg summation. Let us also assume that there are n available receptors on a leg. Then we ascertain by experiment which concentration of sucrose applied to one leg will elicit proboscis response from 50% of a sample of flies. This is the median acceptance threshold, and at this concentration 50% of the flies must have x or more receptors acting. The probability that a fly ($= x$ or more active receptors) will be in the half of the population that is responding is 50%. This probability can be increased either by increasing the concentration of the stimulus or by increasing the number of avail-

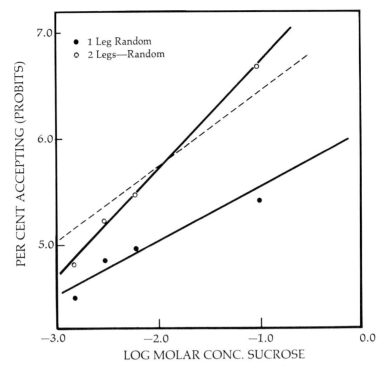

Fig. 61. Comparison of the distribution of acceptance thresholds for sucrose, as a function of concentration, for flies stimulated unilaterally or bilaterally. The broken line represents the theoretical distribution of bilateral thresholds (two-legged flies) calculated from the expression $1 - q^2$ where q equals the fraction of the population of one-legged flies not responding (from Dethier, 1953).

able receptors (n). If n is doubled, the probability of a response is increased. The increase can be calculated from the expression $1 - q^2$, where q is equal to the fraction of receptors not reacting at the median concentration. This being the case, the concentration of sucrose that elicits a response from 50% of the one-legged flies should elicit a response from 75% of two-legged flies if there is no interaction between the two legs. This expectation is realized. The response of the two-legged flies is not greater, except at the highest concentrations, than would be predicted on a probability basis from the behavior of one-legged flies. The difference between the expected line (calculated from the line for one-legged individuals from the expression $1 - q^2$) and the line describing experimentally-determined two-legged thresholds is not significant (Fig. 61). Thus the data do not constitute evidence for true contralateral neural summation but can be accounted for on a probability basis—the supplementation of Smith and Licklider (1949). Nevertheless, from the point of view of the fly the results represent a behavioral summation of no little importance. Similar results were obtained in studies of olfactory responses (Dethier, 1952b).

Another advantage of large numbers of receptors is the possibility that if not all adapt simultaneously they may ensure that the organism will receive a longer continued flow of information. This would provide one means of retaining sensitivity without fatiguing individual receptors. At the same time, the organism could obtain relief from a surfeit of information by providing for adaptation at higher levels. The existence of several levels of adaptation was first demonstrated by behavioral measurements of acceptance thresholds (Dethier, 1952a). The measurement of thresholds was a powerful technique before the advent of electrophysiology, and in the hands of skilled investigators yielded up many secrets of the insect sensory system. The employment of thresholds for the study of summation has already been discussed. In order to appreciate how it was used to investigate adaptation, a more detailed description is necessary.

Various techniques for measuring thresholds were employed over the years, but all of them involved ingestion, and sensitivity changed as the hunger state of the animal changed. Thresholds were calculated from observations of the initial acceptance of rejection of solutions, from measurements of crop loads at the conclusion of feeding (von Frisch, 1934), from the duration of feeding (Kunze, 1927; Weis, 1930), and from a method of mixtures. The last mentioned was developed by von Frisch (1930, 1934) for application to honeybees. Sugars that showed no stimulating effectiveness when tested alone were mixed with a concentration of sucrose to which only a small percent of all bees tested responded. If acceptance of the mixture exceeded that of the sucrose alone, a value of stimulating effectiveness (threshold) could be calculated. This technique did not take into account the effect of synergism, which had not yet been demonstrated. The technique that has yielded the lowest behavioral threshold values is the two-bottle choice test, but it too is not free of the influences of changing nutritional states.

Minnich's (1921) discovery of tarsal sensitivity introduced a new and powerful technique for measuring and comparing the stimulating effectiveness of many compounds. Since animals were prevented from ingesting test solutions, the hunger state could be held constant. Admittedly, the technique was valueless insofar as oral sensitivity was concerned; however, with it a great deal was learned about tarsal receptors that was applicable, as subsequent events proved, to oral receptors (Minnich, 1931). Among the phenomena investigated were summation and adaptation.

The standard technique for measuring acceptance thresholds consists in preparing in petri dishes a series of solutions made up in doubling concentration steps. A sample of 200 flies of both sexes are standardized with respect to age, state of food, and water deprivation. For most tests the flies employed are two and one half days old; they are given 0.1 M sucrose ad libitum until 24 hours before testing, are attached by the wings to wax-tipped sticks 24 hours before testing in order to accustom them to restraint, and are finally given water to repletion immediately before testing. If the compounds to be used are volatile, the olfactory receptors are removed at the time of mounting by amputating the antennae and terminal segments of the proboscis. Preliminary experiments had shown that this operation does not affect behavioral responses to tarsal stimulation (Dethier and Chadwick, 1947).

A test consists of lowering the feet of a fly gently onto the surface of water. If the proboscis is extended, the fly is allowed to drink (the surgery did not impede drinking) until no further extension of the proboscis occurs. The fly is then lifted and lowered over the dish containing the lowest concentration of the compound being used. Many tests showed that it was not necessary to wash and wipe the legs after each stimulation. The fly is tested on each successively higher concentration until extension of the proboscis occurs. Although threshold lies somewhere between the concentration that elicits extension and the one in the series immediately below it, the higher of the two is arbitrarily designated as threshold. Since the aim of practically all experiments was to obtain data for comparative analysis, this fiction was acceptable.

Analysis of second and third tests with 17,000 individual flies have revealed that 80% responded to the same concentration each time. Nearly all of the rest were within two and one half doubling-concentration steps. Two other levels of variability were encountered: interfly variability and intersample variability. The distribution of thresholds within a sample followed a normal distribution with respect to the logarithm of concentration. Intersample variation was controlled, but the underlying causes were never satisfactorily explained.

Presentation of solutions in increasing order of concentrations is only one of several possible procedures. A descending order can be presented and that concentration noted at which flies retract the proboscis; or solutions can be presented in random order. In the latter instance a different sample of 20 flies is exposed to each concentration. Each fly is stimulated only once in this procedure. The data from all three procedures can be treated similarly and compared. Not unexpectedly, in retrospect, each procedure yields a different median acceptance threshold (Fig. 62). Random testing yields the lowest, and descending testing gives a threshold determined by the starting point in the series of concentrations. In other words, individuals subjected

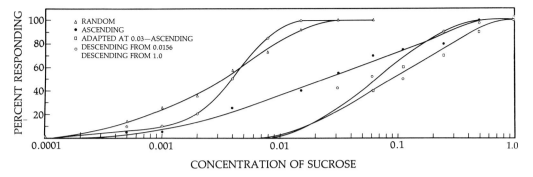

Fig. 62. Change in acceptance threshold to sucrose brought about by altering the order of presentation of different concentrations (from Dethier, 1952a).

to ascending or descending testing fail to respond to solutions that are known to evoke responses if presented directly after water.

Obviously the process of gradually increasing the intensity of the stimulus by small increments from a subliminal value results in a decrease in sensitivity. The site of adaptation could be peripheral, central, or both. In an attempt to separate the alternatives, Dethier (1952a) adapted one prothoracic leg of the fly and tested the contralateral one. Although the results did not rule out the possibility of peripheral adaptation, they indicated that adaptation also occurs centrally. It now appears that there can be at least three levels of adaptation: (1) sensory adaptation in the primary neuron; (2) adaptation (habituation) at the first interneural synapse before the primary neurons converge, as Getting has shown; and (3) a deeper central adaptation whereby the fly becomes refractory (or habituated).

It has already been mentioned that individual labellar hairs can be adapted and that the phenomenon is peripheral. More will be said about this in the following chapter. Arab (1959) adapted simultaneously a large number of hairs on one lobe of the labellum and could still elicit a proboscis extension when a few unadapted hairs on the opposite lobe were stimulated. He argued from this that if central adaptation did in fact occur, massive stimulation of one lobe should have produced it and subsequent stimulation of a few unadapted hairs should not have been able to override it. His conclusion was that central adaptation was not demonstrable *with labellar hairs* and that adaptation is peripheral. There is, of course, no doubt that peripheral adaptation occurs and occurs rapidly. On the other hand, the establishment of central adaptation is not easy. Long periods of stimulation are required. Under appropriate conditions, repeated stimulation of the labellum over a period of 3 to 5 minutes will result finally in failure to respond. This refractoriness persists beyond the period required for disadaptation of the receptors themselves.

The time course of peripheral adaptation varies with the duration of stimulation and the intensity of the stimulus. For example, flies were exposed to 0.001 M sucrose (to which there is no proboscis response) for different periods of time and then tested for threshold by the random procedure. At zero time (no previous stimulation) the threshold was 0.00356 M; at 1 second, 0.0116 M; at 10 seconds, 0.0168 M. No further increase could be demonstrated up to and including 30 seconds. It would appear that for 0.001 M sucrose adaptation was complete after one second. At different concentrations the time necessary for complete adaptation, as measured by time to retraction of proboscis, increases as the logarithm of the adapting concentration (Fig. 63).

Even though it has not been possible by behavioral methods to construct adaptation curves for *Phormia*, the fact that a correlation with adapting concentration can be demonstrated permits one to compare certain aspects of the phenomenon in *Phormia* and in man. The forms of curves of man's adaptation to a variety of substances at different concentrations differ considerably, but two facts are worthy of mention: adaptation is complete in anywhere from 10 to 30 seconds; the time required for total adaptation increases as the concentration is increased (Hahn, 1933; Hahn, Kuckulies, and Bissar, 1940). The situation in *Phormia* is remarkably similar. Adaptation times vary from 1 to 13 seconds; the total time required increases as the concentration is increased. Over the range of lower concentrations, 0.003125 to 0.05 M, the time required is proportional to the logarithm of molar concentration. From 0.05 M on there is no significant change in time with increasing concentration.

Continuous stimulation at constant intensity for long periods of time results in bursts of activity. For example, a fly with its feet exposed to 0.05 M sucrose will continue to extend its proboscis for a period of 13 seconds on the average. Thereupon, the proboscis is retracted. After a varying period of time, extension recommences; then it ceases again. Such rhythmic activity has been observed over periods of continuous stimulation lasting as long as 3 hours. No correlation between frequency of activity cycles and concentration has ever been observed. These cyclic responses could be a consequence of proprioceptive or tactile input related to "discomfort," to injury arising from deleterious effects of long immersion, or to normal chemosensory input as different sugar receptors adapt and disadapt out of phase, especially if the proboscis moves in a way to plunge hairs in and out of the solution.

As a final point, one other possible advantage of large populations of receptors may be mentioned. The precision of individual sensory neurons is very modest; that is, there is considerable variance in in-

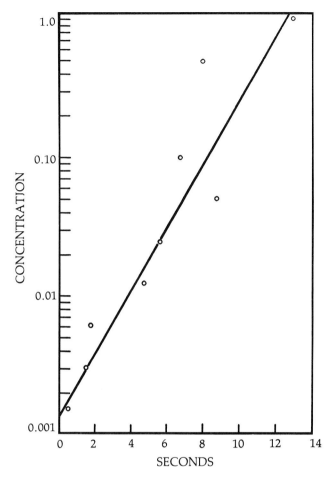

Fig. 63. The relation between the concentration of sugar employed as a stimulus and the time required for the cessation of response (from Dethier, 1952a).

terspike intervals. As Knight (1972a, 1972b) has pointed out, precise results arise from the functioning of a collection of components that have considerable noise in them. As an example, the perfect pitch that some persons possess comes from a population of neurons that individually exhibit very ragged firing patterns. In most systems—the vertebrate visual, auditory, and olfactory, for example, and the visual system of the horseshoe crab, *Limulus*—there are successive convergences in the neural circuitry giving rise to neural responses at successive levels of abstraction. In short, imperfections in individual units can be compensated for by having a large number of such units; for a detailed discussion, consult Knight (1972a, 1972b).

These considerations raise many questions concerning individual receptor units and it is now time to inquire more carefully into the character and physiology of these units.

Literature Cited

Abbott, C. E. 1928. The tarsal chemical sense of the screwworm fly, *Cochliomyia macellaria* Fab. *Psyche* 35:201–204.

Abbott, C. E. 1932. The proboscis response of insects, with special reference to blowflies. *Ann. Ent. Soc. Amer.* 25:241–244.

Adams, J. R. 1961. The location and histology of the contact chemoreceptors of the stable fly, *Stomoxys calcitrans* L. Ph. D. dissertation, Rutgers University, New Brunswick, N.J.

Adams, J. R., Holbert, P. E., and Forgash, A. J. 1965. Electronmicroscopy of the contact chemoreceptors of the stable fly, *Stomoxys calcitrans* (Diptera: Muscidae). *Ann. Ent. Soc. Amer.* 58:909–917.

Arab, Y. M. 1957. A study of some aspects of contact chemoreception in the blowfly. Ph. D. dissertation, The Johns Hopkins University, Baltimore, Md.

Arab, Y. M. 1959. Some chemosensory mechanisms in the blowfly. *Bull. Coll. Sci.*, University of Baghdad, 4:77–85.

Bárány, E. 1946a. A theory of binocular visual acuity and an analysis of the variability of visual acuity. *Acta Ophthalmol.* 24:63–92.

Bárány, E. 1946b. Some statistical observations on the methods in threshold determination in general with particular regard to determination of visual acuity and subliminal addition. *Acta Ophthalmol.* 24:113–127.

Barrows, W. M. 1907. The reactions of the pomace fly, *Drosophila ampelophila* Loew, to odorous substances. *J. Exp. Zool.* 4:515–537.

Bennett-Clark, H. C. 1963. The control of meal size in the blood-sucking bug, *Rhodnius prolixus*. *J. Exp. Biol.* 40:741–750.

Boeckh, J., Kaissling, K.-E., and Schneider, D. 1965. Insect olfactory receptors. Cold Spring Harbor Symp. Quant. Biol., 30:263–280.

Crow, S. 1932. The sensitivity of the legs of certain Calliphoridae to saccharose. *Physiol. Zool.* 5:16–35.

Deonier, C. C. 1938a. The gustatory nature of the chemotarsal stimulations in the housefly, *Musca domestica* L. *J. Exp. Zool.* 79:489–500.

Deonier, C. C. 1938b. Effects of some common poisons in sucrose solutions on the chemoreceptors of the housefly, *Musca domestica* L. *J. Econ. Ent.* 31:742–745.

Deonier, C. C. 1939a. Effects of toxic compounds on the gustatory chemoreceptors in certain Diptera. *Iowa State College J. Sci.* 14:22–23.

Deonier, C. C. 1939b. Responses of the blowflies *Cochliomyia americana* C. and P. and *Phormia regina* Meigen, to stimulation of the tarsal receptors. *Ann. Ent. Soc. Amer.* 32:526–532.

Deonier, C. C., and Richardson, C. H. 1935. The tarsal chemoreceptor response of the housefly, *Musca domestica* L., to sucrose and levulose. *Ann. Ent. Soc. Amer.* 28:467–474.

Dethier, V. G. 1950. Central summation following contralateral stimulation of tarsal chemoreceptors. *Proceedings, Fed. Amer. Soc. Exp. Biol.* 9:31–32.

Dethier, V. G. 1952a. Adaptation to chemical stimulation of the tarsal receptors of the blowfly. *Biol. Bull.* 103:178–189.

Dethier, V. G. 1952b. The relation between olfactory response and receptor population in the blowfly. *Biol. Bull.* 102:111–117.

Dethier, V. G. 1953. Summation and inhibition following contralateral stimulation of the tarsal chemoreceptors of the blowfly. *Biol. Bull.* 105:257–268.

Dethier, V. G. 1954. Olfactory responses of blowflies to aliphatic aldehydes. *J. Gen. Physiol.* 37:743–751.

Dethier, V. G. 1955. The physiology and histology of the contact chemoreceptors of the blowfly. *Quart. Rev. Biol.* 30:348–371.

Dethier, V. G. 1959. The nerves and muscles of the proboscis of the blowfly *Phormia regina* Meigen in relation to feeding responses. *Smithson. Misc. Coll.* 137:157–174.

Dethier, V. G. 1961. The role of olfaction in alcohol ingestion by the blowfly. *J. Insect Physiol.* 6:222–230.

Dethier, V. G. 1968. Chemosensory input and taste discrimination in the blowfly. *Science* 161:389–391.

Dethier, V. G. 1971. A surfeit of stimuli: a paucity of receptors. *Amer. Sci.* 59:706–715.

Dethier, V. G., and Chadwick, L. E. 1947. Rejection thresholds of the blowfly for a series of aliphatic alcohols. *J. Gen. Physiol.* 30:247–253.

Dethier, V. G., and Chadwick, L. E. 1948. Chemoreception in insects. *Physiol. Rev.* 28:220–254.

Dethier, V. G., Hackley, B. E., and Wagner-Jauregg, T. 1952. Attraction of flies by iso-valeraldehyde. *Science* 115:141–142.

Dethier, V. G., Larsen, J. R., and Adams, J. R. 1963. The fine structure of the olfactory receptors of the blowfly. In *Olfaction and Taste* I, Y. Zotterman, ed., pp. 105–114. Pergamon Press, Oxford.

Eltringham, H. 1933. On the tarsal sense organs of Lepidoptera. *Trans. Roy. Ent. Soc. London* 81:33–36.

Ernst, K-D. 1969. Die Feinstruktur von Riechsensillen auf der Antenne des Aaskäfers *Necrophorus* (Coleoptera). *Zeit. Zellforsch. Mikroskop. Anat.* 94:72–102.

Fredman, S. M. 1975. Peripheral and central interactions between sugar, water, and salt receptors of the blowfly, *Phormia regina*. *J. Insect Physiol.* 21:265–280.

Fredman, S. M., and Steinhardt, R. A. 1972. Mechanism of inhibiting action by salts in the feeding behavior of the blowfly, *Phormia regina*. *J. Insect Physiol.* 19:781–790.

Frisch, K. von 1930. Versuche über den Geschmackssinn der Bienen. *Naturwiss.* 18:169–174.

Frisch, K. von 1934. Über den Geschmackssinn der Bienen. *Zeit. vergl. Physiol.* 21:1–156.

Getting, P. A. 1971. The sensory control of motor output in fly proboscis extension. *Zeit. vergl. Physiol.* 74:103–120.

Getting, P. A., and Steinhardt, R. A. 1972. The interaction of external and internal recptors on the feeding behaviour of the blowfly, *Phormia regina*. *J. Insect Physiol.* 18:1673–1681.

Grabowski, C. T., and Dethier, V. G. 1954. The structure of the tarsal chemoreceptors of the blowfly, *Phormia regina* Meigen, *J. Morph.* 94:1–20.

Graham-Smith, G. S. 1930. Further observations on the anatomy and function of the proboscis of the blow-fly, *Calliphora erythrocephala* L. *Parasitology* 22:47–115.

Hahn, M. 1933. Die Adaptation des Geschmackssinnes. *Zeit. vergl. Physiol.* 53:169–177.

Hahn, H., Kuckulies, G., and Bissar, A. 1940. Eine systematische Untersuchung der Geschmacksschwellen. *Zeit. Sinnesphysiol.* 68:185–260.

Haslinger, F. 1935. Über den Geschmackssinn von *Calliphora erythrocephala*

Meigen und über die Verwertung von Zuckern und Zuckeralkoholen durch dies Fliege. *Zeit. vergl. Physiol.* 22:614–639.

Hayes, W. P., and Liu, Y-S. 1947. Tarsal chemoreceptors of the housefly and their possible relationship to DDT toxicity. *Ann. Ent. Soc. Amer.* 40:401–416.

Hertwick, M. 1931. Anatomie and Variabilitat des Nervensystems und der Sinnesorgane von *Drosophila melanogaster* (Meigen). *Zeit. Wiss. Zool.* 139:559–663.

Kaib, M. 1974. Die Fleisch-und-Blumenduftrezeptoren auf der Antenne der Schmeissfliege *Calliphora vicina*. *J. Comp. Physiol.* 95:105–121.

Kay, R. E., Eichner, J. T., and Gelvin, D. E. 1967. Quantitative studies on the olfactory potentials of *Lucilia sericata*. *Amer. J. Physiol.* 13:1–10.

Knight, B. W. 1972a. Dynamics of encoding in a population of neurons. *J. Gen. Physiol.* 59:734–766.

Knight, B. W. 1972b. The relationship between the firing rate of a single neuron and the level of activity in a population of neurons. Experimental evidence for resonant enhancement in the population response. *J. Gen. Physiol.* 59:767–778.

Krijgsman, B. J. 1930. Reizphysiologische Untersuchungen am blutsaugenden Arthropoden im Zusammenhang mit ihrer Nahrugswahl. I. Stomoxys calcitrans. *Zeit. vergl. Physiol.* 11:702–729.

Künckel d'Herculais, J. 1879. Terminaisons nerveuses tactiles et gustatives de la trompe des diptères. *C. R. Assoc. Franc. Avanc. Sci.*, 1878:771–773.

Kunze, G. 1927. Einige Versuche über den Geschmackssinn der Honigbiene. *Zool. Jahrb., Abt. allg. Zool. Physiol.* 44:287–314.

Larsen, J. R. 1962a. The fine structure of the labellar chemosensory hairs of the blowfly, *Phormia regina* Meigen. *J. Insect Physiol.* 8:683–691.

Larsen, J. R. 1962b. Fine structure of the interpseudotracheal papillae of the blowfly. *Science* 139:347.

Larsen, J., and Dethier, V. G. 1965. The fine structure of the labellar and antennal chemoreceptors of the blowfly, *Phormia regina*. *Proc. XVI Int. Congr. Zool.* Washington, 3:81–83.

Lewis, C. T. 1954. Studies concerning the uptake of contact insecticides. I. The anatomy of the tarsi of certain Diptera of medical importance. *Bull. Ent. Res.* 45:711–722.

Leibermann, A. 1926. Correlation zwischen den antennalen Geruchsorganen und der Biologie der Musciden. *Zeit. Morph. Ökol. Tiere* 4:1–97.

McCutchan, M. C. 1969. Behavioral and electrophysiological responses of the blowfly, *Phormia regina* Meigen, to acids. *Zeit. vergl. Physiol.* 65:131–152.

Minnich, D. E. 1921. An experimental study of the tarsal chemoreceptors of two nymphalid butterflies. *J. Exp. Zool.* 33:173–203.

Minnich, D. E. 1926a. The chemical sensitivity of the tarsi of certain muscid flies. *Biol. Bull.* 51:166–178.

Minnich, D. E. 1926b. The organs of taste on the proboscis of the blowfly, *Phormia regina* Meigen. *Anat. Rec.* 34:126.

Minnich, D. E. 1929. The chemical sensitivity of the legs of the blowfly, *Calliphora vomitoria* Linn., to various sugars. *Zeit. vergl. Physiol.* 11:1–55.

Minnich, D. E. 1931. The sensitivity of the oral lobes of the proboscis of the blowfly, *Calliphora vomitoria* Linn., to various sugars. *J. Exp. Zool.* 60:121–139.

Morita, H. and Takeda, K. 1957. The electrical resistance of the tarsal chemosensory hair of the fly. *Mem. Fac. Sci.*, Kyushu University, Ser. E, 3:81–87.

Moulins, M. 1967. Less sensilles de l'organe hypopharyngien de *Blabera craniifer* Burm. (Insecta, Dictyoptera). *J. Ultrastructure Res.* 21:474–513.

Peters, W. 1961. Die Zahl der Sinneszellen von Marginalborsten und das Verkommen multipolarer Nervenzellen in den Labellen von *Calliphora erythrocephala* Mg. (Diptera). *Naturwiss.* 48:412–413.

Peters, W. 1962. Die propriorezeptiven Organe am Prosternum und an den Labellen von *Calliphora erythrocephala* Mg. (Diptera). *Zeit. Morph. Ökol. Tiere* 51:211–226.

Peters, W. 1963. Die Sinnesorgane an den Labellen von *Calliphora erythrocephala* Mg. (Diptera). *Zeit. Morph. Ökol. Tiere* 55:259–320.

Peters, W., and Richter, S. 1965. Morphological investigations on the sense organs of the labella of the blowfly, *Calliphora erythrocephala* Mg. *Proc. XVI Int. Congr. Zool.* Washington, 3:89–92.

Pierenne, H. 1943. Binocular and monocular threshold of vision. *Nature* 152:689–699.

Rees, C. J. C. 1968. The effect of aqueous solutions of some 1:1 electrolytes on the electrical response of the type 1 ("salt") chemoreceptor cell in the labella of *Phormia*. *J. Insect Physiol.* 14:1331–1364.

Rees, C. J. C. 1970. Age dependency of response in an insect chemoreceptor sensillum. *Nature* 227:740–742.

Rice, M. J. 1970a. A study of the innervation, structure and function of the anterior alimentary canal of the adult tsetse fly (*Glossina austini*) and other Diptera. Ph.D. dissertation, University of Birmingham, England.

Rice, M. J. 1970b. Cibarial stretch receptors in the tsetse fly (*Glossina austeni*) and the blowfly (*Calliphora erythrocephala*). *J. Insect Physiol.* 16:277–289.

Rice, M. J., and Finlayson, L. H. 1972. Response of blowfly cibarial pump receptors to sinusoidal stimulation. *J. Insect Physiol.* 18:841–846.

Shiraishi, A., and Tanabe, Y. 1974. The proboscis extension response and tarsal and labellar chemosensory hairs in the blowfly. *J. Comp. Physiol.* 92:161–179.

Slifer, E. H. 1970. The structure of arthropod chemoreceptors. *Annu. Rev. Ent.* 15:121–142.

Slifer, E. H. and Sekhon, S. S. 1964. Fine structure of the sense organs in the antennal flagellum of a flesh fly *Sarcophaga argyrostoma* R.-D. (Diptera Sarcophagidae). *J. Morph.* 114:185–208.

Slifer, E. H. and Sekhon, S. S. 1969a. Some evidence for the continuity of ciliary fibrils and microtubules in the insect sensory dendrite. *J. Cell. Sci.* 4:527–540.

Slifer, E. H. and Sekhon, S. S. 1969b. Nodes on insect sensory dendrites. *27th Annu. Proc. Electron. Microsc. Soc. Amer.* 242–243.

Smith, M. H., and Licklider, J. C. R. 1949. Statistical bias in comparisons of monaural and binaural thresholds: binaural summation or binaural supplementation. *Psychol. Bull.* 46:278–284.

Stoffolano, J. G. 1973. Effect of age and diapause on the mean impulse frequency and failure to generate impulses in labellar chemoreceptor sensilla of *Phormia regina*. *J. Geront.* 28:35–39.

Sturckow, B. 1962. Ein Beitrag zur Morphologie der labellaren Marginalborste der Fleigen *Calliphora* und *Phormia*. *Zeit. Zellforsch.* 57:627–647.

Stürckow, B. 1965. Electrophysological studies of a single taste hair of the fly during stimulation by a flowing system. *Proc. XVI Int. Congr. Zool.*, Washington, D. C., 1963, 3:102–104.

Stürckow, B. 1967. Occurrence of a viscous substance at the tip of a labellar taste hair of the blowfly. In *Olfaction and Taste, II,* T. Hayashi, ed. pp. 707–720. Pergamon Press, Oxford.

Stürckow, B. 1970. Responses of olfactory and gustatory receptor cells in insects. *Adv. Chemoreception* 1:107–160.

Stürckow, B. 1971. Electrical impedance of the labellar taste hair of the blowfly, *Calliphora erythrocephala* Mg. *Zeit. vergl. Physiol.* 72:131–143.

Stürckow, B., Adams, J. R., and Wilcox, T. A. 1967. The neurons in the labellar nerve of the blowfly. *Zeit. vergl. Physiol.* 54:268–289.

Stürckow, B., Holbert, P. E., and Adams, J. R. 1967. Fine structure of the tip of chemosensitive hairs in two blowflies and the stable fly. *Experientia* 23:780–782.

Stürckow, B., Holbert, P. E., Adams J. R., and Anstead, R. J. 1973. Fine structure of the tip of the labellar taste hair of the blowflies, *Phormia regina* (Mg.) and *Calliphora vicina* R.-D. (Diptera, Calliphoridae). *Zeit. Morph. Tiere* 75:87–109.

Tateda, H., and Morita, H. 1959. Initiation of spike potentials in contact chemosensory hairs of insects. I. Generating site of the recorded spike potentials. *J. Cell. Comp. Physiol.* 54:171–176.

Thurm, U., 1965. An insect mechanoreceptor. *Cold Spring Harbor Symposia on Quant. Biol.,* 30:75–82.

Tinbergen, L. 1939. Über den Bau der Geschmacksorgane auf den Proboscislippen und den Beinen von *Calliphora erythrocephala* Meig. *Archiv. Néerland. Zool.* 4:82–92.

Tominga, Y., Kabuta, H., and Kuwabara, M. 1969. The fine structure of the interpseudotracheal papilla of a fleshfly. *Annotationes Zool. Jap.* 42:91–104.

Verlaine, L. 1927. Le déterminisme du déroulement de la trompe et la physiologie du gout chez les lépidoptères (*Pieris rapae* Linn.). *Ann. Bull. Soc. Ent. Belg.* 67:147–182.

Wallis, D. I. 1962. The sense organs on the ovipositor of the blowfly, *Phormia regina* Meigen. *J. Insect Physiol.* 8:453–467.

Weis, I. 1930. Versuche über die Geschmacksrezeption durch die Tarsen des Admirals, *Pyrameis atalanta* L. *Zeit. vergl. Physiol.* 12:206–246.

Wilczek, M. 1967. The distribution and neuroanatomy of the labellar sense organs of the blowfly *Phormia regina* Meigen. *J. Morph.* 122:175–201.

Wolbarsht, M. L., and Dethier, V. G. 1958. Electrical activity in the chemoreceptors of the blowfly. I. Responses to chemical and mechanical stimulation. *J. Gen. Physiol.* 42:393–412.

Zacharuk, R. Y., and Blue, S. G., 1971. Ultrastructure of the peg and hair sensilla on the antenna of larvae of *Aedes aegypti* (L.), *J. Morph.* 135:433–456.

Zacharuk, R. Y., Yin, L. R., and Blue, S. G., 1971. Fine structure of the antenna and its sensory cone in larvae of *Aedes aegypti* (L.). *J. Morph.* 135:273–298.

5
Detecting the Unpalatable and the Dangerous

> *. . . and aching Pleasure nigh,*
> *Turning to poison while the bee-mouth sips.*
> Keats, *Ode on Melancholy*

To eat or not to eat when in a state of hunger is a question decided on the basis of information provided by contact chemoreceptors. In the simplest of all worlds the minimum requirement would be a receptor or collection of receptors to signal acceptability. What was not acceptable would pass undetected. This arrangement would suffice were all foodstuffs pure, unspoiled, unadulterated, and always equally acceptable. Because in nature these absolutes do not exist, it is as desirable to sense the unpalatable as the palatable. To the possession of receptors that mediate acceptance, therefore, must be added some that can mediate rejection. Since there are undoubtedly more unacceptable substances in the world than acceptable, the sensory elements detecting them either must be very broad in their sensitivities or must have multiple sensing capacities.

The detection of acceptable foods is only one of many functions delegated to the chemical senses of insects. They are also involved in courting, mating, locating oviposition sites, regulating certain prey-predator relationships, following trails, homing, identifying nest mates and queens, and regulating many aspects of social life. In addition they serve to detect potentially dangerous noxious materials, whether these emanate from the defense glands of other arthropods, the leaves of plants, or the laboratories of industry. They cannot concern themselves exclusively with food; therefore, collectively they must be sensitive to a wide spectrum of substances and be able to send unambiguous information about the different ones. How many different kinds of receptors are required and by what mechanisms do they interact with the environment?

The primary insectan chemoreceptor is a unique cell which exists simultaneously in two contrasting environments. One of its surfaces is presented to the changing world in which the insect moves, while its opposite pole makes contact with the comparatively stable environment of the central nervous system. The essence of this cell is the capacity to generate nerve impulses when specific chemical changes in the external environment act upon its outer surface. This capacity embraces three fundamental functions: selective employment of some fraction of the potential energy of chemicals to release large amounts of endogenous cellular energy (the first level of transduction); transduction of cellular energy to electrical energy; and transmission of an electrical code to the central nervous system.

When the contact chemoreceptors of the tarsi and labellum of the blowfly were finally identified there was no knowledge of how they performed the essential functions of chemoreceptors, nor was it known how many different kinds there were. It was not even known which part of the hair constituted the sensitive area. That only the tip was sensitive (in contrast to many olfactory receptors where the entire surface is involved) was first demonstrated by an experiment in which a minute drop of solution containing sucrose, water, and propanol was transferred from a needle to the base of a labellar hair (Dethier, 1955). The drop was carefully coaxed up the shaft of the hair. The purpose of the propanol was to provide a drop large enough to handle initially but later on to be free of solvent. As the drop was moved along, evaporation of the propanol decreased the size of the drop and concentrated the sucrose. With judicious rolling the reduction in size of the drop could be accommodated to the decreasing diameter of the shaft of the hair. During the maneuver watch was kept for extension of the proboscis. Only when the drop surmounted the tip did extension occur. Today the whole demonstration can be accomplished with great ease by electrophysiological techniques.

The demonstration that only the tip was sensitive, coupled with the histological observation that four dendrites terminated there and the behavioral observation that the fly could differentiate sugar, water, and salt, raised the question of receptor specificity and modalities. Was there one hair for each modality or was each hair sensitive to all? And, if the latter, were all the cells equally sensitive or were they differentially sensitive? The fly in fact showed only two kinds of behavior: it extended its proboscis to some compounds, notably water and certain carbohydrates, and it rejected a vast array of other compounds.

Behavioral experiments provided the first answers. A single labellar hair of a water- and food-deprived fly was prodded gently with a dry needle. In response to this mechanical stimulation the fly extended its proboscis. Stimulation was continued until no further responses were obtained. Failure of response could have been due to sensory adaptation, central adaptation, or muscle fatigue. A neighboring hair was then bent. An immediate response ensued, indicating that cessation had been due to sensory adaptation. Now, while the hair was still refractory to mechanical stimulation, a drop of water was applied to the tip. Response was immediate. This was interpreted as indicating that mechanoreception and reception of water were subserved by two different cells. The hair was next adapted to water and the same test for adaptation as before was applied. While the hair was adapted to touch and to water, a drop of sucrose was applied. Again there was a response. The hair could also be adapted to this concentration of sucrose after which a higher concentration would again initiate proboscis extension. Because of this result the series water, dilute sucrose, concentrated sucrose, was thought of as a continuum. On the basis of this result alone it was not possible to separate water and sugar receptors, and only one receptor was postulated.

The final experiment of this series consisted in adding sodium chloride to a solution of sucrose that was demonstrably effective. With the addition of enough salt the sugar failed to elicit a response. Failure could have been due to some interference with the sugar receptor by the salt. As a check against this possibility, a high concentration of salt was applied to one hair while another was being stimulated with sucrose. No proboscis extension occurred.

It was concluded from these experiments that each labellar hair contained a mechanoreceptor, a receptor that mediated acceptance (and since the test substance was sucrose, the receptor was termed a "sugar" receptor), and a receptor that mediated rejection, designated as a "salt" receptor because one of the most effective stimuli was sodium chloride. On the basis of geometry and position the mechanical function was assigned to the cell whose dendrite terminated in the socket. Additional support for this interpretation came from the observation that amputation of the tip of the hair abolished behavioral responses to chemicals but not to bending. No function was ascribed to the remaining two of the five cells.

It was still believed that there were basically two kinds of receptors, one mediating proboscis extension and the other, retraction. The sugar receptor was believed to be highly specific in its sensitivity and the salt receptor broadly sensitive. But was unacceptability a unitary modality? Theoretically a solution could be rejected because it failed to stimulate the sugar receptor, because it stimulated the salt receptor, because it inhibited activity of the sugar receptor, or because it caused a receptor to send a message different from normal. Different mechanisms of action might conceivably have different central interpretations. The search turned to mechanism.

The first indication that more than one mechanism might be involved in rejection emerged from behavioral measurements, not of rejection thresholds, but of acceptance thresholds of mixtures of sugars (Dethier, Evans, and Rhoades, 1956). It was already known that the effectiveness of sugars in eliciting extension of the proboscis ranged all the way from fructose and maltose, for which tarsal thresholds were 0.0058 M and 0.0043 M respectively, to lactose and L-rhamnose, which are non-stimulating at all concentrations (Table 8). Mixtures did not behave as would have been predicted from the threshold values of their constituents, however. For example, the median acceptance threshold for fructose is 0.0058 M, for glucose 0.132 M; for an equimolar mixture of the two, 0.0078 M. In other words, the concentration at which the mixture is stimulating represents 0.0039 M glucose and 0.0039 M fructose. Even if the two sugars were simply additive, they would not be expected to stimulate at the combined concentration. The fact that they do stimulate implies synergism, an enhancement of one sugar by the other. Mannose, on the other hand, when added to

Table 8. Effectiveness of Various Carbohydrates in Stimulating Single Labellar Hairs and Groups of Tarsal Hairs in *Phormia regina.*[a]

Compound	Single hair	Tarsi	Compound	Single hair	Tarsi
Triose			idose	−	−
DL-glyceraldehyde	−	−	D-altrose	−	−
Tetroses			Heptoses		
D-erythrose	−	−	D-α-glucoheptose	−	−
L-erythrose	−	−	D-gluco-D-guloheptose	−	−
Pentoses			Octose		
L-fucose	+	0.087*	D-gluco-L-gala-octose	−	−
D-arabinose	+	0.144*	Disaccharides		
L-arabinose	+	0.536*	sucrose	+	0.0098*
D-xylose	+	0.440*	turanose	+	0.011
L-xylose	+	0.337*	D-maltose	+	0.0043*
D-ribose	−	8.99*	D-trehalose	+	0.133*
D-xylose‡	−	42.27*	cellobiose‡	−	5.01*
ribulose	−	−	lactose	−	−
L-rhamnose	−	−	melibiose	−	−
2-desoxyribose	−	−	invert sugar	+	0.0062
Hexoses			equimolar glucose-		
D-fructose	+	0.0058*	fructose mixture	+	0.0078
D-glucose	+	0.132*	Trisaccharides		
L-sorbose	+	0.140*	melezitose	+	0.064*
D-galactose	+	0.50*	raffinose	+	0.20*
D-mannose‡	−	7.59*	Polysaccharides		
D-tagatose	−	−	levo-glucosan	−	−
D-gulose	−	−	xylan	−	−

From Dethier, 1955.

[a] Compounds that elicit proboscis extension when applied to single hairs are designated (+). For compounds that are effective on entire tarsi, the molar concentration that causes 50% of the flies to accept is given, when known. Starred values are from Hassett, Dethier, and Gans (1950). Compounds designated (−) are nonstimulating at all concentrations; those designated (‡) are effective on groups of labellar hairs but not on single hairs.

fructose inhibits it, that is, causes a ten-fold rise in the fructose threshold. This effect cannot be attributed to stimulation of the rejection receptor because in a preference test mannose is preferred to water at all concentrations above threshold. Furthermore, mannose has no effect on glucose, sucrose, or maltose (compare also Tables 9 and 10).

That the effects observed represent inhibition (of the acceptance receptor) rather than deterrence (stimulation of the rejection receptor) is further confirmed by the action of sorbose. Sorbose is stimulating in its own right, yet causes an increase in the thresholds of glucose and fructose when mixed with them. Its action is revealed clearly in the representative result shown in Table 11, in which the percent response of a sample of flies to various concentrations of glucose and sorbose is compared with their response to a series of solutions that contain additionally 0.5 mole of sorbose. In this series the stimulating effect of the mixture at low glucose concentrations stems entirely from the sorbose that is present. At higher concentrations of glucose when the same amount of sorbose is present there is little change in the stimulating effectiveness. Not only do the two sugars fail to add, but also

Compound	Single hair	Tarsi	Compound	Single hair	Tarsi
glycogen	−	−	NO_2-β-glucose	±	
hydroxycellulose	−	−	diacetone glucose	+	
Polyhydric alcohols			α-pentaacetyl glucose	−	−
sorbitol	−	−	diacetone nitrobenzoyl		
glycerol	−	−	glucose	−	−
dulcitol	−	−	1,2,3,4-acetyl-6-		
m-erythritol	−	−	trimethyl glucose	−	−
penta-erythritol			β-NO_2-tetraacetyl		
mannitol	−	−	glucose	−	−
L-arabitol	−	−	α-D-glucose-1-phos-		
inositol	+	0.194*	phate (dipotassium)	−	0.101
Glycosides			α-D-fructose-6-phos-		
α-D-methyl glucoside	+	0.069*	phate (barium salt)	−	−
β-D-methyl glucoside	−	−	glucose-6-phosphate		
NO_2-benzyl glucoside	−	−	(barium)	−	0.083
p-aminophenyl glucoside	+		α-D-fructose-1,6-		
monoacetyl NO_2-β-			diphosphate		
glucoside	−	−	(calcium)	−	−
phenyl β-glucoside			(barium)	−	−
tetraacetate	−	−	(magnesium)	−	−
α-D-methyl mannoside	+		NO_2-maltose	+	
p-aminophenyl			NO_2-7-acetyl maltose	−	−
β-maltoside	+		NO_2-cellobiose	?	?
nitrophenyl maltoside	+		sucrose octaacetate	−	−
7-acetyl p-nitrophenol			diacetyl glucuron	−	−
cellobioside	−	−	β-triacetyl glucuron	−	−
Substituted sugars			glycuronic acid mono-		
n-acetyl glucosamine	−	−	benzoate	−	−
glucosamine hydro-			gulonic lactone	−	−
chloride	+	0.204	α-glucoheptonic lactone	−	−
NO_2-α-glucose	+				

Table 9. Intake of Mixed Solutions Compared With That of Water or Single Sugars in a Two-choice Test.

Concentration of each sugar in mixture	Vol. consumed (ml/fly/24 hr.)	Water or sugar	Vol. consumed (ml./fly/24 hr.)
0.05 M fucose and 0.05 M sorbose	0.0125	water	0.0019
0.5 M fucose and 0.5 M sorbose	.0116	water	.0030
0.05 M fucose and 0.05 M mannose	.0213	water	.0025
0.5 M mannose and 0.5 M sorbose	.0184	water	.0018
0.1 M fructose and 0.1 M mannose	.0234	0.1 M fructose	.0130
0.05 M glucose and 0.05 M mannose	.0090	.1 M glucose	.0030
0.1 M glucose and 0.1 M mannose	.0228	.1 M glucose	.0090
0.05 M glucose and 0.05 M sorbose	.0099	.1 M glucose	.0162
0.05 M fructose and 0.05 M sorbose	.0220	.05 M fructose	.0160
0.05 M glucose and 0.05 M rhamnose	.0260	.05 M glucose	.0130
0.1 M glucose and 0.1 M rhamnose	.0160	.1 M glucose	.0070
0.05 M fructose and 0.05 M rhamnose	.0120	.05 M fructose	.0130
0.1 M fructose and 0.1 M rhamnose	.0170	.1 M fructose	.0140
0.05 M sucrose and 0.05 M rhamnose	.0180	.05 M sucrose	.0230
0.1 M glucose and 0.1 M D-arabinose	.0210	.1 M glucose	.0080
0.05 M glucose and 0.05 M D-arabinose	.0130	.1 M glucose	.0160

From Dethier, Evans, and Rhoades, 1956.

Table 10. Examples of Inhibition Revealed by Ascertaining the Effects of Sugar Mixtures on Tarsal Thresholds.

Sugar	Effect	Sugar affected
Mannose	inhibits	fructose
	does not affect	glucose, sucrose, fucose, maltose
Sorbose	inhibits	glucose, fructose
Fucose	does not affect	glucose, fructose
Rhamnose	does not affect	fucose, glucose
	inhibits	fructose
D-arabinose	does not affect	glucose
Mannitol	does not affect	fructose

From Dethier, Evans, and Rhoades, 1956.

Table 11. Effect of Sorbose on Glucose Threshold.

Molar concentration of glucose solutions	0.0625	0.125	0.25	0.50	1.0
Per cent response	0	5	15	50	80
Molar concentration of glucose solutions containing 0.5 M sorbose	0.0625	0.125	0.25	0.50	1.0
Per cent response	45	50	60	50	65
Molar concentration of sorbose solutions	0.0625	0.125	0.25	0.50	1.0
Per cent response	5	20	45	60	80

From Dethier, Evans, and Rhoades, 1956.

the stimulating effect to be expected of the high concentrations of glucose is absent. Therefore, when a smaller volume of a mixture of sugars is ingested than of either of the constituents alone, the result cannot always be ascribed to deterrence, especially when in other tests both constituents can be shown to be preferred to water. Experiments involving the quantity of fluid ingested by houseflies (*Musca domestica* L.) led to the conclusion that D-xylose, L-arabinose, ribose, rhamnose, and sorbose were "repellent" (Galun, 1955). This conclusion, however, was based on the fact that the addition of those sugars to an acceptable sugar solution lowered the intake. Unless the sugars can be shown to be unacceptable when compared with water, the possibility of inhibition cannot be overlooked.

The concept of inhibition was supported by another series of experiments, which tested, not thresholds, but volume consumed in one-choice and two-choice situations (Dethier, Evans, and Rhoades, 1956). For example, the consumption of a mixture of mannose and glucose was greater than that of glucose alone, as would be expected (Table 9). In contrast, the volume ingested of a mannose fructose mixture, as compared with fructose alone, was not so great as would be expected. Similarly, with mixtures containing rhamnose there was a large increase in consumption whenever the other sugar was glucose, but no significant increase when the other sugar was fructose or sucrose (Table 9). These results agreed with threshold data (Table 10) indicating that mannose and rhamnose inhibit fructose but not glucose.

The situation with regard to sorbose is not clear; there is a tendency for the consumption of sorbose mixtures to be less than that predicted on a purely additive basis. This conclusion is in agreement with the postulated inhibitory effect of sorbose on glucose and fructose. Since the effects of carbohydrate mixtures on the oral papillae is not known, the results obtained in preference tests cannot always be equated with demonstrated inhibition of tarsal and labellar receptors.

The only comparable study of mixtures on another insect was that of Wykes (1952), who measured ingestion of single sugars and mixtures of sugars by honeybees. The experiment tested the hypothesis, although not explicitly stated, that the volume ingested of the four sugars examined singly and in mixtures was related to concentration by the formula $V = a + C$, where V is the volume ingested at concentration C. For all four sugars there was assumed to be a linear relationship between volume ingested and concentration, with a slope of unity and an intercept characteristic of the sugar tested. However, since the units of volume employed were arbitrarily established and apparently were changed from one concentration to the next, this hypothesis was not tested directly; it was implicitly assumed in the analysis of the ingestion of mixtures. The experiments with mixtures consisted in measuring the volume ingested of a solution containing equal proportions by weight of two to four sugars with a total sugar concentration of $x\%$, and testing the significance of the difference between this value and the average of the volumes ingested of each of the component sugars at $x\%$. For example, the volume ingested of a solution containing 8.5% sucrose and 8.5% glucose was compared with one-half the sum of the volumes ingested of 17% glucose and 17% sucrose. Rather surprisingly, the calculated and measured figures did not differ significantly; hence, volume and concentration are linearly related, with a slope of unity for sucrose and glucose within the concentration range 17.1–51.3%. The same relationship was found to be true of maltose. Furthermore, with one exception, these sugars in mixtures are nearly additive insofar as ingestion is concerned. The one exception was the glucose-sucrose-fructose mixture, of which more was ingested than predicted (i.e., there was synergism).

The data for *Phormia* relating volume and concentration, whether measured by single feedings or by preference-aversion techniques, never presented so simple a picture as do Wyke's results. The only similarity may be the striking parallelism (with the exception of fructose) of volume intake from low to optimum concentrations on a semilog plot of ingestion at a single feeding (Fig. 64). Preference-aversion experiments on ingestion of mixtures probably are not comparable to ingestion as measured by Wykes; clearly in the former case simple additivity of sugars in a mixture is not the rule.

By the early 1950s, behavioral analyses of chemoreceptor function in insects had apparently been pushed to the limit. Each labellar hair and each tarsal hair was believed to have a mechanoreceptor, a salt receptor, and a sugar receptor. The obvious next step was to revert to electrophysiological techniques. This approach had in fact been attempted in 1939 on the olfactory receptors of caterpillars by Dethier and Prosser (Dethier, 1941), and by recording summed spontaneous activity in antennal nerves of adult insects (Boistel and Coraboeuf, 1953; Boistel, Lecompte, and Coraboeuf, 1956; Roys, 1954; Smyth and Roys, 1955; Schneider, 1955; 1957a, 1957b; Schneider and Hecker, 1956). None of these pioneering efforts were very successful, chiefly because the chemosensory fibers are extremely small (picking out chemoreceptive impulses from whole nerves is still well-nigh impossible) and the equipment used at that time was still primitive. The breakthrough came from two widely separated laboratories, Tufts University and Kyushu University, and was due as much to the creativity of the experimenters as to the availability of modern apparatus.

Hodgson, Lettvin, and Roeder (1955) invented an ingenious new technique for detecting the nerve impulses generated by labellar chem-

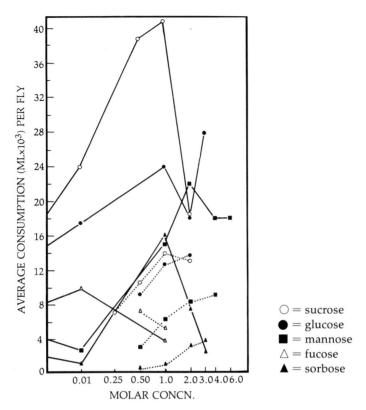

Fig. 64. Comparison of ingestion measured by single feeding and by preference-aversion intake during the first 24 hours of feeding. Solid lines, preference-aversion; dotted lines, single feeding (from Dethier, Evans, and Rhoades, 1956).

oreceptors in response to various stimulating solutions. The technique was as follows. The decapitated head of a fly was impaled on a silver microelectrode, which was to serve as the indifferent electrode. The recording electrode consisted of a fine glass pipette, approximately 10 μ in diameter at its tip, filled with saline (0.1 M NaCl), and connected to a silver-silver chloride wire. Leads from the two electrodes passed to a cathode follower, thence to an amplifier, and finally to an oscilloscope for display and for storage on tape or film. Recording commenced when the glass capillary electrode was placed on the tip of a hair (Fig. 65). The capillary performed the dual function of recording electrode and container for the stimulus. For example, sodium chloride provided a salt bridge for conducting electrical activity from the open tip of the hair where the dendrites lay to the silver-silver chloride wire. At the same time it stimulated the receptors. The stimulating effect of other solutions could be assessed by adding them to the salt. The identical technique was discovered independently by Morita et al. (1957).

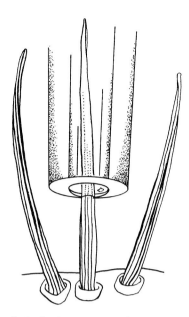

Fig. 65. Position of electrode with reference to hair during tip recording.

Tip recording, as the technique came to be called, was a tremendous technical advance and a powerful tool. It was not without its handicaps. One was the necessity of having an electrolyte present in the recording capillary at all times. Because of this, compounds that were immiscible in water could not be investigated nor could the possibility of interaction between the electrolyte and added stimuli be eliminated or assessed accurately. A second drawback was that all four dendrites were exposed to stimulation simultaneously. There was no way of isolating them, and the possibility of receptor interaction was a constant threat.

Once discovered, however, the technique was almost immediately improved upon and is constantly undergoing refinement. The inability to deal with receptors individually remains. The other major handicap posed by tip recording was eliminated by a spectacular modification introduced by Morita (1959). He succeeded in making a small crack in the wall of the hair against which he could place the recording electrode, thus leaving the tip free for the application of any kind of a stimulus. The crack was made by the "leather-punch" principle; a glass capillary of rather large diameter (15 μ) was placed against one side of the hair, and another capillary of smaller diameter, or a needle, was pressed against the other side of the hair and directly opposite the first (Fig. 66). Pressure was applied until a crack had been made. The recording electrode was placed either directly in contact with the crack or into the large end of a saline-filled glass capillary (tip diameter 30

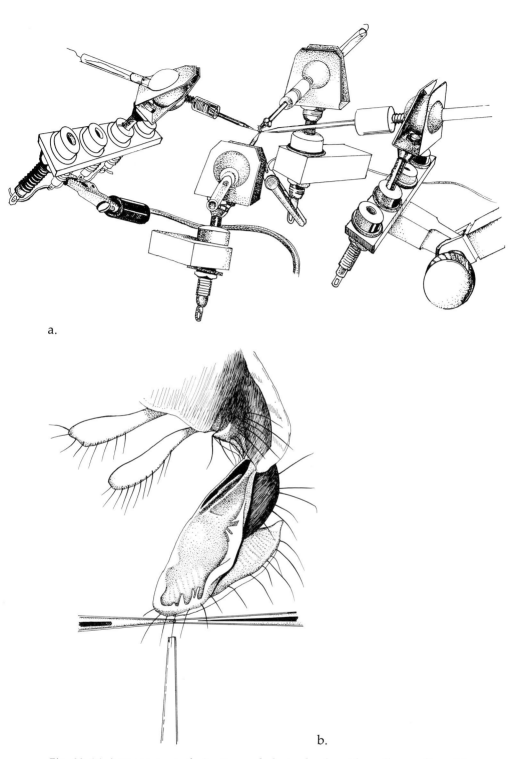

Fig. 66. (a) Arrangement of pipettes and electrodes for side-wall recording; (b) detail of arrangement.

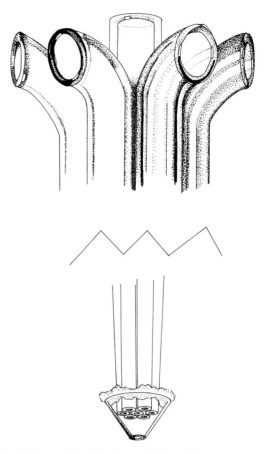

Fig. 67. Method of recording stimulation with a flowing solution (after Stürckow, 1970).

μ), which was itself in touch with the crack. This latter technique compensated for evaporation at the point of juncture with the crack by constantly drawing fluid from the electrode. Another technique, which permits the use of constantly flowing solutions and the uninterrupted interchange of stimuli, was devised by Stürckow (1965, 1970) and is illustrated in Figure 67.

Despite the many advantages, "side-wall" recording is not without its drawbacks. For example, the degree of trauma that the electrode inflicts influences the amount of neural activity that is present in the absence of stimulation at the tip. If the electrode is applied with the utmost delicacy to a labellar hair, only one cell fires (the salt cell). As the pressure of the electrode is increased, more cells fire. It is impossible, therefore, to assert with confidence that any of the cells are or are not "spontaneously" active (*cf.* Evans anf Mellon, 1962a; Rees, 1968). A more serious difficulty arises from the fact that the level of so-

Fig. 68. Response of labellar hair "largest 7" to 0.5 M NaCl plus 0.1 M sucrose. Record begins 500 msec. after onset of stimulation; w = water receptor; 5 = fifth cell; c = sugar receptor; s = salt receptor (from Dethier and Hanson, 1968).

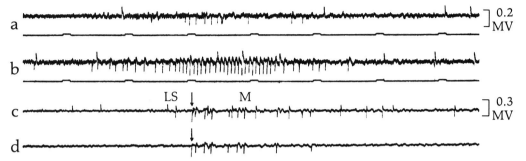

Fig. 69. Electrical response to a labellar hair of *Phormia* to stimulation by salt, fructose, and bending. (a) Intact hair stimulated with 0.01 M NaCl and bending; (b) same hair with top removed responding to the same stimuli; (c) cut hair stimulated by 0.01 M NaCl + 0.1 M fructose; at arrow the hair is moved into a more concentrated solution at the tip of the pipette; (d) same hair responding to motion after adaptation to chemicals. Time, 0.2 sec. L = salt cell, S = sugar cell, M = mechanoreceptor. Positive potential at recording electrode is down (from Wolbarsht and Dethier, 1958).

called spontaneous activity in the absence of stimulation at the tip affects the sensitivity of receptor to normal stimulation (Dethier, 1972). For example, if the base level activity of a cell is high, a solution applied at the tip is more likely to stimulate than if the base level is low. This behavior tends to confound results. Furthermore, with sidewall recording, as with tip recording, the spike height of a cell does not always remain constant. This inconstancy adds to the difficulty of identifying the activity of individual cells when more than one is active.

The first task that electrophysiology set itself was to clarify the functional identity of hairs. Hodgson, Lettvin, and Roeder (1955) immediately confirmed the behavioral conclusion that one of the four dendrites in the labellar hair responded to sodium chloride and another to sugar (Fig. 68). In rapid succession Wolbarsht (1957) discovered that another neuron responded to water, and Wolbarsht and Dethier (1958) demonstrated that the dendrite inserted in the socket of the hair was a mechanoreceptor (Fig. 69). Mellon and Evans (1961) then studied the water receptor in greater detail (Fig. 70a–d).

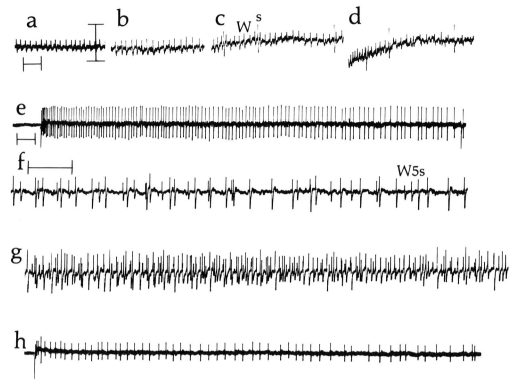

Fig. 70. Response of labellar hair "largest 9" to (a) 0.05 M NaCl in H_2O (water spike), (b) 0.05 M NaCl in D_2O (water spike), (c) 0.1 M NaCl in water (w = water spike, s = salt spike), (d) 0.1 M NaCl in D_2O (water and salt spike). Vertical bar equals 1.5 mv. Horizontal bar equals 100 msec. Records begin at onset of stimulation. (e) Response of tarsal hair D_2 to 1.0 M NaCl (salt spike). (f) Response of labellar hair "largest 7" to 0.1 M NaCl. Record begins 500 msec. after stimulation. w = water cell; 5 = fifth cell; s = salt receptor. (g) Response of labellar hair "largest 2" to 1.0 M sodium valerate (water, salt, and fifth-cell spikes). (h) Response of tarsal hair D_5 to 1.0 M NaCl plus 1.0 M sodium propionate (salt spike) (from Dethier and Hanson, 1968).

Throughout all of these studies the ghost of the fifth receptor still lurked in the background. Periodically, reports appeared stating that two cells responded to sodium chloride (Sturckow, 1960, 1965; den Otter and van der Poel, 1965; Steinhardt, 1965; and Gillary, 1966a). They were seen in *Calliphora* and in *Phormia*. It is difficult to interpret the earliest data gathered with *Calliphora,* and it is risky trying to equate them with those of *Phormia* because there are differences in the details of electrical responses. Nevertheless, it is obvious that another cell giving a spike with an amplitude close to that of the sugar receptor (in *Phormia*) occurs moderately frequently when sodium chloride is the stimulus (Fig. 70e–f). It arises from the cell that Steinhardt has called the anion receptor, the "fifth" cell of Dethier and Hanson (1968).

In an attempt to learn more about this fifth cell Dethier and Hanson (1968) stimulated labellar hairs with the sodium salts of homologous (C_1 to C_{14}) fatty acids. In some respects the salts from C_1 to C_{14} evoked the same kinds of response as NaCl. At concentrations below 0.1 M only the water cell was stimulated. At 0.1 M one or more additional spikes appeared. With the C_1 to C_4 salts the usual response was the large spike alone, the classical salt spike. No additional spike occurred with the addition of NaCl. On the other hand, there was a marked tendency for cells other than the salt cell to respond more often and with greater frequency to fatty acid salts than to any concentration of NaCl (Fig. 70g). In these cases the responses to mixtures sometimes resembled the response to NaCl alone (Fig. 70h), in that the supernumerary spikes disappeared. Furthermore, when the response to mixtures was compared with the response to either component alone, the frequency of salt spikes was lower in the mixture.

Salts with a chain length of C_5—C_{14} invariably caused two or three cells to respond, even in hairs that normally gave only one kind of spike in response to NaCl. The addition of sucrose (0.1 or 1.0 M) to the solution caused an additional cell to respond. In mixtures containing one fatty acid and NaCl the response sometimes resembled that of the fatty acid salt alone (Fig. 71a) and sometimes that of NaCl alone (Fig. 71b).

As exposure was prolonged, the longer-chain salts began to impose reversible inhibition. Sodium valerate (C_5) inhibited the salt cell (Fig. 70g). Sodium laurate (C_{12}) (0.1 M) markedly inhibited the salt cell but had less effect on the sugar cell and least on the water cell (Fig. 72a–g). Salts lower in the series (C_1–C_4) did not give these effects.

Salts with chain lengths C_5–C_{14} also caused the salt cell to burst into violent activity (Fig. 72h–j). An exposure as short as two seconds frequently constituted sufficient stimulation. The incidence of volleying was directly related to concentration and duration of exposure. Valerate caused the salt to volley (Fig. 72h–j); caprylate (C_8), laurate, and myristate caused volleying of both the salt cell and another cell that was not the water receptor (Fig. 72i, j).

Even if the salt cell did not volley, it sometimes ceased normal firing after one or two seconds, while the remaining cells continued their activity. The salt cell was clearly the first to be inhibited (Fig. 72j). When the concentration of stimulus was low and the duration short, recovery occurred after 20–60 minutes. The degree of recovery from inhibition was inversely related to the duration of stimulation and of volleying. With sodium laurate the water cell recovered first, the sugar cell next, the salt cell last (Fig. 72b–g).

These studies agree with many others (den Otter and van der Poel, 1965; Steinhardt, 1965; Gillary, 1966a; Rees, 1968) in showing that the

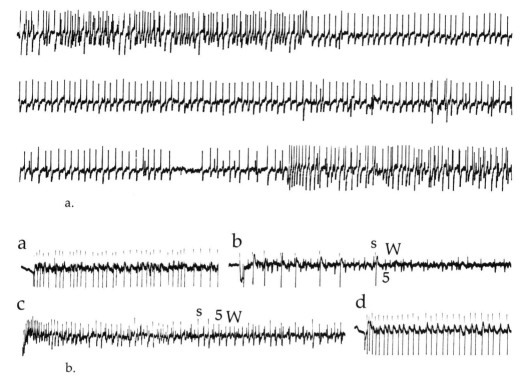

Fig. 71. (a) Continuous record of response of labellar hair "largest 7" to 0.5 M NaCl plus 0.5 M sodium valerate. Note the inhibition of the salt spike beginning in the first line and its volleying beginning in the last line. (b) Response of labellar hair "intermediate 3" to (a) 1.0 M NaCl, (b) 0.5 M sodium caprylate, (c) 1.0 M sodium caprylate, (d) 0.5 NaCl plus 0.5 M sodium caprylate (salt spike) (from Dethier and Hanson, 1968).

anion does affect the action of salt on the salt receptor. This is especially noticeable at C_5 and above. With these long-chain salts the frequency of the fifth cell increases. There is clear evidence that the anion inhibits the salt cell in this range. A strong suspicion exists that chloride inhibits the fifth cell. Throughout, however, the fifth cell responds with a low frequency so that there is doubt that salt is actually its adequate stimulus. Clearly more investigation is required.

The studies with the fatty acid salts are especially interesting in a behavorial context. Rejection thresholds of these salts mixed with 0.5 M sucrose are essentially the same for C_1 through C_4 (Fig. 73) (Dethier, 1956). From C_6 on, thresholds are considerably lower, and decrease with increasing chain length in a manner reminiscent of the homologous alcohols. Two different modes of action are suggested. Furthermore, the change of threshold of each member as the ambient sugar concentration is changed follows a different pattern for C_1–C_4 than for C_6 and above (Fig. 74) (Dethier, 1955). It has been inferred that behav-

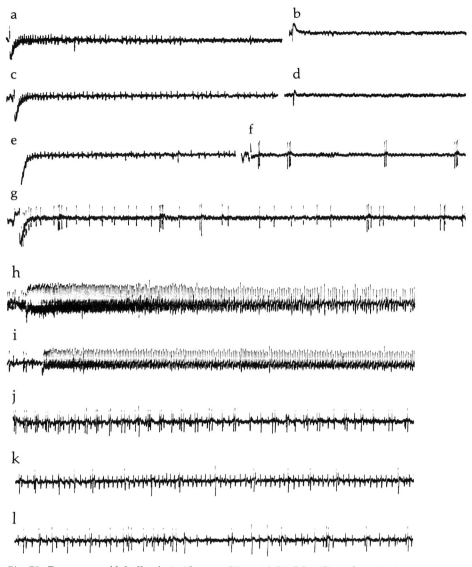

Fig. 72. Response of labellar hair "largest 3" to (a) 0.1 M sodium laurate (water spikes and one salt spike), (b) 1.0 M NaCl applied 60 seconds after sodium laurate (no activity), (c) 0.05 NaCl applied 60 seconds after sodium laurate (water spike only), (d) 0.1 M NaCl + 0.1 M sucrose applied after sodium laurate (no activity), (e) 0.05 M NaCl applied 15 minutes after sodium laurate (water spike), (f) 1.0 M NaCl applied 15 minutes after sodium laurate (salt spike triplets), (g) 0.1 M NaCl + 0.1 M sucrose applied 15 minutes after sodium laurate (salt and sugar spikes). (h) Volleying of salt receptor of labellar hair "largest 4" after 2.5 minutes of stimulation by 0.1 M sodium laurate. (i) One minute later the salt receptor ceases firing and the fifth cell begins volleying. (j–l) Continuous record beginning 500 msec. after onset of stimulation of labellar hair "largest 1" by 1.0 M sodium caprylate. Note the reduced frequency of the salt spike and the increased frequency of the spike of the fifth cell (from Dethier and Hanson, 1968).

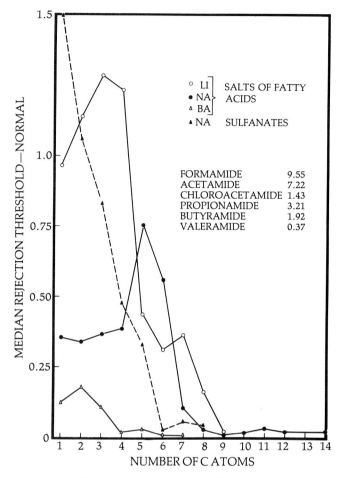

Fig. 73. Behavioral response of *Phormia* (rejection thresholds) to series of sodium salts in sucrose applied to the tarsi (from Dethier, 1956).

ioral rejection of the short-chain members is related to activity of the salt cell, as with sodium chloride, while rejection of the long-chain members is related to activity of the fifth cell.

Insofar as the labellar hairs of *Phormia* are concerned, all of these early studies, behavioral and electrophysiological, supported the hypothesis that each hair is equipped with a mechanoreceptor, a receptor that responds most effectively to water, one that responds most effectively to certain sugars, and two that respond most effectively to salts. Beyond this little was known of the extent of the specificity (although very narrow specificity was assumed in each instance) and knowledge of the tarsal hairs was scanty.

Behavioral and electrophysiological studies also pointed to multiple modes of action resulting in behavioral "rejection." Such studies

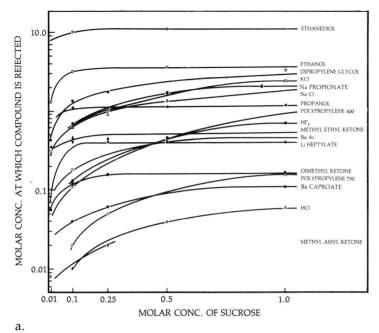

a.

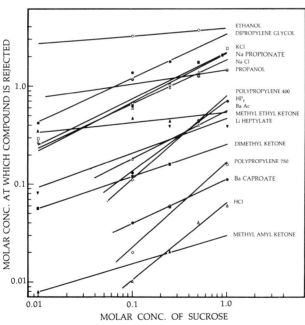

b.

Fig. 74. (a) Relation between the concentration at which a compound is rejected and the concentration of sugar with which it is mixed. (b) Same data plotted logarithmically to show two categories of unacceptable compounds, those for which rejection thresholds change very slightly with change in sucrose concentration, and those for which thresholds change greatly (from Dethier, 1955).

showed that compounds may stimulate the salt receptor, they may stimulate the fifth cell, they can cause complex inhibitory interactions, they can inhibit the water cell or the sugar cell. The implication is that the fly can differentiate among various kinds of unacceptable materials and may behave in a significantly different manner toward each. There was a possibility, then, that its chemoreceptive world might not be simply a matter of "acceptable" and "unacceptable," but might involve many more subtle taste distinctions. However, proof was lacking.

Electrophysiology also set itself the task of trying to fathom where the action potentials were generated and how the supposed depolarization at the tip could manage to travel the distance it did (more than 300 μ in long hairs) without becoming so attenuated as to become ineffective at the site of spike generation.

The electrical events that ensue after stimulation of a chemoreceptive cell follow the classical pattern. It had been suggested that stimuli depolarize the membrane of the end of the dendrite at the tip of the hair, that this depolarization travels down the dendrite to the cell body, and that action potentials are generated somewhere in this region (Dethier, 1956). The observation that alteration of the temperature at the tip of the hair in the range 2° to 41°C caused no change in the behavioral threshold (Dethier, 1956) nor any change in the frequency of impulses (Hodgson, 1956; Dethier and Arab, 1958) was construed as support for this idea. When the temperature of the whole preparation, and hence the cell bodies, was changed, marked changes in the frequency of action potentials did occur (Hodgson and Roeder, 1956; Hodgson, 1956). The ever-present hazard in trying to infer mechanisms from behavioral data proved to be very real in this case, when Gillary (1966b) showed by carefully controlled experiments that there is indeed a temperature dependence, and Uehara and Morita (1972) proved beyond all doubt that the rates of rise of the receptor potentials of the salt and sugar receptor do increase with temperature. No temperature effect on the water receptor has been demonstrated thus far.

An early attempt to find a slow potential associated with stimulation revealed a resting potential of about 60 mV when one electrode was placed on the tip of the hair and the other elsewhere in the head or labellum (Wolbarsht, 1958). Oddly enough, this did not change when the hair was stimulated and obviously was not the sought-after receptor potential. Working with the mechanoreceptive component of the hair, and with simple mechanoreceptors elsewhere on the body, Wolbarsht did find a graded slow potential that increased in negativity at the recording electrode when the hair was bent. This is a receptor potential. It varies directly with the magnitude of the stimulus, varies smoothly, has to reach a critical level before any impulses are generated, and always occurs prior to the initiation of impulses. It is located at a site distad of the site of impulse generation, and is not invaded by the subsequent propagated impulses.

Conclusive pin-pointing of the receptor potential of the chemoreceptors was provided by Wolbarsht (1958), by Morita (1959) and by Wolbarsht and Hanson (1965). By recording through the side wall of a hair Morita found that a slow d.c. potential appeared when the tip of the hair was stimulated. The closer the recording electrode was removed to the tip, the greater the potential. It was negative in response to sugar and sodium chloride, positive in response to quinine. Morita concluded that this potential travelled to the base of the hair and initiated impulses there.

Caution must be exercised, however, in relating any observed shifts of slow potential to a particular stimulus or cell. As Wolbarsht and Hanson (1967) have pointed out, the receptor potential is the summed potential from several dendrites because all terminals are exposed to the same extracellular ions. They may be affected differently with respect to ionic flux leaving across their membranes. For example, water hyperpolarizes the salt fiber, and there is a post-inhibitory rebound. Since no d.c. shift occurs in the positive direction in this instance, one must assume that it is counterbalanced by a negative potential from the water fiber; therefore, the "receptor potential" is an algebraic sum. Impulse stimulation would be independent of the direction of this potential.

The exact site of impulse generation was made even clearer by work carried out on the flesh fly *Lucilia caeser* and the butterfly *Vanessa indica* by Tateda and Morita (1959), Morita and Takeda (1959), Morita (1959), and Morita and Yamashita (1959a, 1959b). First they proved that spikes never occurred at the tip of an amputated hair. They then confirmed data from *Phormia* which showed that anodal stimulation elicited spikes and cathodal current blocked them. This finding had been interpreted to indicate that the dendrites act as rather poorly insulated extensions of the recording pipette into the cell so that recording was in effect intracellular (Wolbarsht, 1958). This would have explained the puzzling fact that in tip recording the initial component of the spike is positive and not negative as would be expected (Morita, Takeda and Kuwabara 1957; Wolbarsht, 1958; Hodgson, 1958). Behavioral responses had been evoked by cathodal stimulation and did not occur with anodal stimulation (Arab, 1957, 1959), but this apparent discrepancy was reconciled by the finding that the behavioral response was evoked by current on the neurons at the bases of hairs adjacent to the one on which the electrodes had been placed. In short, the polarizing current was extracellular; hence, cathodal stimulation was excitatory and anodal stimulation was blocking (Wolbarsht, 1958).

Analyses of changes in the shape of the spike (tip recording) when different parts of the hair were cooled, first the middle, then the base, showed that cooling the middle affected predominantly the falling phase of the spike, whereas cooling the base affected both rising and falling phases (Morita, 1959). This finding was construed as indicating that the spike was generated at the base of the hair and fired in both directions, that is, back along the dendrite to the tip and centripetally along the axon to the central nervous system. It was also concluded, on the basis of analyses of simultaneous tip and side-wall recordings, that impulses are recorded extracellularly and not intracellularly as Wolbarsht (1958) had concluded (Morita and Yamashita, 1959b). A further contribution to our understanding of spike generation came from studies of the effects of special narcotics on the salt receptor; e.g., xylocaine, cocaine, tetrodotoxin (Wolbarsht and Hanson, 1965). These narcotics block the negative phase of the diphasic action potential; strong salts increase the negative phase and reduce the positive phase. Records taken simultaneously from the tip and the side of the hair revealed that both phases of the spike occurred first at the side. The conclusion, in agreement with others was that the impulses are generated in or near the cell body and pass antidromically up the dendrite to the tip.

All of these results, together with information about the structure of chemosensory hairs and new data, enabled Rees (1968) to construct an electrical analogue for the labellar hair complex that represents our present concepts of this sensory system (Fig. 75). An electrically conducting, pathway is formed by the series-connected resistances of the loop 6, 1, 3, 4, 5 — receptor membrane, dendrite interior, membrane separating dendrite contents from trichogen contents, lumen of large cavity which is continuous with the trichogen and tormogen, sinuses and a communicating channel reported to connect the large lumen with the chamber of the pore at the tip (Hori and Morita, cited in Rees, 1968). The wall separating the two lumina is probably a nearly perfect insulator, so that there will be no flow across it except for the presumed connection at the tip. There is some reasonable doubt however, about the existence of this connection. Electronmicrographs by Stürckow, Holbert, and Adams (1967) show a spongy area at the tip of the large lumen and a termination of similar appearance at the small lumen; continuity of the two is suggested, but the pictures are not convincing. More suggestive evidence comes in the form of impedance measurements; however, the issue is still in doubt (Stürckow, 1971). It has been proposed that the resultant differences in electrical potential between the two electromotive forces at 3 and 6 may drive a current around this loop, which would alter the state of polarization of the generator membrane (3), promoting or inhibiting the discharge of action potentials (Rees, 1967).

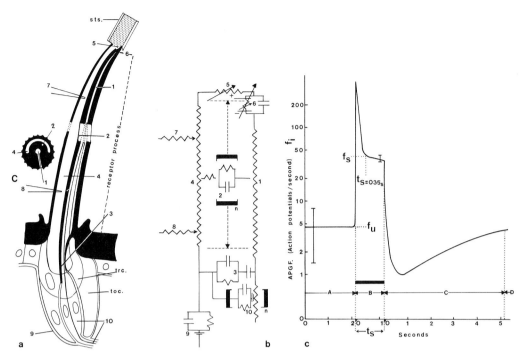

Fig. 75. (a) Diagrammatic representation of a chemoreceptor sensillum. 1, dendrite; 2, insulating cuticular wall; 3, generator membrane; 4, large lumen; 5, tip chamber; 6, receptor membrane; 7, recording electrode; 8, indifferent electrode; 9, envelope membrane (continuation of basement membrane); 10, neuron cell body and axon; trc, trichogen cell, toc, tormogen cell; dotted connections, foreshortening; sts, stimulating solution. (b) Electrical analogue of (a). n, element repeated many times (e.g., 2 extends for the whole length of the cuticular insulating wall). (c) Diagram showing the time-course of generation frequency of action potentials (APGF) of the salt receptor of *Phormia*. Solid line, response to 0.4 M KCl solution. Stimulus applied for a period B (black bar). A and D are periods of unstimulated discharge frequency (f_u). The changes of frequency during adaptation may be seen during period B. The adapted discharge frequency (f_s) is defined here at $t_s = 0.35$ sec. The post-stimulation response is shown during period C. Vertical bars show 95 per cent confidence limits about mean frequencies where APGF is constant enough to calculate them (from Rees, 1968).

The important membrane sites are those at the receptive tip of the dendrite and those at the base. The ionic composition on the inside is probably the same at both locations, as is probably true also of the composition on the outside at both points; however, the resting potentials need not be the same because relative permeability to ions may be different at the tip and the base (Rees, 1968). The constant resting potential of about 60 mV (Wolbarsht, 1958) is probably maintained by the enveloping membrane (Fig. 75a, b), presumably has nothing whatsoever to do with stimulation, and may, according to Rees (1968), act as a shield protecting the whole receptor complex from the effects of major fluctuations in the ionic composition of the haemolymph, which might otherwise disturb the state of polarization of the generator membrane. In this connection, it is worth noting that some substances injected into the haemocoele can enter the cellular complex of the sensillum and exert effects there. Tritiated water injected into the haemolymph can be detected at the tip of a labellar hair in amounts proportional to the time that has elapsed after injection (Hodgson, 1968). If the tip of the hair is amputated, more radioactivity can be detected in the hair, indicating that the pore has a restrictive effect. Further experiments involving the injection of epinephrine and dopamine revealed increases in firing of the salt cell when these hormones were present (Hodgson, Ishibashi, and Wright, cited in Hodgson, 1968).

The picture, then, is the following. A stimulating solution impinges on the tip of the hair. Here it mixes with the fluid in the terminal pore. Different latencies have been recorded, depending upon the solutions and concentrations employed as well as the particular event selected as a marker. A latency of 100 msec. from stimulation of the legs to extension of the proboscis was recorded cinematographically for tarsal sugar receptors (Dethier, 1955). Much of this delay can be attributed to the long conduction pathway from the tarsi through the thoracic ganglion to the brain, and to inertial properties of the proboscis and its muscles. A latency of 80–100 msec. from stimulation of a labellar hair with sugar to contraction of the muscles of the proboscis was recorded electrophysiologically by Dethier, Solomon, and Turner (1965). The same technique revealed a latency of 100 msec. when the stimulating solution was 0.2 M sodium chloride, and 64 msec. (to retraction of the proboscis) when the solution was 0.5 M sodium chloride (Dethier, 1968). Getting (1971) found a latency of 43 msec. to the appearance of the first motor spike (small unit), and 54 msec. (large unit) in the efferent fibers to the extensor muscles of the haustellum. The difference between latencies for spike generation in the hair and the appearance of motor spikes might represent synaptic delays and conduction times from the hairs to the extensor muscles. Latencies in the receptors themselves represent the time required for the stimulating solution to attain effective concentration in the pore, to initiate a receptor potential, and in turn to initiate action potentials.

Other electrophysiological techniques employed to measure the latency between the application of a stimulus to the tip of the hair and the appearance of the first spike gave the following values: for 0.01 M NaCl, 1 msec (Hodgson and Barton Browne, 1960); for water, 6 msec. (Evans and Mellon, 1962b).

The generation of spikes that follows is characterized by the curve in Figure 75c (Rees, 1968). There is first a very rapid rise of the rate of action potentials to a maximum, followed by a precipitous decline. Until about 0.35 sec. from time zero there is a slower decline, almost a steady state. After removal of the stimulus there is a poststimulatory decline, followed by a slow recovery after the next five seconds. Although adaptation (of the salt receptors) is generally considered to be complete by 0.35 sec., slow decline does actually continue, as Rees (1968) has indicated. Furthermore, it must be remembered that adaptation time varies with the concentration of the stimulating solution. The fact that the times for complete behavioral adaptation are considerably longer suggests that the receptor is still sending useful information to the brain even when in the steady (tonic) phase; or that, as some hairs adapt, others disadapt.

The nature of the primary event, how a stimulus initiates a receptor potential, is still a matter for lively discussion and currently the subject of a great deal of experimentation. Effort has divided itself principally into a search for the mechanism of stimulation of the salt receptor and a search for the mechanism underlying stimulation of the sugar receptor by carbohydrates. Most studies have been based on conditions in the "completely" adapted or steady-state phase of the response.

Nearly two decades ago Beidler (1954) had advanced a hypothesis to explain stimulation of the taste receptors of the rat by electrolytes. Beidler started with the following assumptions: (1) the stimulus reacts with a receptor substance; (2) the reaction obeys the law of mass action; the stimulus is more likely in equilibrium than in a steady state; (3) the magnitude of response is directly related to the number of ions or molecules that have reacted with the receptor. An equation was derived that relates the magnitude of response to the concentration of the stimulating solution: $c/R = (c/R_m) + (1/KR_m)$, where c is the concentration, R the magnitude of response, R_m the magnitude of maximum response, and K the equilibrium constant. The validity of the equation was tested by measuring c and R. This was done by recording and measuring total integrated responses in the form of action potentials from bundles of fibers in the chorda tympani. A linear relationship between c and R was found. Now, knowing the equilibrium constant, Beidler calculated the free energies of reaction from the expression $\Delta F = RT \ln K$, where ΔF is the change of free energy, R the gas constant, and T the absolute temperature. The low value found for ΔF (−1.22 to −1.37 cal/mole) for a series of sodium salts was interpreted as indicating that physical rather than chemical forces are involved in the initial interaction between the stimulus and the receptor. It is highly unlikely that enzymatic forces are involved in this stage.

The low values of ΔF together with the small temperature dependence suggested a reaction similar to those that occur with ion binding by proteins and natural polyelectrolytes. The absence of a pH effect over the wide range 3.0–11.0 suggested that the molecules of the receptor substance are strong acidic radicals. The implication of relatively weak carboxyl radicals of proteins is unlikely. Phosphate and sulphate radicals of such natural polyelectrolytes as nucleic acids and some polysaccharides are able to bind cations in a manner consistent with the properties just outlined.

In a more extensive exposition of his hypothesis Beidler (1961, 1963, 1967) related it to a model of a typical membrane. The membrane is composed of many charged amino acid side-chains that could bind ions. There are also the PO_4 groups of phospholipids. Various experiments have shown that the amino acids are not involved in binding; it is assumed, therefore, that most cations are bound to PO_4. Some sites are better for sodium, some for potassium, etc., and the kinds and numbers are different in different species of animals. Nonelectrolytes are presumed to react with protein by hydrogen bonding where there are many available sites. Steric configuration is important. Anions inhibit cations and are bound at other sites independently. The net result of all this activity is presumed to be a change in configuration of macromolecules in the receptor surface, with the result that a hole is formed. Potassium would leak out of this hole and initiate the electrical events that constitute the first measurable response.

Evans and Mellon (1962a) tested the applicability of Beidler's theory to the salt receptor of *Phormia*. The great advantage of the blowfly over the rat as an experimental organism is that response can be measured more directly and accurately because of the possibility of recording from a single primary receptor. The same assumptions and expressions that Beidler had made were applied, with the difference that thermodynamic activity was substituted for concentration, and number of sites filled for magnitude of response. Again the relation between magnitude of response and concentration of stimulus was found experimentally to be linear. A value of -0.06 to -0.73 kcal/mole was calculated for ΔF. Evans and Mellon concluded that the Beidler hypothesis was valid for *Phormia* and that anionic and strongly acidic radicals in the receptor substrate were involved. They suggested PO_4 or SO_4 groups.

Recently, theories that have been based upon steady-state conditions (e.g. Beidler, 1954; Evans and Mellon, 1962a) have been questioned on the grounds that the relevant part of the response of a receptor is the phasic segment rather than the tonic. Indeed it is true that the rat is able to make a discrimination within 250 msec. of the onset of stimulation (Halpern and Tapper, 1971). *Phormia* also acts upon information contained in the first few milliseconds of spike discharge (Dethier, 1968; Getting, 1971). Nevertheless, even though the phasic part of the response is instrumental in initiating action (be it acceptance or rejection), information contained in the tonic phase apparently also is used. It need not necessarily be required to maintain feeding but it can be conceived of as monitoring feeding, in the sense that whatever behavior is initiated continues as long as there is no change in the input but that any change in input brings about a reaction. In any case, it is reasonable to demand that a theory of mechanism of action be able to explain both phasic and tonic phases of the response (Faull and Halpern, 1972).

The various hypotheses attempted to offer an explanation of the nature of the initial reaction between stimulus and receptor, but they did not explain how depolarization of the dendritic membrane would result. A model to explain this step in the sequence of events in *Phormia* was proposed by Rees (1968). This model permits predictions of the a.c. and d.c. response-concentration curves following stimulation by NaCl, KCl, KBr, and KI. It is proposed that the driving currents of the electromotive forces through the loop result from the difference between potentials established across receptor (tip) and generator membranes. Normally one would not expect electrotonic transmission of d.c. displacements from the receptor membrane over so great a distance as a 300 μ hair; however, the wall of the small lumen acts as a very effective insulator between two longitudinal conductors (dendrite and large lumen contents) and prevents great cable losses.

This displacement in potential across the receptor membrane of the salt receptor during stimulation by NaCl, etc., can be explained by applying a derivative of the Hodgkin and Katz (1949) equation to the probable conditions of ionic activity on either side of the receptor membrane. The only ions within the dendrite assumed to play a part are Na^+, K^+, and Cl^- (Rees, 1967).

All monavalent salts probably act alike. The salt cell responds to a whole range of alkali chlorides and sodium halides in a manner indistinguishable from that to sodium chloride (Gillary, 1966c; Rees and Hori, 1968; den Otter, 1972a). Mixing and cross adaptation showed that all act the same way (Gillary, 1966c). Most workers have found that the stimulating effectiveness of cations is greatest for potassium and declines progressively as atomic number either increases or decreases (Gillary, 1966c). This finding is in general agreement with behavioral observations. For the tarsi of the horsefly *Tabanus sulcifrons,* for example, the order of increasing effectiveness is (Frings and O'Neal, 1946)

$$Li^+ < Na^+ < Mg^{++} < Ca^{++} = Sr^{++} < K^+ \leqq Cs^+ = Rb^+ < NH_4^+ < < < H^+.$$

The hierarchy is stable at all molarities. However, there are minor species differences (in vertebrates as well), and it is risky to generalize. Roughly speaking, the order is the same as that of partition coefficients, which parallel ionic mobilities. An intensive electrophysiological study employing the flesh fly *Calliphora vicina* yielded somewhat different results (den Otter, 1972a). The receptors involved were one (the salt receptor) in tarsal hairs and sometimes two (the salt cell and the "fifth" cell) in labellar hairs. For these, the order of increasing effectiveness generally was $Li^+ < Na^+ < Cs^+ < Rb^+ < K^+$; however, in some hairs the order was different. Den Otter also discovered that nonsalt receptors frequently responded to these salts but were insensitive to changes in concentration. The order of effectiveness of cations, as far as could be told, was the same as for salt receptors. The order is not a function of atomic weights or ionic mobilities (see also Gillary, 1966c). The irregular sequence was interpreted as indicating a reaction with highly polarizable negative groups, probably phosphates, in the dendritic membrane (den Otter, 1972a).

The order of effectiveness of anions is generally $I > NO_3 > Cl > F$ (Steinhardt, 1965) and $I > Br > Cl$ (Rees, 1968). Their effectiveness increases monotonically with atomic numbers (Gillary, 1966c). Clearly, contrary to the beliefs of Evans and Mellon (1962b), both the cation and the anion are very important in determining the stimulating effectiveness of a salt.

Divalent salts proved difficult for investigators to understand. Behavioral studies had shown that they, like monovalent salts, caused rejection. With some exceptions, notably salts of heavy metals, the behavioral response was normal. Salts of heavy metals were different in that they inhibited or damaged the sugar receptor for long periods (1 to 5 hours) even when applied briefly at millimolar concentrations (Deonier, 1938; Dethier, 1955). Other divalent salts were said to behave anomalously in so far as electrophysiological responses were concerned (Evans and Mellon, 1962b). Calcium and magnesium chloride failed to stimulate the salt receptor (Evans and Mellon, 1962b). Calcium hyperpolarized the salt receptor (Morita, 1959) and $CaCl_2$ and $MgCl_2$ inhibited the salt receptor and the water receptor (Morita and Yamashita, 1966). The oral papillae respond quite differently; calcium evokes intense activity from one of the four receptors (Dethier and Hanson, 1965).

It was proposed (Rees and Hori, 1968) that calcium produces three effects: (1) it dilutes the solution normally present in the chamber of the pore of the hair, thus increasing electrical resistance of this component of the loop-conductor; (2) if the membrane is impermeable to calcium, the activity of Cl^- will increase as the concentration of the $CaCl_2$ in the tip increases and there will be hyperpolarization due to the increased Cl^-; (3) the divalent cation combines with some component of the receptor membrane, reduces its permeability to K^+, Na^+, and Cl^- and possibly results in an accumulation of ions on the inside of the membrane. Upon removal of the $CaCl_2$ there is an after-discharge from the salt cell. Normally, in the resting receptor the concentration of permanent ions inside at the receptor tip is presumed to be less than it

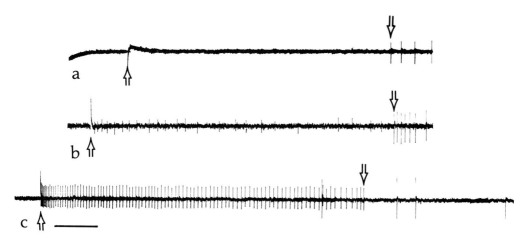

Fig. 76. Records illustrating the discharge of the salt cell following removal of a water stimulus. Arrows pointing up indicate the onset of stimulation by water. Arrows pointing down indicate the removal of water. In (a) (labellar hair "largest" no. 10), water did not elicit discharges from the water cell; nevertheless there was an after-discharge from the salt cell. In (b) (labellar hair "largest" no. 9) the salt cell begins to respond as soon as water is removed. In (c) (labellar hair "largest" no. 1) there is a delay between the removal of water and the response of the salt cell. All recordings are side-wall. Horizontal bar = 100 msec.

is in the more proximal regions of the dendrite because of membrane leakage at the tip. External concentrations are probably lower than internal. If calcium blocks leakage at the receptor tip, K^+, Na^+, and Cl^- will accumulate there internally. After removal of $CaCl_2$, Ca^+ ions at the tip will diffuse away into the large chamber. Reduction of external calcium will result in a desorption of calcium from the receptor tip membrane. Then the K^+, Na^+, and Cl^- accumulated inside will reestablish their normal gradients across the membrane and may carry a depolarizing current, resulting in an after-discharge. No after-discharge followed water; however, we have found that this does occur (Fig. 76).

Of the various divalent ions that depress spike activity, namely Ba^{++}, Sr^{++}, Ca^{++}, Mg^{++}, and Be^{++}, calcium is the most effective, and effectiveness decreases as atomic numbers diverge from that of calcium. Calcium nitrate, iodide, and acetate resemble the chloride in their action.

Gathering together all of the available data relating to stimulation of the salt receptor by electrolytes, den Otter (1972a, 1972b, 1972c) advanced a new hypothesis to explain the interaction between ions and the dendritic membrane of the salt receptor. Having assessed earlier hypotheses, he concluded that Rees's is an oversimplification. If no chemisorptive process is involved in transduction (Rees, 1968), the generator potential should occur as a mere effect of change in external ionic concentration, and dendritic membranes should have constant relative permeability to different ions. Beidler's hypothesis and the use of it made by Evans and Mellon are rejected insofar as their applicability to the receptors of blowflies is concerned, on two counts: (1) they do not account for the influence that anions exert in stimulation; (2) in *Calliphora* there is no linear relationship between concentration of stimulating solution and magnitude of response.

The hypothesis proposed by den Otter may be summarized as follows. Antagonism between calcium ions in the membrane and ions of the applied salts plays a role in the process of stimulation. Calcium holds the molecules of the membrane together by forming linkages between anionic sites. Anions in stimulating solutions may screen calcium ions in the membrane, leading to breakage of linkages and thus increasing the number of freely moving anionic groups. This change will open (depolarize) the membrane. Fixation of cations to negative groups in the membrane will increase compactness and may hyperpolarize the membrane.

For a detailed discussion of ion selectivity of biological membranes the review of Diamond and Wright (1969) should be consulted.

Although the effort to understand stimulus/receptor interaction has been directed more toward electrolytes than toward any other class of compounds, there is ample behavioral evidence that nonelectrolytes can also initiate rejection. The rejection thresholds of more than two hundred aliphatic compounds were measured; the flies used had been rendered anosmic by removal of antennae and labella (Dethier and Chadwick, 1947, 1948, 1950; Chadwick and Dethier, 1947, 1949; Dethier, 1951). The relationships revealed by these data solved some of the puzzle as to how compounds with limited or no solubility in water were just as able as electrolytes to cause rejection.

Within any homologous series, exceptionally high correlations were found between the stimulating power of the compounds and such properties as boiling point, molecular area, oil-water partition coefficients, molecular moments, vapor pressures, and activity coefficients. Molecular weight, number of carbon atoms, and osmotic pressure were eliminated from consideration, in part by the use of isomers, which proved usually to have different thresholds. Also the correlation with vapor pressures was inverse. Because of the paucity of data relating to the various chemical and physical properties enumerated above, it became convenient to plot stimulative efficiency against the chain length of the molecule. Graphs so constructed (Fig. 77) show clearly that, for all series of homologous aliphatic compounds studied, the members of the series are rejected at logarithmically decreasing concentrations as the carbon chain length is increased. It is apparent

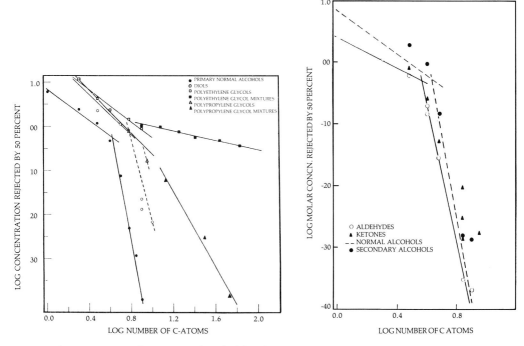

Fig. 77. (a) Tarsal rejection thresholds of glycols and alcohols by *Phormia*. (b) Same for aldehydes and ketones (from Dethier and Chadwick, 1948).

from Figure 77 that this relation is not a continuous one. For each series the curve shows a sharp break in the region of a definite chain-length characteristic for each series.

Taking a saturated straight-chain hydrocarbon as a starting point, one may, without altering appreciably the arrangement of the carbon linkages, substitute various kinds of polar groups for one or more of the hydrogen atoms. Whereas all substitutions raise the molecular weight and the boiling point, the effect on solubility is variable. The different polar groups may be arranged in increasing order of solubility as follows: $Br < Cl < CH_3 < CHO < C = O < OH$. The CH_3 radical represents the unsubstituted compound. This series also represents in reverse order the relative stimulating efficiencies of these polar groups. When the number of substitutions of the groups to the right of the CH_3 (in the above representation) is increased, the stimulating efficiency decreases. Thus, glycols are less effective than alcohols, and diketones are as a rule less effective than ketones. Increase of the number of substitutions of those polar groups to the left of the CH_3 tends to increase stimulating effectiveness. Thus, dibromo compounds are more effective than monobromo homologues.

The effects of the positions of the functional groups are illustrated

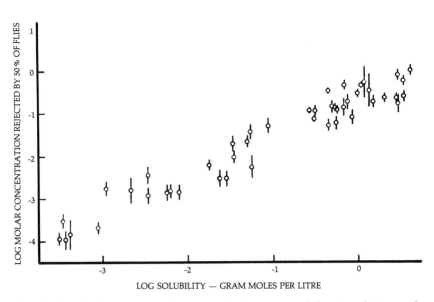

Fig. 78. Correlation between the concentration required for stimulation and the solubility of compounds at 25–27°C in water. Each of the 46 points represents a different aliphatic compound (from Dethier and Chadwick, 1950).

by experiments with the glycols. The following rules are found to hold. (1) Juxtaposition of two hydroxyl groups in a short molecule (e.g., 1, 2-butanediol) makes for a high threshold; this effect is reduced as chains become longer. (2) In a chain with no terminal OH groups, i.e., subterminal but not adjacent (e.g., 1, 3-butanediol) the thresholds are low. (3) If both OH groups are terminal, thresholds are intermediate. (4) Branching tends to raise the threshold, other factors being equal.

Taken all together, the foregoing results state essentially that the length of the free alkyl group largely determines the stimulating effectiveness and that its power is modified to varying degrees by the nature of the attached polar groups. The length of the alkyl group is also of prime importance in determining the solubility characteristics of compounds of this type; hence the same structural characteristics that decrease water solubility likewise decrease threshold.

There is then only one molecular property, for which data are available, that brings all the data from the different series into a single homogeneous system. This is water solubility (Fig. 78). The order of stimulative efficiencies follows the inverse of the order of water solubilities with fewer contradictions than appear in most of the other comparisons attempted. Additional evidence that solubility is of importance in this connection has been presented by Dethier (1951), who showed that the thresholds of alcohols are altered as the alcohol is presented as an aqueous solution, a glycol solution, or a mineral oil solution (Fig. 79).

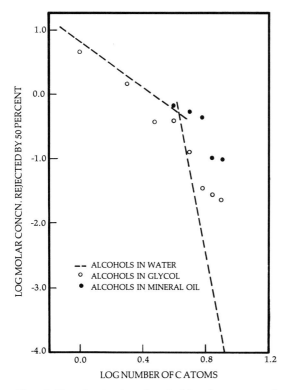

Fig. 79. Tarsal rejection thresholds of aqueous, glycol, and oil solutions of primary alcohols by *Phormia* (from Dethier, 1951).

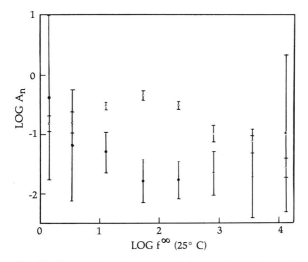

Fig. 80. Comparison in terms of thermodynamic activity of the effectiveness of the first eight normal alcohols acting in aqueous solution (open circles) on tarsal chemoreceptors and as gases (solid circles) on olfactory receptors of *Phormia*. In each case the value represents a threshold of rejection. The vertical lines represent 2.575 standard errors for aqueous solutions and 2 for vapors (from Dethier and Yost, 1952).

When threshold values are expressed as thermodynamic activities rather than as moles, the differences between successive homologues of a series are not so marked, but a plot of the logarithm of these values against the logarithm of activity coefficients (Fig. 80) does not produce a straight line of the sort that one is accustomed to expect from parallel experiments on narcosis; for a complete discussion consult Ferguson (1951), Brink and Posternak (1948), Dethier (1954).

In spite of the generally good correspondence between low solubility in water and high stimulating power, it seems likely, from the data on oil-water partition coefficients, that plots of threshold values against water solubilities would also yield a smaller slope for the lower than for the higher range of compounds in each series, if such an analysis could be made. This type of relationship seems to have no counterpart in any of the tabulated values for the physical properties, and its consistent recurrence prompted Chadwick and Dethier (1949) to consider the possibility that different forces may be of primary importance in stimulation by the lower and higher members of each of the types investigated. This amounts to postulating at least a two-phase system for the limiting mechanism in contact chemoreception. The hypothesis that small molecules gain access to the receptors in part through an aqueous phase, whereas the larger aliphatic molecules penetrate chiefly through (or accumulate in) a lipoid phase, would appear to offer a basis for reconciling most of the contradictions encountered when an attempt is made to fit the facts into a single-phase system.

Movement of the smaller molecules through an aqueous medium should occur at rates related inversely to the molecular weight, which would help to account for their being more stimulating than is expected from the relationship found for the higher members of the series, although it is doubtful that the entire difference can be explained in this way. It may be noted also that the inflections in the curves relating thresholds to molecular size occur at increasing chain lengths in passing from the less to the more water-soluble species. At the same time, the predominant importance of lipoid affinity is suggested by the logarithmically increasing stimulating power of both lower and higher members of all series, as well as by the inverse relationship between water solubility and stimulating effectiveness in comparisons of the several series with each other.

Further analyses of the data of Dethier and Chadwick (1948, 1950) suggest that the important factor in stimulation by nonelectrolytes is adsorption (Davies and Taylor, 1959). When the logarithm of rejection thresholds (molar concentration) is plotted against the logarithm of the adsorption constant for an oil/water interface, the slope of the resulting line is very near that expected for adsorption from solution onto a pure lipid membrane.

These results are in striking agreement with a wealth of data from experiments dealing with the effect of nonelectrolytes in a wide variety of biological membranes. Diamond and Wright (1969) have summarized nonelectrolyte selectivity of biological membranes by observing that its basis lies in differences in intermolecular forces between solute:water and solute:lipid as operating in a bulk lipid phase. Exceptions are attributed to the more organized structure of lipids in cell membranes. Furthermore, there are some predominantly polar regions in membranes that bypass lipids and offer parallel access to small polar solutes. The most important forces in addition to differences in water: solute intermolecular forces are hydrogen bonds, van der Waals forces, and entropy effects in hydrocarbon:water interactions.

Considering specifically the fly, den Otter (1972c) noted that the plasma membrane of the receptor undoubtedly has as its framework a bimolecular leaflet composed mainly of phospholipids but possibly also helical segments of protein. Changes in the permeability of the membrane may be assumed to result from opening or closing of "organized pores" and "statistical pores," the latter arising as a consequence of thermal agitation of molecules in the membrane. The effects of substances dissolved in water on the packing of molecules in the membrane are assumed to depend on the size and shape of these substances and on their distribution between the water and lipid bilayer of the system, and within the bilayer itself.

In the foregoing discussion the expression "stimulating power" has been employed. On the other hand, the chemical/physiological relationships observed are strikingly similar to narcosis throughout. If the compounds do indeed act as narcotics, the expression "stimulating power" can be applied correctly only to the behavioral event. One experiment, however, stood in the way of accepting the idea of narcosis. It was discovered that while sugar stimulation of one leg caused extension of the proboscis, simultaneous stimulation of the contralateral leg with propanol caused rejection (Dethier, 1950). From this result it was argued that some positive message (rather than an absence of messages due to narcosis) did reach the central nervous system from that leg, and that the arrival of this message in the central nervous system caused contralateral inhibition there of excitatory input from the sugar receptors of the other leg. Behavioral rejection resulted. In retrospect, with the electrophysiological evidence now available, it is likely that there were indeed messages from the leg in propanol but that these were injury potentials, such as Hanson (1965), Hodgson and Steinhardt (1967), and Steinhardt, Morita, and Hodgson (1966) have shown to occur, rather than stimulation in the normal sense. Although injury potentials can indeed mediate behavioral rejection, it is more likely that the usually observed rejection occurs as a result of narcosis. Low concentrations of alcohols were found (by sidewall recording) to inhibit reversibly the salt, sugar, and water receptors in that order. The effects on the salt and water receptors resembled narcosis of nerve; the effect on the sugar receptor was different.

Hanson (1965) proposed that the contradiction between the results of contralateral rejection and electrophysiologically demonstrated narcosis could be resolved by considering the fact that excitation responsible for proboscis extension is the algebraic sum of all sensory inputs, of which water is one even if the animal is water-satiated. Propanol inhibits the water cell. In the control behavioral experiment, where water and sucrose were applied to opposite legs, both water and sucrose provided excitatory input to the central nervous system. When alcohol and sucrose were applied to opposite legs, excitation from the water cell was greatly reduced because of narcosis; therefore, the central nervous system received excitation from the sucrose alone. Thus there was a smaller *total* excitation and a lower probability of response. A similar explanation would apply to the contralateral experiments involving other alcohol-sucrose-water combinations.

Whatever theory ultimately triumphs and explains the mode of action of compounds that are behaviorally unacceptable, the fact remains that more than one mechanism of action occurs. The existence of multiple mechanisms provides for the detection of many unrelated compounds that could be toxic, nonnutritious, or token stimuli associated with potential danger. To what extent these different modes result in the generation of messages with different meanings for the central nervous system cannot be assessed until we are able to read the neural messages and correlate them with behavior. This is the subject matter of Chapter 7.

Literature Cited

Arab, Y. M. 1957. A study of some aspects of contact chemoreception in the blowfly. Doctoral dissertation, Johns Hopkins University, Baltimore.

Arab, Y. M. 1959. Some chemosensory mechanisms in the blowfly. *Bull. Coll. Sci.*, University of Baghdad, 4:77–85.

Beidler, L. M. 1954. A theory of taste stimulation. *J. Gen. Psysiol.* 38:133–139.

Beidler, L. M. 1961. Taste receptor stimulation. In *Progress in Biophysics and Biophysical Chemistry*, Vol. 12, J. A. V. Butler, H. E. Huxley, and R. E. Zirkle eds., pp. 107–151. Pergamon Press, Oxford.

Beidler, L. M. 1963. Dynamics of taste cells. In *Olfaction and Taste*, I, Y. Zotterman ed., pp. 133–148. Pergamon Press, Oxford.

Beidler, L. M. 1967. Anion influences on taste receptor response. In *Olfaction and Taste*, II, T. Hayashi ed., pp. 509–534. Pergamon Press, Oxford.

Boistel, J., and Coraboeuf, E. 1953. L'activité électrique dans l'antenne isolée de Lépidoptère au cours de l'étude de l'olfaction. *C. R. Soc. Biol.*, Paris, 147:1172–1175.

Boistel, J., Lecompte, J., and Coraboeuf, E. 1956. Quelques aspects de l'étude électrophysiologique des récepteurs sensoriels des antennes d'hymenoptères. *Insectes Sociaux* 3:25–31.

Brink, F., and Posternak, J. M. 1948. Thermodynamic analysis of the relative effectiveness of narcotics, *J. Comp. Cell. Physiol.* 32:211–233.

Chadwick, L. E., and Dethier, V. G. 1947. The relationship between chemical structure and the response of blowflies to tarsal stimulation by aliphatic acids. *J. Gen. Physiol.* 30:255–262.

Chadwick, L. E., and Dethier, V. G. 1949. Stimulation of tarsal receptors of the blowfly by aliphatic aldehydes and ketones. *J. Gen. Physiol.* 32:445–452.

Davies, J. T., and Taylor, F. H. 1959. The role of adsorption and molecular morphology in olfaction: the calculation of olfactory thresholds. *Biol. Bull.* 117:222–238.

den Otter, C. J. 1972a. Differential sensitivity of insect chemoreceptors to alkali cations. *J. Insect Physiol.* 18:109–131.

den Otter, C. J. 1972b. Interactions between ions and receptor membrane in insect taste cells. *J. Insect Physiol.* 18:389–402.

den Otter, C. J. 1972c. Mechanism of stimulation of insect taste cells by organic substances. *J. Insect Physiol.* 18:615–625.

den Otter, C. J., and van der Poel, A. M. 1965. Stimulation of three receptors in labellar chemosensory hairs of *Calliphora erythrocephala* Mg. by monovalent salts. *Nature* 206:31–32.

Deonier, C. C. 1938. Effects of some common poisons in sucrose solutions on the chemoreceptors of the housefly, *Musca domestica* L. *J. Econ. Ent.* 31:742–745.

Dethier, V. G. 1941. The function of the antennal receptors in lepidopterous larvae. *Biol. Bull.* 80:403–414.

Dethier, V. G. 1950. Central summation following contralateral stimulation of tarsal chemoreceptors. *Proceedings, Fed. Amer. Soc. Exp. Biol.* 9:31–32.

Dethier, V. G. 1951. The limiting mechanism in tarsal chemoreception. *J. Gen. Physiol.* 35:55–65.

Dethier, V. G. 1954. The physiology of olfaction in insects. *Ann. N.Y. Acad. Sci.* 58:139–157.

Dethier, V. G. 1955. The physiology and histology of the contact chemoreceptors of the blowfly. *Quart. Rev. Biol.* 30:348–371.

Dethier, V. G. 1956. Chemoreceptor mechanisms. In *Molecular Structure and Functional Activity of Nerve Cells*, R. G. Grenell and L. J. Mullins, eds., pp. 1–30. A. I. B. S., Washington, D.C.

Dethier, V. G. 1968. Chemosensory input and taste discrimination in the blowfly. Science, 161:389–391.

Dethier, V. G. 1972. Sensitivity of the contact chemoreceptors of the blowfly to vapors. *Proc. Nat. Acad. Sci. U.S.A.* 69:2189–2192.

Dethier, V. G., and Arab, Y. M. 1958. Effect of temperature on the contact chemoreceptors of the blowfly. *J. Insect Physiol.* 2:153–161.

Dethier, V. G., and Chadwick, L. E. 1947. Rejection thresholds of the blowfly for a series of aliphatic alcohols. *J. Gen. Physiol.* 30:247–253.

Dethier, V. G., and Chadwick, L. E. 1948. The stimulating effect of glycols and their polymers on the tarsal receptors of blowflies. *J. Gen. Physiol.* 32:139–151.

Dethier, V. G., and Chadwick, L. E. 1950. An analysis of the relationship between solubility and stimulating effect in tarsal chemoreception. *J. Gen. Physiol.* 33:589–599.

Dethier, V. G., Evans, D. R., and Rhoades, M. V. 1956. Some factors controlling the ingestion of carbohydrates by the blowfly. *Biol. Bull.* 111:204–222.

Dethier, V. G., and Hanson, F. E. 1965. Taste papillae of the blowfly. *J. Cell. Comp. Physiol.* 65:93–100.

Dethier, V. G., and Hanson, F. E. 1968. Electrophysiological responses of the blowfly to sodium salts of fatty acids. *Proc. Nat. Acad. Sci. U.S.A.* 60:1296–1303.

Dethier, V. G., Solomon, R. L., and Turner, L. H. 1965. Sensory input and central excitation and inhibition in the blowfly. *J. Comp. Physiol. Phychol.* 60:303–313.

Dethier, V. G., and Yost, M. T. 1952. Olfactory stimulation of blowflies by homologous alcohols. *J. Gen. Physiol.* 35:823–839.

Diamond, J. M., and Wright, E. M. 1969. Biological membranes: The physical basis of ion and nonelectrolyte selectivity. *Annu. Rev. Physiol.* 31:581–646.

Evans, D. R., and Mellon, De F. 1962a. Stimulation of a primary taste receptor by salt. *J. Gen. Physiol.* 45:651–661.

Evans, D. R., and Mellon, De F. 1962b. Electrophysiological studies of a water receptor associated with the taste sensilla of the blowfly. *J. Gen. Physiol.* 45:487–500.

Faull, J. R., and Halpern, B. P. 1972. Taste stimuli: time course of peripheral nerve response and theoretical models. *Science* 178:73–75.

Ferguson, J. 1951. Relations between thermodynamic indices of narcotic potency and the molecular structure of narcotics. In *Mechanisme de la Narcose*, 26:25–39. Coll. Internat. Centre Nat. Rech. Sci.

Frings, H., and O'Neal, B. R. 1946. The loci and thresholds of contact chemoreceptors in females of the horsefly, Tabanus sulcifrons Macq. *J. Exp. Zool.* 103:61–80.

Galun, R. 1955. Physiological responses of three nutritionally diverse dipterous insects to selected carbohydrates. Doctoral dissertation, University of Illinois, Urbana.

Getting, P. A. 1971. The sensory control of motor output in fly proboscis extension. *Zeit. vergl. Physiol.* 74:103–120.

Gillary, H. L. 1966a. Stimulation of the salt receptor of the blowfly. I. NaCl. *J. Gen. Physiol.* 50:337–350.

Gillary, H. L. 1966b. Stimulation of the salt receptor of the blowfly. II. Temperature. *J. Gen. Physiol.* 50:351–357.

Gillary, H. L. 1966c. Stimulation of the salt receptor of the blowfly. III. The alkali halides. *J. Gen. Physiol.* 50:359–368.

Grabowski, C. T., and Dethier, V. G. 1954. The structure of the tarsal chemoreceptors of the blowfly *Phormia regina* Meigen. *J. Morph.* 94:1–20.

Halpern, B. P., and Tapper, D. N. 1971. Taste stimuli: quality coding time. *Science* 171:1256–1258.

Hanson, F. E. 1965. Electrophysiological studies on chemoreceptors of the blowfly. Doctoral dissertation, University of Pennsylvania, Philadelphia.

Hassett, C. C., Dethier, V. G., and Gans, J. 1950. A comparison of nutritive values and taste thresholds of carbohydrates for the blowfly. *Biol. Bull.* 99:446–453.

Hodgkin, A. L., and Katz, B. 1949. The effect of sodium ions on the electrical activity of the giant axon of the squid. *J. Physiol.* 108:37–77.

Hodgson, E. S. 1956. Temperature sensitivity of primary chemoreceptors of insects. *Anat. Rec.* 125:560–561.

Hodgson, E. S. 1958. Electrophysiological studies of arthropod chemoreception. III. Chemoreceptors of terrestrial and fresh water arthropods. *Biol. Bull.* 115:114–125.

Hodgson, E. S. 1968. Taste receptors of arthropods. *Symp. Zool. Soc. London* (1968) 23:269–277.

Hodgson, E. S., and Barton Browne, L. 1960. Electrophysiology of blowfly taste receptors. *Anat. Rec.* 137:365.

Hodgson, E. S., Lettvin, J. Y., and Roeder, K. D. 1955. Physiology of a primary chemoreceptor unit. *Science* 122:417–418.

Hodgson, E. S., and Roeder, K. D. 1956. Electrophysiological studies of arthropod chemoreception. I. General properties of the labellar chemoreceptors of Diptera. *J. Cell. Comp. Physiol.* 48:51–76.

Hodgson, E. S., and Steinhardt, R. A. 1967. Hydrocarbon inhibition of primary chemoreceptor cells. In *Olfaction and Taste,* IV, T. Hayashi ed., pp. 739–748. Pergamon Press, Oxford.

Mellon, De F., and Evans, D. R. 1961. Electrophysiological evidence that water stimulates a fourth sensory cell in the blowfly taste receptor. *Amer. Zool.* 1:372.

Morita, M. 1959. Initiation of spike potentials in contact chemosensory hairs of insects. III. D.C. stimulation and generator potential of labellar chemoreceptor of Calliphora. *J. Cell. Comp. Physiol.* 54:189–204.

Morita, H., Doira, S., Takeda, K., and Kuwabara, M. 1957. Electrical responses of contact chemoreceptor on tarsus of the butterfly, *Vanessa indica. Mem. Fac. Sci. Kyushu Univ.* E2:119–139.

Morita, M., and Takeda, K. 1959. Initiation of spike potentials in contact chemosensory hairs of insects. II. The effect of electric current on tarsal chemosensory hairs of Vanessa. *J. Cell. Comp. Physiol.* 54:177–187.

Morita, H., and Yamashita, S. 1959a. Generator potential of insect chemoreceptor. *Science* 130:922.

Morita, H., and Yamashita, S. 1959b. The back-firing of impulses in a labellar chemosensory hair of the fly. *Mem. Fac. Sci. Kyushu Univ.* Ser. E (Biol.), 3:81–87.

Morita, H., and Yamashita, S. 1966. Further studies on the receptor potential of the blowfly. *Mem. Fac. Sci. Kyushu Univ.* E4:83–93.

Rees, C. J. C. 1967. Transmission of receptor potential in dipteran chemoreceptors. *Nature* 215:301–302.

Rees, C. J. C. 1968. The effect of aqueous solutions of some 1:1 electrolytes on the electrical response of the type 1 ("salt") chemoreceptor cell in the labella of *Phormia*. *J. Insect Physiol.* 14:1331–1364.

Rees, C. J. C., and Hori, N. 1968. The effect of electrolytes of the general formula XCl_2 on the response of the type 1 labellar chemoreceptor of the blowfly *Phormia*. *J. Insect Physiol.* 14:1499–1513.

Roys, C. 1954. Olfactory nerve potentials a direct measure of chemoreception in insects. *Ann. N.Y. Acad. Sci.* 58:250–255.

Schneider, D. 1955. Mikro-Elektroden registrieren die elektrischen Impulse einzelner Sinnes-nervenzellen der Schmetterlingsantenne. *Industrie-Elektronik* (Elektro-Spezial, Hamburg) 3:3–7.

Schneider, D. 1957a. Electrophysiological investigation on the antennal receptors of the Silk Moth during chemical and mechanical stimulation. *Experientia* 13:89–91.

Schneider, D. 1957b. Elektrophysiologische Untersuchungen von Chemo-und Mechanorezeptoren der Antenne des Seidenspinners *Bombyx mori* L. *Zeit. vergl. Physiol.* 40:8–41.

Schneider, D., and Hecker, E. 1956. Elektrophysiologie der Antenne des Seidenspinners *Bombyx mori* bei Reizung mit angereicherten Extrakten des Sexuallockstoffes. *Zeit. Naturf.* 11b:121–124.

Smyth, T., and Roys, C. C. 1955. Insect chemoreception and the action of DDT. *Biol. Bull.* 108:66–76.

Steinhart, R. A. 1965. Cation and anion stimulation of electrolyte receptors of the blowfly, *Phormia regina*. *Amer. Zool.* 5:651.

Steinhardt, R. A., Morita, H., and Hodgson, E. S. 1966. Mode of action of straight chain hydrocarbons on primary chemoreceptors of the blowfly, *Phormia regina*. *J. Cell. Physiol.* 67:53–62.

Stürckow, B. 1960. Elektrophysiologische Untersuchungen am Chemorezeptor von *Calliphora erythrocephala* Meigen. *Zeit. vergl. Physiol.* 43:141–148.

Stürckow, B. 1965. Electrophysiological studies of a single taste hair of the fly during stimulation by a flowing system. *Proc. XVI Int. Congr. Zool.*, Washington, D.C. 1963, Vol. 3:102–104.

Stürckow, B. 1970. Responses of olfactory and gustatory cells in insects. In *Communication By Chemical Signals,* Vol. 1, J. W. Johnston, D. G. Moulton, and A. Turk, eds., pp. 107–159. Appleton-Century-Crofts, New York.

Stürckow, B. 1971. Electrical impedance of the labellar taste hairs of the blowfly, *Calliphora erythrocephala* Mg. *Zeit. vergl. Physiol.* 72:131–143.

Stürckow, B., Holbert, P. E., and Adams, J. R. 1967. Fine structure of the tip of chemosensitive hairs in two blow flies and the stable fly. *Experientia* 23/9:1–7.

Tateda, H., and Morita, H. 1959. Initiation of spike potentials in contact chemosensory hairs of insects. I. Generating site of the recorded spike potentials. *J. Cell. Comp. Physiol.* 54:171–176.

Uehara, S., and Morita, H. 1972. The effect of temperature on the labellar chemoreceptors of the blowfly. *J. Gen. Physiol.* 59:213–226.

Wolbarsht, M. L. 1957. Water taste in *Phormia*. *Science* 125:1248.

Wolbarsht, M. L. 1958. Electrical activity in the chemoreceptors of the blowfly. II. Responses to electrical stimulation. *J. Gen. Physiol.* 42:413–428.

Wolbarsht, M. L., and Dethier, V. G. 1958. Electrical activity in the chemoreceptors of the blowfly. I. Responses to chemical and mechanical stimulation. *J. Gen. Physiol.* 42:393–412.

Wolbarsht, M. L., and Hanson, F. E. 1965. Electrical activity in the chemoreceptors of the blowfly. III. Dendritic action potentials. *J. Gen. Physiol.* 48:673–683.

Wolbarsht, M. L., and Hanson, F. E. 1967. Electrical and behavioral response to amino acid stimulation in the blowfly. In *Olfaction and Taste,* II, T. Hayashi ed., pp. 749–760. Pergamon Press, Oxford.

Wykes, G. R. 1952. The preference of honeybees for solutions of various sugars which occur in nectar. *J. Exp. Biol.* 29:511–519.

6
The Sweet Tooth and the Freshness of Water

And a perpetual feast of nectar'd sweets,
Where no crude surfeit reigns.

Milton, *Comus*

Next to water, sucrose is the most universally acceptable compound. Considering that sucrose is an ideal source of instant energy and is generally abundant and ubiquitous, this popularity is not surprising. Acceptance of sugar is an eminently reasonable evolutionary adaptation especially with respect to herbivores. On the other hand, with certain exceptions (e.g., aphids, corn borers, etc.), herbivores do not select their food on the basis of its sugar content (Dethier, 1973; Drickamer, 1972), nor indeed is any unusually great quantity of sugar detectable in most plant material (notable exceptions being sugar cane, sugar beet, carrot, and similar plants with high sugar concentrations). Another curious aspect of sensitivity to sugar is that it may be as well developed in nonherbivores as, for example, the rat. Even more interesting is the powerful behavioral effect that sugar exerts. Animals will strive mightily to obtain sugar whether there is a metabolic need or not, and they will eat it to the point of becoming ill. Such a driving influence is matched in normal animals only by that produced by water and salt. In contrast to sugar, water and salt evoke obsessive behavior only when there is serious deprivation. It is only the thirsty animal that fights for water; it is only the salt-deprived animal that consistently comes to the salt lick. In human cultures whole civilizations, patterns of commerce, and patterns of conquest have been built on the need for water and salt. In this respect the craving for sugar is anomalous. Hedonistic factors appear to be more influential than metabolic ones.

In human experience sugar is identified with the sensation "sweet." The same sensation is evoked to varying degrees by a limited odd assortment of totally unrelated compounds, chief among them being some L-amino acids, certain synthetic aromatic compounds such as saccharin (*o*-sulphobenzimide), dulcin (*p*-ethoxphenyl urea), cyclamates (Na and Ca cyclohexylsulphamate), chloroform, glycerine, etc., as well as lead acetate, beryllium salts, the protein from the African fruit *Thaumatoccus daniellii* Benth (van der Wel, 1972), the protein monellin from the African plant *Dioscoreophyllum cumminsii* (Stapf) (Cagan, 1973), and the glycoprotein (molecular weight $\pm$ 44,000) miraculin from the African "miracle fruit," *Synsepalum dulcificum* (Schum. and Thonn.), which is actually a taste modifier in that it renders the taste of sour solutions sweet.

It has always been assumed that a single receptor subserves the sensation "sweet." Insofar as nonhuman vertebrates are concerned, it has been a working principle that substances that evoke a response similar to that elicited by sucrose belong in the same behavioral category, and the modality is equivalent to "sweet." This working hypothesis is checked by making comparisons against a basic response to sucrose or glucose and by cross-adaptation. Additionally, if a given concentration of one substance can be matched by some other concentration of a second substance so that an animal cannot distinguish one from the other, the gustatory modality can be assumed to be the same for both. In short, our approach to the study of the "sweet taste" experienced by other animals has been to test their responses to substances that are sweet to us. This course of action is safest as long as one explores among the carbohydrates; it becomes more hazardous as totally unrelated compounds are investigated. The best insurance against error is provided by electrophysiological evidence that the physiologically unknown compounds stimulate the same receptor that responds to sugar.

Against the background of human experience, the earliest studies of the "sweet tooth" of insects aimed at a comparison of the relative stimulating effectiveness of sugars and closely related carbohydrates. Although the motivation underlying these investigations varied (sometimes the purpose was to discover a better bait) analyses of the data usually ended as attempts to relate effectiveness to chemical structure. These comparisons laid the groundwork for a characterization of the sugar receptor.

Two of the most extensive early studies were those of von Frisch (1935) with the honeybee and Dethier (1955a) with the black blowfly. They were based on the assumption, supported by training experiments (with the honeybee), tests with mixtures, and tests for cross adaptation, that a single kind of receptor was involved. The criterion for judging the comparative stimulating effectiveness of the test compounds was some measure of acceptability (extension of the proboscis, duration of feeding, crop-load, two-choice tests, etc.). The two studies were in agreement in concluding that the stimulating quality of sugars simply does not find expression in the structural formula of the compounds. With the blowfly the following conclusions were made:

(1) Molecular size is important. Sugars shorter than pentoses and longer than hexoses are ineffective. Some mono-, di-, and trisaccharides do stimulate. Oligosaccharides larger than trisaccharides are nonstimulating. (2) Linear polyols (e.g., D-glucitol) are ineffective or at best weakly stimulating; Haslinger (1935) found that *Calliphora,* when starved, responded to polyhydric alcohols, and Evans (1963) was able to elicit responses from *Phormia* to saturated solutions of mannitol and glucitol by starving his flies. (3) The carbonyl group is not required for stimulation. Myo-inositol and nonreducing glycosides are effective. (4) *Alpha* glucoside derivatives are more stimulating than their *beta* isomers. (5) The D isomers tend to be more effective than the L. D-arabinose gives a lower threshold value than L-arabinose.

The concept of a single receptor could not be reconciled with known chemical configurations, and no further studies of this sort were undertaken until the research took a new direction. A hint of a solution to the puzzle was offered by the discovery that synergism and competitive inhibition occurred between certain sugars, especially that fructose was specifically inhibited by mannose (Dethier, 1955a; Dethier, Evans, and Rhoades, 1956). On the basis of this finding and other apparently unrelated bits of information, Evans (1963) proposed that the sugar receptor had to have multiple combining sites with different structural requirements in order to account for the extreme structural specificity on the one hand and the odd array of effective molecules on the other. The idea of multiple sites had been developed earlier by Beidler (1961a, 1961b, 1963) as a result of investigations of the sense of taste of rats. A taste receptor was conceived of as having multiple specific sites at the molecular level. The numbers and kinds vary from receptor to receptor and species to species. A chemical can react with two different kinds of sites, and several ions can compete for the same site.

In the fly, the selective inhibition by mannose suggested that there was a glucose site and a fructose site. Evans reevaluated the earlier data of Hassett, Dethier, and Gans (1950) and Dethier (1955a). He then proceeded to analyze in detail the structural requirements for the receptor site for glucose. By comparing the effectiveness relative to glucose of twenty-eight closely related carbohydrates in the D series he came to the conclusion that the hydroxyl groups on C_3 and C_4 alone are responsible for stimulation, that furanose forms of glucose cannot stimulate (the reverse of Dethier's conclusion), and that linear polyols can stimulate weakly because they are free to rotate around single C—C bonds and all conformations are potentially rapidly interconvertible.

Table 12. Relative Stimulating Effectiveness (Related to Sucrose = 1) of Oligosaccharides and Glycosides.[a]

Compound	Structure	Stimulating effectiveness
p-Nitrophenyl-α-glucopyranoside	Glc α-φ NO$_2$	approx. 10
Sucrose	Glc α 1-2 βFruf	approx. 1
Turanose	Glc α 1-3 Frup	approx. 1
Palatinose	Glc α 1-6 Fru	approx. 1
Maltose	Glc α 1-4 Glc	approx. 1
Maltotriose	Glc α 1-4 Glc α 1-4 Glc	approx. 1
Maltopentaose	Glc α 1-(4 Glc α 1)$_3$-4 Glc	approx. 1
Melezitose	Glc α 1-3 Fruf β 2-1 α Glc	approx. 0.5
Methyl-α-glucopyranoside	Glc α 1-CH$_3$	approx. 0.4
Trehalose	Glc α 1-1 α Glc	0.05 –0.1
Gentianose	Glc β 1-6 Glc α 1-2 β Fruf	0.05 –0.1
Raffinose	Gal α 1-6 Glc α 1-2 β Fruf	0.05 –0.1
Methyl-β-glucopyranoside	Glc β 1-CH$_3$	0.05 –0.1
Cellobiose	Glc β 1-4 Glc	0.02 –0.1
p-Nitrophenyl-β-glucopyranoside	Glc β 1-φ NO$_2$	0.02 –0.05
Panose	Glc α 1-6 Glc α 1-4 Glc	0.02 –0.05
Planteose	Glc α 1-2 β Fruf 6-1 α Gal	0.01 –0.05
Stachyose	Gal α 1-6 Gal α 1-6 Glc α 1-2 β Fruf	0.01 –0.05
Gentiobiose	Glc β 1-6 Glc	0.003–0.01
Melibiose	Gal α 1-6 Glc	0.003–0.01
p-Nitrophenyl-α-mannopyranoside	Man α 1-φ NO$_2$	<0.01
Lactose	Gal β 1-4 Glc	0
Methyl-α-mannopyranoside	Man α -CH$_3$	0
Phenyl-α-mannopyranoside	Man α 1-φ	0
Methyl-α-rhamnopyranoside	Rha α 1-CH$_3$	0

From Pflumm, 1972.

[a] With the exception of methyl-α-rhamnopyranoside all compounds are of the D-series.

Now that the sugar receptor was conceived of as being molecularly heterogeneous insofar as its acceptor was concerned, further analyses of structural requirements for stimulation seemed profitable. Kijima (1970) reviewed the conformational characteristics of molecules in substances that are "sweet" for insects. He classified the effectively stimulating monosaccharides into the fructopyranoside type of 1C conformation and the glucopyranoside type of C1 conformation. He did not consider furanose structures because of reports that methyl-furanosides are ineffective even though p-fructopyranose does stimulate. More tests with additional compounds (Table 12) on *Phormia terraenovae* by Pflumm (1971, 1972) showed agreement with earlier work but presented evidence that there is indeed a furanose site in addition to the glucose and fructose sites. Fructose was envisioned as combining with both the fructose and the furanose sites. Sucrose was presumed to combine at two sites, the glucose site and the furanose site (Pflumm, 1972).

Analyses of concentration-response curves, estimates of affinities for the sugar site, measurements of maximum responses obtained, and calculation of the Hill coefficient for a number of glycosides acting on the labellum of *B. peregrina* led Hanamori, et al. (1972) to the following conclusions: (1) a polar group such as OH both in α and β positions is not necessary for stimulation; (2) α-glucose has the same affinity as β-glucose but a larger maximum response and a smaller Hill coefficient; (3) a polar group like the ethylene glycol residue in the α position is favorable for both affinity and maximum response; (4) nonpolar groups (methyl, ethyl, p-nitrophenyl) in the α position are favorable for affinity but not for maximum response, whereas in the β position they are unfavorable in both respects; (5) derivatives with β configuration show a tendency to give concentration-response curves with a higher Hill coefficient than those with a β configuration (Hanamori et al., 1972). Although a polar group is not necessary on C1, the nature of the group there affects the value of the maximum response (e.g., a hydrophilic group increases affinity but decreases maximum response). In tests with two glycosides consisting of aglycons of the same size but different polar properties (e.g., ethyl α-glucopyranoside and ethylene glycol α-glucopyranoside) the *relative* concentration-response curves were the same, the affinities almost identical, but the maximum response was greater for ethylene glycol α-glucoside. All of these results confirm the results of Dethier (1955a) and Evans (1963).

Structural similarities between sugars and cyclitols suggested to Jakinovich and Agranoff (1972) that these chemically stable compounds would be model substances for the study of stereospecificity on the fly's sugar receptor. Of the nine isomeric cyclitols only two stimulated "largest" labellar hairs, namely, *myo*-inositol, as previously reported (Dethier, 1955a; Hodgson, 1957), and D-chiroinositol.

An extensive electrophysiological study by Morita and Shiraishi (1968) and Shiraishi and Morita (1969) of stimulation of "largest" labellar hairs of the fleshfly *Boettcherisca peregrina* by sugars provided additional evidence for molecular heterogeneity and added details to the characterization of the sites. The starting assumption was that the magnitude of the response of the receptor is proportional to the

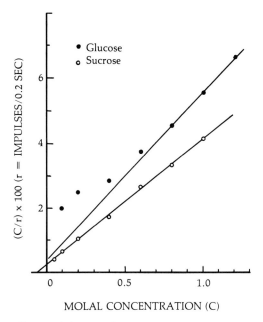

Fig. 81. Relation of response to sugar concentration plotted after the equations of Lineweaver and Burk (1934) and Beidler (1954). Note the deviation of glucose from a straight line (from Morita and Shiraishi, 1968).

number of receptor sites occupied (Morita, Hidaka, and Shiraishi, 1966; Morita and Shiraishi, 1968). Each site was assumed to be occupied by one molecule of sugar. This is the basis of Beidler's hypothesis, which is modeled on the Michaelis-Menton theory of enzyme reactions. Morita and Shiraishi then measured these relations experimentally and used Lineweaver's and Burk's (1934) modification of the Michaelis-Menton equation to analyze the results. The outcome of this analysis is depicted in Figure 81. They tested sucrose, maltose, glucose, fructose, and mannose. As the figure shows, data for sucrose give a straight line, except at low concentrations, whereas those for glucose do not. For disaccharides the results are in agreement with the hypothesis that one molecule combines with one site. The model for stimulation by monosaccharides requires two molecules at a site.

Hodgson (1957) had been unable to detect synergism or inhibition by mannose electrophysiologically. He did report inhibition by ribose and 2-desoxyribose. Morita and Shiraishi found that at most concentrations glucose and fructose were simply additive. They concluded that if there were two sites, glucose could occupy the fructose site as well as its own and that differentiation between the two sites was poor. On the other hand, mannose does indeed inhibit fructose and has a synergistic effect with glucose and sucrose; therefore, mannose must occupy one of the sites that two molecules of fructose normally

Table 13. Effect of Stimulation of the Sugar Receptor by Mixtures as Compared With Individual Components.

		Mixture		
Sugar 1	Concentration (M)	Sugar 2	Concentration (M)	p^a
Mannose	1.0	Fructose	0.1	0.005
	0.75	Fructose	0.25	0.005
	0.30	Fructose	0.10	0.005
	0.05	Fructose	0.05	0.025
	0.75	Sorbose	0.25	0.01
	0.75	Glucose	0.25	0.005
	0.75	D-arabinose	0.25	0.005
	0.75	Fucose	0.25	0.01
	0.75	Fucose	0.25	>0.05
	0.75	Trehalose	0.25	0.025
	0.75	Trehalose	0.25	>0.05
	0.75	Alpha-D-methyl-glucoside	0.25	0.005
	1.0	Maltose	0.10	0.005
	0.75	Maltose	0.25	>0.05
	1.0	Sucrose	0.10	>0.05
	0.75	Sucrose	0.25	>0.05
	0.30	Sucrose	0.10	>0.05
	0.75	Melezitose	0.25	>0.05
	1.0	Turanose	0.10	0.005
	0.75	Turanose	0.25	0.005
	0.75	Turanose	0.25	0.005
Fructose	0.10	Glucose	0.10	>0.05
	0.05	Glucose	0.05	0.005
	0.025	Glucose	0.025	>0.05
	0.0039	Glucose	0.0039	>0.05
	0.0039	Glucose	0.0039	0.005
	0.05	Sorbose	0.05	0.005
	0.05	Sorbose	0.0167	0.005
	0.05	Sorbose	0.005	0.005

From Omand and Dethier, 1969.

The individual components were at the same concentration in the mixtures as when used separately. The effect was synergistic when responses to mixtures were greater than the sum of responses to individual components. When responses to mixtures were less than those to the more effective component, the effect was termed inhibition.

[a] p values determined by the application of the Wilcoxon matched-pair test.
[b] One animal also tested with mannose and sorbose at the same concentrations.
[c] One animal also tested with fructose plus mannose at the same concentrations.
[d] This animal also tested with 0.05 M glucose plus 0.05 M fructose and with 0.1 M glucose plus 0.1 M fructose.
[e] This animal also tested with 0.05 M glucose plus 0.05 M fructose.

(Table 13. Continued)

Effect	Number of hairs	Number of flies tested	Age (days)
Inhib.	23	2	2, 3
Inhib.	53	5	2, 3, 4
Inhib.	30	3	1
Inhib.	11	1	1
Inhib.	22	2	2, 4
Syn.	22	2^b	4, 4
Syn.	10	1	2
Syn.	9	1^c	3
None	13	1^c	3
Syn.	26	3^c	3, 4
None	7	1^c	4
Syn.	21	2	2, 4
Syn.	24	2	3
None	15	1	2
None	24	2	2, 3
None	18	2	1
None	19	2^c	1
None	17	2	2
Syn.	27	2	2, 3
Syn.	15	1^c	2
Inhib.	13	1	4
None	27	3	1
Syn.	38	4	1, 2
None	8	1^d	1
None	18	2	1
Syn.	9	1	3
Inhib.	56	5	1, 2
Inhib.	12	1^e	1
Inhib.	12		

occupy. The simplest explanation of these results is that there is one type of receptor site rather than two on the sugar cell and that it has two subunits that must be occupied simultaneously for excitation to occur. Monosaccharides accomplish this with one molecule; disaccharides require two. Fructose has the highest affinity of all sugars but has very weak competitive effects on other sugars and is deeply depressed by low concentrations of mannose. If we assume that one receptor site can be shared by each sucrose and fructose molecule in a mixture of the two sugars and that this shared complex is more effective than the two molecules per site of fructose alone, then only slight fructose inhibition should be observed in mixtures. This assumption predicts that stimulation by a fructose-glucose mixture forms the complex, fructose-glucose-receptor site, and that there should be synergism at low concentrations. Dethier et al. had found it behaviorally. Morita and Shiraishi found it once electrophysiologically. Omand and Dethier (1969) also found it electrophysiologically.

Independently, Omand and Dethier (1969) were working along similar lines. Their experiments fell into three categories: (1) stimulation with mixtures and their individual components; (2) stimulation before and immediately after prolonged exposure to mannose; (3) stimulation with individual sugars over a wide range of concentrations. The results from the first series are summarized in Table 13, which shows that behavioral indications of synergism and inhibition are confirmed (Fig. 82). The second series showed that the inhibiting sugar, in this case mannose, need not be present simultaneously with the stimulating sugar for inhibition to occur. Exposure to mannose for two minutes greatly reduced subsequent activity in response to fructose (Table 14). Prior exposure reduced activity in response to sucrose and maltose, whereas in mixtures there was no synergism or other change (cf. Table 13). The reduction was less (62% and 36%), however, than fructose (80%) and may have been due to a nonspecific deleterious effect of mannose. Concentration-response relationships of the different sugars, experimental series 3, revealed marked differences (Fig. 83). For all sugars tested response was proportional to the logarithm of concentration except at very low concentrations. The response of the sugar receptor to the ketose sugars fructose and sorbose was affected less by changes in concentration than were those to glucose, D-arabinose, or maltose. The curves of responses to sucrose had lower slopes than that relating to glucose. The x intercepts for fructose and sorbose had lower values than those for either glucose or D-arabinose and were about the same as those for the disaccharides sucrose and maltose (Table 15). The values of the x intercepts, with the exception of that for sorbose, were not significantly different from the thresholds of tarsi determined behaviorally (Hassett, Dethier, and Gans, 1950). The x intercept for sorbose was lower than anticipated.

The sugars may be divided into two, and possibly three, groups on the basis of the slopes and intercepts of concentration-response curves and of responses to mixtures. The sugar receptor is almost exactly three times more sensitive to changes in the concentration of aldose and aldose-derived sugars than it is to changes in the concentration of ketoses. The response curves relating to aldoses intersect the x axis at higher concentrations than those relating to the ketose sugars tested. The addition of mannose enhances responses to the aldose and aldose-derived sugars and consistently reduces responses to the ketoses. Ad-

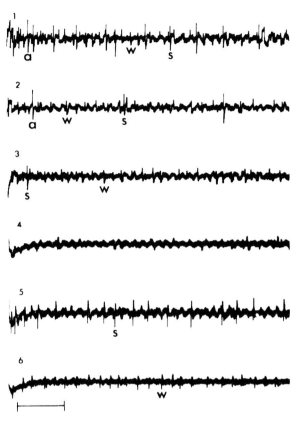

Fig. 82. Electrophysiological records of inhibition (1–3) and synergism (4–6). Stimulating solutions were: (1) 0.2 M fructose; (2) 0.2 M fructose plus 1.6 M mannose; (3) 1.6 M mannose, all on the same hair; (4) 0.25 M turanose; (5) 0.25 M turanose plus 0.75 M mannose; and (6) 0.75 M mannose, all on the same hair. Note that the frequency of discharge of the water receptor was markedly reduced by turanose. The horizontal bar represents 0.1 sec. The symbols indicate impulses of: (w) the water receptor; (s) the sugar receptor; (a) the water and the sugar receptor occurring simultaneously (from Omand and Dethier, 1969).

Table 14. Effect of Two-minute Exposure to 1.0 M Mannose on Responses to Sugars Applied One Second Later.

Sugar	Per cent of pretreatment response	Number of hairs	Number of flies tested
Fructose (0.1 M)	12 ± 9	32	8
Maltose (0.1 M)	38 ± 18	13	3
Sucrose (0.1 M)	64 ± 9	11	3

From Omand and Dethier, 1969.
Confidence intervals (99%) are indicated.

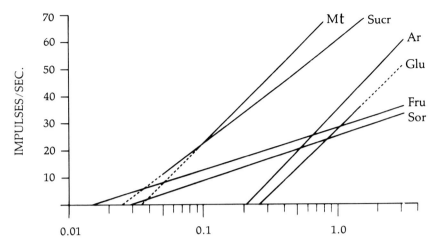

Fig. 83. Concentration-response curves of the sugar receptor for six sugars; maltose (Mt), sucrose (Sucr), D-arabinose (Ar), glucose (Glu), fructose (Fru), and sorbose (Sor). Solid lines indicate ranges of concentrations tested (from Omand and Dethier, 1969).

Table 15. Relation of Response to Concentration.

	Electrophysiological responses of labellar hairs					
				Number of		
Sugar	Slope[a]	x intercept (molar conc.)	flies	hairs	concentrations	Behavioral responses[b]
Fructose	16 ± 2	0.015 ± 0.01	11	111	4–10	0.0058
Sorbose	17 ± 3	0.029 ± 0.02	6	63	4–6	0.140
Glucose	48 ± 2	0.264 ± 0.12	11	111	4–10	0.132
D-arabinose	52 ± 7	0.210 ± 0.09	3	30	4	0.144
Maltose	50 ± 13	0.035 ± 0.02	3	30	3	0.0043
Sucrose	38 ± 7	0.025 ± 0.02	6	62	3–5	0.0098

From Omand and Dethier, 1969.

[a] The slope is expressed as impulses per second per logarithm unit.

[b] From Hassett, Dethier, and Gans (1950). Tarsal thresholds are expressed as molar concentrations. Differences representing not less than 1.5 doubling concentration steps are significant at the 1% level.

ditionally, responses to the ketose fructose are enhanced by the presence of the aldose, glucose, and are reduced by the presence of the ketose, sorbose.

Disaccharides are anomalous insofar as this categorization is concerned. Maltose (aldose-derived) approximates the aldose sugars in that the slope of the response curve is identical to that representing glucose. Mixtures of maltose and mannose show synergism. The x intercept for the curve for maltose is lower than those for glucose and D-arabinose. The curve relating sucrose has a slope intermediate in relation to aldoses and ketoses. It also gives a low intercept. Melezitose, which is also made up of aldose (glucose) and ketose (fructose) units (2G-1F) does not interact with mannose. Turanose (1G-1F) mixed with mannose gave equivocal results: synergism with younger flies, inhibition in the oldest animal tested.

The low x intercepts for maltose and sucrose gave rise to the suggestion that the disaccharides may have some features in common in their mode of action; however, it is also possible that disaccharides containing only glucose units and those containing both glucose and fructose may act in different ways. This possibility, together with the different stimulating characteristics of ketose and aldose sugars, suggested that the sugar receptor may be sensitive to both furanose and pyranose forms of sugars. Taken together, all of the data are in agreement with the early behavioral data and the results of Morita and Shiraishi (1968).

Although it is now agreed that the sugar receptor is composed of a number of different sites, their nature and number is still an unsettled issue. Evans (1963) and Omand and Dethier (1969) suggested a glucose and a fructose site. Morita and Shiraishi (1968) postulated for *Phormia regina* and *Boettcherisca peregrina* one site with two subunits. Each had to be occupied simultaneously, so that two molecules of monosaccharide but only one of disaccharide were required. With sucrose, the fructofuranose ring was assumed to occupy the fructose site. This idea was later modified (Morita, 1972) because of a report that the fructofuranose ring was ineffective. It was then assumed that the furanose ring of sucrose occupies a site for aglycons which is not highly specific; that is, sucrose acts merely like an α-glucoside. On the other hand, Pflumm (1972), using *Phormia terraenovae,* and Hansen and Kühner (1972) (on *Sarcophaga bullata*) found that fructopyranosides were ineffective and suggested that fructose may stimulate in its open chain structure.

Shimada et al. (1974) have presented direct evidence of the existence of two sites, for which they propose the terms pyranose site and furanose site. In their view D-fructose, D-fucose, and D-galactose react "chiefly" with the furanose site, whereas D-glucose, D-arabinose, L-fucose, L-arabinose, L-sorbose, D-xylose, and L-glucose react principally with the pyranose site.

However many sites there may be, little attention had been paid to the probable events that take place between the fitting of a proper molecule to a specific site and the initiation of an electrical event. The sugar receptor is unique in that it is stimulated by uncharged molecules, whereas almost every other kind of excitable cell is stimulated by charged particles or ions. How the uncharged molecules manage to trigger electrical events is a mystery that has excited the interest of many investigators and so far been solved by none. Morita (1967) had proposed a mechanism based upon the effect of salt on the sugar response as studied in *B. peregrina*. All investigators have observed that salt depresses the frequency of impulses from the sugar receptor. It is generally agreed that the effect is initiated at a place and time antecedent to the generation of spikes. At some concentrations, however, salts increase the response to sugar in the increasing order of effectiveness Cs^+, $NH_4^+ > K^+ = Na^+ > Li^+ >> Ca^{++}$. Morita concluded that Cl^- has no special importance in producing the membrane current but that cations do and that the membrane is nonselectively permeable to them. A clue to the possible mechanism of action came from the observation that deionized distilled water stimulated both the salt cell and the water cell. It was concluded that water applied to the tip of the hair caused cations in the extracellular fluid elsewhere in the hair to diffuse to the region immediately outside of the receptor membrane. Sugar was presumed to do the same thing, at which point the cations passed into the cell activated by the sugar, and a membrane current was initiated.

It had early been shown that such metabolic inhibitors as azide, fluoride, cyanide, and iodoacetate did not block stimulation; therefore, it was concluded that enzymes involved in the glycolytic cycle were not necessary for stimulation (Dethier, 1955a). Furthermore, in behavioral tests sugars were still effective over a wide range of pH and over a temperature range of 0–35°C (Dethier and Arab, 1958). Frings and Cox (1954) reported that high (37–39°) and low (19–21°) temperatures elevated the tarsal threshold of *Sarcophaga bullata* to sucrose; however, the entire fly was subjected to these temperature changes so that little can be concluded about the effect of temperature specifically on the receptors. When electrophysiological techniques became available, many of these stimulus variables were reexamined and new ones studied. In particular, attention was paid to the effects of pH, temperature, and salts. Independence of pH was confirmed for the range pH 3–10. Above and below these values, however, inhibition occurred (Shiraishi and Morita, 1969). At very high values the sugar receptor is excited in the absence of sugar. Analyses of these findings led to a hypothesis that there are at least five ionizable groups at the receptor site, but that these are not directly coupled with the response to sugars.

The story with respect to temperature was different. Although no behavioral effect had been observed, Uehara and Morita (1972) demonstrated that the stationary amplitudes of potentials from the salt and sugar (but not water) receptors decreased at 12°C to 50–80% of the val-

Fig. 84. Various trapping arrangements employed in the study of the mode of action of sugar-baited traps. (a) Baited with dry sucrose; (b) with dry sucrose beneath a wire screen partition; (c) with dry sucrose and living flies beneath a single screen partition; (d) with dry sucrose and living flies beneath a double partition; (e) empty trap with two funnels, the top one of which was soiled by flies (from Dethier, 1955b).

ues measured at 28°C. In effect, the maximum rate of rise of the receptor potential strongly increased with rising temperature. These results were interpreted as favoring Morita's (1969) regulator model over the "complex" one. Morita and Shiraishi (1968) had proposed, tentatively, a model in which the receptor site is composed of allosteric macromolecules. In support of that proposition are the facts that: (a) the specificity of particular sugars can be attributed only to macromolecular structure; (b) uncharged molecules like sugars may induce electrical charges in membranes only through structural changes in the receptor sites and these changes should be related to or synonymous with allosteric transitions; (c) the allosteric model agrees better with experimental facts, i.e., the relation between response magnitude and the number of sites, than other models, e.g., Beidler's (1954, 1970) model for salt.

The next clue directing research into the nature of the sugar receptor came from unexpected sources. In the process of screening candidate fly baits for the United States Army, Barnhard and Chadwick (1953) discovered that bait attended by flies was rendered more attractive than untrod bait. In one experiment half of a petri dish containing a bait composed of malt extract was protected from flies while the other half was exposed. A few minutes later all flies were chased away and the entire bait exposed. Almost all of the returning flies ignored the formerly protected side (Fig. 22). Interest in the "fly-factor" prompted a detailed investigation that employed glass fly traps (Fig. 84) and confirmed the conclusion that flies made surfaces attractive by regurgitating and defecating on the substrate. Dethier (1955b) sealed the mouth and anus of each of a number of flies and stood them for one hour in a solution of sucrose. Analysis of the sucrose then revealed that it had been hydrolyzed. A series of controls confirmed the conclusion that there was an enzyme on or in the legs that did not come from contamination with saliva or feces and exceeded by many orders of magnitude the quantity found elsewhere on the body. Tests with several different kinds of sugars indicated that the enzyme in question was an α-glucosidase that reduced sucrose, melezitose, and raffinose in decreasing order of effectiveness. Tests with glucose, fructose, and lactose failed to reveal the presence of other enzymes. This finding was later confirmed by Wiesmann (1960).

The discovery was picked up by Hansen (1968, 1969), who undertook a quantitative topographical study. By comparing the quantity of enzyme on the different tarsal segments and femur and tibia of each leg with the number of chemoreceptive sensilla on the respective segments he was able to demonstrate a positive correlation between concentration and a number of hairs. He made the following additional conclusions: (1) Only substrates of the glucosidase give low behavioral thresholds, for example, sucrose (Glc-1-α-2-β-Frut), maltose (Glc-1-α-4-Glc), and nitrophenyl-α-glucoside; other nonhydrolizable sugars give higher thresholds (cellobiose, Glc-1-β-4-Glc) or are not accepted at all (lactose, Gal-1-β-4-Glc). (2) The proportionality between the affinity constants of the enzyme (apparent Michaelis constants) and the behavioral thresholds is approximately constant for all substrates. (3) Higher alkylamines and alkylalcohols (C_6-C_8) inhibit the enzyme and the receptor (Steinhardt, Morita, and Hodgson, 1966) in the same allosteric way.

Hansen proposed that the glucosidase was part of the receptor membrane and acted as an acceptor protein for α-glucosidic disaccharides and α-glucosides. Membrane permeability, and hence the frequency of the action potentials, would be controlled by the percentage of reversible glucosidase-substrate complexes, being itself a function of the surrounding substrate (stimulus) concentration. This mechanism would be described by Beidler's (1954) equation and by the Michaelis equation derived for the fly's sugar receptor (Morita, Hidaka, and Shiraishi, 1966; Morita and Shiraishi, 1968).

The obvious next move was to attempt to extract the enzyme in a purified state. When crude extracts of 400 tarsi of *Phormia terraenovae* were fractionated by gel chromatography, three peaks of activity were obtained (Hansen and Kühner, 1972). Peaks II and III exhibited the greatest activity and differed in a number of respects. The substrates (sugars) were split with different maximal velocities by the two fractions; the highest rate was with sucrose followed by p-nitrophenyl-α-glucoside. Maltose and isomaltose were split better by III than by II. Turanose was split better by II and melezitose solely by II. The apparent Michaelis constants for the same substrate were different in II and III; the pH optimum of III was slightly more acidic than that for II; for II the Michaelis law was obeyed at low substrate concentrations only, and there was an indication that glucosidase II had more than one substrate site.

Whether one or both enzymes are involved in stimulation by sugars, and if one, which one, is uncertain. Enzyme III shows higher activity in the distal four tarsomeres than in the first but can be washed off partially. Significant amounts of enzyme II are found in regions richly clothed with chemoreceptive hairs. In crude homogenates the proportions of the Michaelis constants for sucrose, turanose, melizitose, and p-nitrophenyl-α-glucoside are approximately the same as the proportions of the corresponding behavioral thresholds. After fractionation the constants that correlated with thresholds for sucrose and maltose belonged to two different enzymes. Either there are two glucosidases in each of the four types of tarsal hairs or each type contains its own enzyme.

Neither α-methylglucoside nor trehalose (Glc-1-α-Glc) are split by the enzymes even though they possess α-glucosidic linkages. A specific trehalase is responsible for hydrolyzing trehalose. Both sugars inhibit enzyme II, indicating that they are bound by it. Since both exhibit higher thresholds than sucrose, Hansen and Kühner concluded that affinity constants and thresholds are related only if splitting occurs, and possibly the primary process of reception may involve the entire enzyme reaction rather than the formation of the enzyme-substrate complex only as earlier proposed (Hansen, 1968). Thus the energy of the glucosidic bond would be of paramount importance in reception.

Many complex properties are characteristic of the glucosidases. There is inhibition by glucose, fructose, and α-methylglucoside. The inhibition of hydrolysis of sucrose by fructose provides further evidence for the existence of a fructofuranoid site (Dethier, 1955a; Morita and Shiraishi, 1968; Pflumm, 1972). There are other examples of inhibition and synergism (cf. Dethier, Evans, and Rhoades, 1956; Omand and Dethier, 1969), some of which are competitive and others of which are noncompetitive. Glucosidase II obviously has more than one binding site for monosaccharides. All of these kinetic complexities, apparently common for glucosidases, suggest that these enzymes might also be involved in stimulation by sugars that are not substrates for the enzymes (Hansen and Kühner, 1972).

Hansen's hypothesis stimulated a flurry of additional biochemical work (Morita, 1972; Amakawa et al., 1972; Hanamori et al., 1972; Kawabata et al., 1973; Kijima, Koizumi, and Morita, 1973; Shimada et al., 1974; Koizumi et al., 1973, 1974). Purified isozymes of α-glucosidase were extracted from the labella and tarsi of *Phormia regina* and *Boettcherisca peregrina*. The most likely candidate for sugar reception is enzyme II. An α-glucosidase having the characteristics of enzyme II was demonstrated in the tip of an intact chemosensory hair where it was fixed and never dissolved into the reaction mixture (Kijima et al., 1973). All of the data thus far available strengthen Hansen's hypothesis that α-glucosidase is the sugar receptor molecule.

As the picture now stands there is an implicit assumption that there is reversible complexing between the receptor and the stimulus, which results in a conformational change of the receptor molecule, a change that leads directly or indirectly to a permeability change in the dendrite. This differs from the earliest concept (Dethier, 1956) in that more details are injected into the model. The most likely candidates for the role of receptor proteins are α-glucosidases as proposed by Hansen (1968). However, discrepancies exist between the properties of the enzymes dissolved from the hairs and the properties of the sugar receptor. Koizumi, Kijima, and Morita (1973) have found that there is an α-glucosidase stably localized at the tip of the labellar hair that possesses properties different from those of the enzymes in the soluble form. The properties of the stable enzyme and those of the receptor are more compatible than those of the soluble enzyme and the receptor.

Pursuing the idea that α-glucosidases could be the receptor proteins, Koyama and Kurihara (1971) examined behaviorally the effects of several reagents that react with specific amino acid residues in the protein molecule; and Shimada, Shiraishi, Kijima, and Morita (1972) studied electrophysiologically the effects of some very specific sulphhydryl reagents on the "largest" labellar hairs of *B. peregrina*. It has been known that very low concentrations of the salts of heavy metals interfered with stimulation by sugars (Deonier, 1938; Dethier, 1955a), but the action of these is too nonspecific to give such information about the nature of the receptor. On the other hand, *p*-chloromercuribenzoate (PCMB) is highly specific for sulphydryl groups. Side-wall recording from "largest" hairs of *B. peregrina* revealed that 0.5 mM of PCMB depressed the sugar receptor and the salt receptor, but not the water receptor. The duration of the treatment was critical for recovery. After exposure for 3 minutes, recovery occurred within 30 minutes. There was no recovery after 10 minutes of treatment. Depression is easily reversed by treatment with 5 mM β-mercaptoethanol or 10 mM L-cysteine. Since no change in the shape of the impulse was observed it was concluded that the effects observed relate to the primary process of reception. Depression and spontaneous recovery were also obtained with 1,4-dithiothreitol (DTT). No effect followed application of N-ethylmaleimide (NEM).

Depression by PCMB differs from that by heavy metals not only in being more specific but also in being less rapid. PCMB reacts with sulphydryl groups presumably present in proteins of the dendritic membrane and blocks changes in permeability. It is envisioned as acting in two steps: it is absorbed; it makes bonds. It is noncompetitive in the first step; therefore, it does not act at the sugar site, but elsewhere. One can conclude, therefore, that it does not block the binding of sugar, but instead blocks the change in the membrane normally resulting from binding.

However the sugar receptor may work, any theory of action must take into account the fact that compounds other than sugars can also stimulate it. None of the artificial sweeteners (e.g., saccharin, sucaryl, monellin) do, but some amino acids are effective. Although the sole required diet for the blowfly during most of its adult life is carbohydrate, there are certain times in the reproductive cycle of the female when a protein meal is required. At these times, as Chapter 8 will describe, the female can differentiate between carbohydrate and protein even in the absence of olfactory cues. Knowledge of this behavioral preference prompted search for a protein receptor. None was found (Dethier, 1961). Although there was no evidence at that time that the blowfly could respond to pure amino acids, gustatory attractiveness of the phosphate salts of some amino acids to the housefly (*Musca domestica* L.) had been demonstrated (Robbins et al., 1965). Amino acids were also known to be effective stimuli for many marine invertebrates (Case and Guilliam, 1963; Laverack, 1963), for some phytophagous insects (Thorsteinson, 1960), and, of course, for man. An early attempt to detect electrophysiological responses to a series of DL- and L-amino acids applied to the labellar hairs of *Phormia* failed (Wolbarsht and Hanson, 1967). In no cases were impulses detected; however, three effects on receptor potentials were observed: a negative shift, a positive shift, no change. It was suggested that even though no impulses were generated and hence no behavioral response could be expected, the negative shift might indicate a subthreshold effect of some of the acids on salt and sugar receptors and that this might form the basis for synergism.

A later reinvestigation of stimulation by amino acids, however, did uncover effects. Shiraishi and Kuwabara (1970), testing L-forms instead of racemic mixtures discovered a wide and complex series of responses to these compounds. The conclusions did not solve the problem of protein differentiation, but they did change ideas about the specificity of receptors. On the basis of how amino acids react with labellar receptors they can be divided into four clearly defined categories: (1) completely nonstimulating (glycine, alanine, serine, threonine, cystine, tryosine); (2) inhibiting all cells (aspartic acid, glutamic acid, histidine, arginine, lysine); (3) stimulating the salt cell (proline, hydroxyproline); (4) stimulating the sugar cell (valine, leucine, isoleucine, methionine, phenylalanine, tryptophane). Curiously enough, members of class 4 taste bitter to man, but this is just one more respect in which the fly receptor for sugar differs from that in man (the receptor of *Phormia* is insensitive to artificial sweeteners and is not blocked by the specific sweet inhibitor, gymnemic acid). Extract of *Gymnema* does not affect the sugar receptors of *Sarcophaga* (Larimer and Oakley, 1968). Obviously the sugar receptor is not specific to sugars nor the salt receptor to salts. It is equally clear that there is no amino acid receptor as such, nor do amino acids constitute a functionally homogeneous group of compounds.

These findings were extended and generally confirmed by Goldrich (1973) who combined electrophysiological with behavioral studies.

One more acceptable compound remains to be discussed. Water, the universally acceptable compound, has a receptor all to itself in the fly. How it works is almost as much a mystery as how the sugar receptor works. The most effective stimulus thus far tested is pure water. Aqueous solutions of many nonelectrolytes depress its activity as a linear function of the logarithm of osmotic pressure. At 5-molal sucrose the frequency of impulses decreases to less than half of maximum (Evans and Mellon, 1962). Inorganic electrolytes also inhibit response to water but in a more specific manner. Sodium chloride begins to inhibit at 0.1 molal and completely inhibits at about 0.3 molal. The receptor is not depressed by *p*-chloromercuri-benzoate. Also, it does not show any temperature dependence. It is the most resistant of all cells to narcosis. It is unaffected by acetylcholine, cyanide, dinitrophenol, ATP, and ouabain (Rees, 1970). On the other hand it is greatly stimulated by some narcotics, especially Halothane (2-bromo-2-chloro-1, 1, 1-trifluoroethane) (Dethier, 1972).

One hypothesis regarding how water could stimulate was that proposed by Morita (1967); namely, that water caused the release in the region of the dendrite of cations originating elsewhere in intracellular space, and that the cations actually accomplished the stimulation. In speaking of water responses by vertebrates, Beidler (1967) argued that there is no fundamental difference between salt and water, only a difference in degree. In this view the vertebrate receptor is not responding to water but is spontaneously active because of the relative number of anionic and cationic groups in the receptor membrane. It is the relative effectiveness of these opposing groups that determines response.

Rees (1970) has advanced a hypothesis that electrokinetic streaming potentials are responsible for the receptor potential. A streaming potential is established across a semipermeable porous membrane if permanent ions are present that can be carried through it by the flow of water driven by an osmotic pressure difference, if the membrane is preferentially permeable either to anions or cations, and if the pores possess linings with a net positive or negative surface charge. As the solute concentration of stimulating sollutions decreases, depolarization of the water receptor increases. Ions must be involved in the passage of current through the membrane. With the application of water there is an outward downhill concentration gradient and selective anion egress must occur, influenced by the extracellular concentration of anions. Pores selective for cations (i.e., pores with a net negative surface charge) would permit the entry of cations and thus initiate depolarization.

Support for the hypothesis is derived from experiments that prove that the response is almost independent of concentration, species, and valency of anions but is strongly affected by cationic valency and concentration. The unimportance of anions rules out diffusion potentials as the possible source of the receptor potential. The cations probably govern the streaming potential by modifying the pore-surface zeta potential and the specific conductance of the contents of the pores. Inhibition of the response by pH below 5 and above 11 suggests the existence of negatively charged groups with an isoelectric point at about pH 3.7, lining the transmembrane pores. Observed inhibition by nonelectrolytes probably arises as a result of a reduction in the osmotic pressure difference across the membrane, which in turn reduces the streaming potential. Figure 85 expresses the situation diagrammatically.

Rees (1967) also suggested that most water receptors are in fact generalist ion receptors which may in addition exhibit sensitivity to changes in osmotic pressure. It is therefore the solute that is important rather than the solvent. Unambiguous information about the water content of a stimulus can be given only over the solute concentration range where the response is constant irrespective of concentration; this is below 0.1 mM. In the range between 0.1 mM and 50 mM NaCl, both cells respond. At 50 mM NaCl, the water cell stops responding and the salt cell begins. By employing two receptors with different ranges of ion sensitivity, the fly is the recipient of continuous information about the salt concentration in its environment, and salt, with potassium, is probably the most important ion encountered.

Since the receptor of the fly responds to pure water (in contrast to the water receptor of fishes, which does not) (Konishi and Hidaka, 1969; Konishi, 1965), there must be an external source of ions. Morita (1967) suggested that these were drawn from elsewhere in the hair.

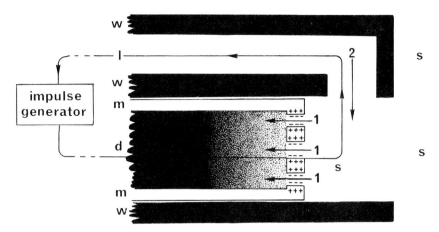

Fig. 85. Simplified diagram of the tip of a chemoreceptor sensillum showing the dendrite of the water receptor with transmembrane pores with charged walls. 1, flow of water and cations into dendrite; 2, communication between large lumen of sensillum and the region of the pores of the sensillum with a possible supply of cations or solution of high osmolarity to "hold off" the receptor in the unstimulated condition. m, dendrite membrane; w, impermeable cuticular walls; d, dendrite contents; l, large lumen; s, stimulating solution. Shading indicates region of lower osmotic pressure near inner mouths of pores. Streaming potential may drive current in pathway shown by arrows via region of the receptor cell, which generates action potentials repetitively (from Rees, 1970).

Rees has suggested that excessive dilution of these by applied water is prevented by the structural characteristics of the pore in the tip of the hair. Since the pore is very small but leads into an enlarged chamber in which the dendrites lie (Fig. 39), it is possible that the chamber provides a reservoir for ions, whereas the pore imposes appreciable delays in diffusion. That this may indeed be a restrictive mechanism is supported by the observation of Hodgson (1968) that the flow of tritiated water from the haemolymph up the shaft of a hair to the tip is greatly accelerated if its tip is amputated.

The specific nature of all five cells of the labellar hair has apparently been established. The evidence that each responds differently is convincing. The situation with respect to the identity and specificity of tarsal hairs is less clear and some contradictions remain unresolved. There is no doubt that the fly can discriminate among water, sugar, and salts applied to the tarsi. Several electrophysiological studies of tarsal hairs led to the conclusion that salt-, sugar-, and mechanoreceptors existed; however, the types and locations of the hairs were not noted in these studies (Wolbarsht and Dethier, 1958; Hanson, 1965; Steinhardt, Morita, and Hodgson, 1966). One study restricted to C and D hairs (Grabowski's and Dethier's nomenclature) reported two cells (in *Phormia*) responsive to salts (Dethier and Hanson, 1968). On the other hand, den Otter and van der Starre (1967) and van der Starre (1970) reported two cells in *Phormia* and *Calliphora* responsive to water.

The first systematic study resulted in the conclusion that the tarsal hairs of *Phormia* were essentially like the labellar hairs except that differences among different types of hairs and among hairs of the same type were found (McCutchan, 1969). In an intensive study of the D hairs of *Calliphora* van der Starre (1972) concluded that one, two, or three cells responded to water. He further concluded, primarily on the basis of concentration-response analyses and cross-adaptation, that there is no specific water receptor and no specific sugar receptor but that the same cell or cells respond to pure water and to aqueous solutions of sucrose.

Studies of the tarsal hairs of *Phormia* by Shiraishi and Tanabe (1974) have revealed marked difference among hairs with respect to salt and sugar, and it is by no means clear which or how many cells are sensitive to which stimulus. Obviously a great deal more work on the tarsal hairs is required before the characteristics of the various receptors are understood.

There is general agreement regarding the identity of the labellar receptors. Common usage has now formalized the designation of the five cells as: mechanoreceptor, classical salt receptor (or simply, salt receptor), sugar receptor, water receptor, anion receptor (although "fifth cell" is to be preferred until more is known about it). At least one objection has been raised to this classification on the grounds that it is teleological, and Rees (1968) prefers to refer to the cells simply by number. In fact, calling a cell a salt receptor is no more teleological than calling it a chemoreceptor, unless, of course, one is referring to the human sensation "salt" rather than the identity of the stimulus.

A more serious objection can be raised, and it is in this connection that the bias of human sensory experience asserts itself. Classically, the human gustatory sense was divided into the basic categories sweet, salt, sour, and bitter; and the standard test stimuli were sucrose, sodium chloride, hydrochloric or acetic acid, and quinine. What more natural than that the same classical stimuli should be applied to insects? The relevance of these stimuli to the life of the insect was never questioned nor, for a very long time, were any other stimuli applied to insect gustatory receptors. Therefore the fiction had arisen that the salt receptor was indeed a salt receptor, the sugar receptor, a sugar receptor, and so forth. When behavioral tests were extended beyond the four classical stimuli, they tended to remain within families of compounds (except for the large series of deterrent compounds tested by Dethier and Chadwick). The development of electrophysiological techniques prompted investigations of a wide range of stimuli, especially those that had relevance of the life of the fly. Included among these were compounds characteristic of proteinaceous foods and fermenting fruits (e.g., alcohols, organic acids, etc.), and the foods themselves. One result of these explorations is the discovery that although the five receptors are indeed different, their specificity is not so narrow as once believed. The breadth of their action spectra assumes great importance when the discussion turns to the nature of the information that the receptors convey to the central nervous system.

Literature Cited

Amakawa, T., Kawabata, K., Kijima, H., and Morita, H. 1972. Isozymes of α-glucosidase in the proboscis and legs of flies. *J. Insect Physiol.* 18:541–553.

Barnhard, C. S., and Chadwick, L. E. 1953. A "fly factor" in attractant studies. *Science* 117:104–105.

Beidler, L. M. 1954. A theory of taste stimulation. *J. Gen. Physiol.* 38:133–139.

Beidler, L. M. 1961a. Taste receptor stimulation. In *Progress in Biophysics and Biophysical Chemistry,* Vol. 12, J. A. V. Butler, H. E. Huxley, and R. E. Zirkle, eds., pp. 107–151. Pergamon Press, Oxford.

Beidler, L. M. 1961b. The chemical senses. *Annu. Rev. Psychol.* 12:363–388.

Beidler, L. M. 1963. Dynamics of taste cells. In *Olfaction and Taste,* I, Y. Zotterman, ed., pp. 133–145. Pergamon Press, Oxford.

Beidler, L. M. 1967. Anion influences on taste receptor response. In *Olfaction and Taste,* II, T. Hayashi, ed., pp. 509–534. Pergamon Press, Oxford.

Beidler, L. M. 1970. Physiological properties of mammalian taste receptors. In *Taste and Smell in Vertebrates,* G. E. W. Wolstenholme and J. Knight, eds., pp. 51–82. J. A. Churchill, London.

Cagan, R. H. 1973. Chemostimulatory protein: A new type of taste stimulus. *Science* 181:32–35.

Case, J., and Guilliam, G. F. 1963. Amino acid detection by marine invertebrates. *Proc. XVI Int. Zool. Congr.,* Washington, D.C., 3:75.

Deonier, C. C. 1938. Effects of some common poisons in sucrose solutions on the chemoreceptors of the housefly, *Musca domestica* L. *J. Econ. Ent.* 31:742–745.

Dethier, V. G. 1955a. The physiology and histology of the contact chemoreceptors of the blowfly. *Quart Rev. Biol.* 30:348–371.

Dethier, V. G. 1955b. Mode of action of sugar-baited fly traps. *J. Econ. Ent.* 48:235–239.

Dethier, V. G. 1956. Chemoreceptor mechanisms. In *Molecular Structure and Functional Activity of Nerve Cells,* R. G. Grenell and L. J. Mullins, eds., pp. 1–33. Amer. Inst. Biol. Sci., Washington, D.C.

Dethier, V. G. 1961. Behavioral aspects of protein ingestion by the blowfly *Phormia regina* Meigen. *Biol. Bull.* 121:456–470.

Dethier, V. G. 1972. Sensitivity of the contact chemoreceptors of the blowfly to vapors. *Proc. Nat. Acad. Sci. U.S.A.* 69:2189–2192.

Dethier, V. G. 1973. Electrophysiological studies of gustation in lepidopterous larvae. II. Taste spectra in relation to food plant discrimination. *J. Comp. Physiol.* 82:103–134.

Dethier, V. G., and Arab, Y. M. 1958. Effect of temperature on the contact chemoreceptors of the blowfly. *J. Insect Physiol.* 2:153–161.

Dethier, V. G., Evans, D. R., and Rhoades, M. V. 1956. Some factors controlling the ingestion of carbohydrates by the blowfly. *Biol. Bull.* 111:204–222.

Dethier, V. G., and Hanson, F. E. 1968. Electrophysiological responses of the chemoreceptors of the blowfly to sodium salts of fatty acids. *Proc. Nat. Acad. Sci. U.S.A.* 60:1296–1303.

Drickamer, L. C. 1972. Food selection in wood mice. *Behaviour* 41:269–287.

Evans, D. R. 1963. Chemical structure and stimulation by carbohydrates. In *Olfaction and Taste,* I, Y. Zotterman, ed., pp. 165–192. Pergamon Press, Oxford.

Evans, D. R., and Mellon, De F. 1962. Stimulation of a primary taste receptor by salts. *J. Gen. Physiol.* 4:651–661.

Frings, H., and Cox, B. L. 1954. The effects of temperature on the sucrose thresholds of the tarsal chemoreceptors of the flesh fly Sarcophage bullata. *Biol. Bull.* 107:360–363.

Frisch, K. von 1935. Über den Geschmackssinn der Biene. *Zeit. vergl. Physiol.* 21:1–156.

Goldrich, N. R. 1973. Behavioral responses of *Phormia regina* (Meigen) to labellar stimulation with amino acids. *J. Gen. Physiol.* 61:74–88.

Hanamori, T., Shiraishi, A., Kijima, H. and Morita, H. 1972. Stimulation of labellar sugar receptor of the fleshfly by glycosides. *Zeit. vergl. Physiol.* 76:115–124.

Hansen, K. 1968. Untersuchungen über den Mechanismus der Zucker-Perzeption bei Fliegen. Habilitationsschrift der Universität Heidelberg.

Hansen, K. 1969. The mechanism of insect sugar reception, a biochemical investigation. In *Olfaction and Taste,* III, C. Pfaffmann, ed., pp. 382–391. Rockefeller University Press, New York.

Hansen, K., and Kühner, J. 1972. Properties of a possible receptor protein of the fly's sugar receptor. In *Olfaction and Taste,* IV, D. Schneider, ed., pp. 350–356. Wissenschaftliche Verlagsgesellschaft MBH, Stuttgart.

Hanson, F. E. 1965. Electrophysiological studies on chemoreceptors of the blowfly, *Phormia regina* Meigen. Doctoral dissertation, University of Pennsylvania.

Haslinger, F. 1935. Über den Geschmackssinn von *Calliphora erythrocephola* Meigen und die Verwertung von Zuckern und Zuckeralkoholen durch diese Fliege. *Zeit. vergl. Physiol.* 22:614–640.

Hassett, C. C., Dethier, V. G., and Gans, J. 1950. A comparison of nutritive values and taste thresholds of carbohydrates for the blowfly. *Biol. Bull.* 99:446–453.

Hodgson, E. S. 1957. Electrophysiological studies of arthropod chemoreception. II. Responses of labellar chemoreceptors of the blowfly to stimulation by carbohydrates. *J. Insect Physiol.* 1:240–247.

Hodgson, E. S. 1968. Taste receptors of arthropods. *Symp. Zool. Soc. London,* 23–269–277.

Jakinovich, W., and Agranoff, B. W. 1972. Taste receptor response to carbohydrates. In *Olfaction and Taste,* IV, D. Schneider, ed., pp. 371–377. Wissenschaftliche Verlagsgesellschaft MBH, Stuttgart.

Kawabata, K., Kijima, H., Shiraishi, A., and Morita, H. 1973. α-Glucosidase isozymes and the labellar sugar receptor of the blowfly. *J. Insect Physiol.* 19:337–348.

Kijima, H. 1970. (In Japanese.) *Kaguka,* 40:523–530.

Kijima, H., Koizumi, O., and Morita, H. 1973. α-Glucosidase at the tip of the contact chemosensory seta of the blowfly, *Phormia regina. J. Insect Physiol.* 19:1351–1362.

Koizumi, O., Kijima, H., Kawabata, K., and Morita, H. 1973. α-Glucosidase activity on the outside of labella and legs of the fly. *Comp. Biochem. Physiol.* 44B:347–356.

Koizumi, O., Kijima, H., and Morita, H. 1974. Characterization of a glucosidase at the tips of the chemosensory setae of the fly, *Phormia regina. J. Insect Physiol.* 20:925–934.

Konishi, J. 1965. Fresh-water fish chemoreceptors responsive to dilute solutions of electrolytes. *J. Gen. Physiol.* 49:1241–1264.

Konishi J., and Hidaka, I. 1969. On the stimulation of fish chemoreceptors by dilute solutions of polyelectrolytes. *Jap. J. Physiol.* 19:315–326.

Koyama, N., and Kurihara, K. 1971. Modification by chemical reagents of proteins in the gustatory and olfactory organs of the fleshfly and cockroach. *J. Insect Physiol.* 17:2435–2440.

Laverack, M. S. 1963. Aspects of chemoreception in crustacea. *Proc. XVI Inter. Zool. Congr.,* Washington, D.C., 3:72–74.

Larimer, J. L., and Oakley, B. 1968. Failure of *Gymnema* extract to inhibit the sugar receptors of two invertebrates. *Comp. Biochem. Physiol.* 25:1091–1097.

Lineweaver, H., and Burk, D. 1934. The determination of enzyme dissociation constants. *J. Amer. Chem. Soc.* 56:658–666.

McCutchan, M. C. 1969. Responses of tarsal chemoreceptive hairs of the blowfly, *Phormia regina. J. Insect Physiol.* 15:2059–2068.

Morita, H. 1967. Effects of salts on the sugar receptor of the fleshfly. In *Olfaction and Taste,* II, T. Hayashi, ed., pp. 787–798. Pergamon Press, Oxford.

Morita, H. 1969. Electrical signs of taste receptor activity. In *Olfaction and Taste,* III, C. Pfaffmann, ed., pp. 370–381. Rockefeller University Press, New York.

Morita, H. 1972. Properties of the sugar receptor site of the blowfly. In *Olfaction and Taste,* IV, D. Schneider, ed., pp. 357–363. Wissenschaftliche Verlagsgesellschaft MBH, Stuttgart.

Morita, H., Hidaka, T., and Shiraishi, A. 1966. Excitatory and inhibitory effects of salts on the sugar receptor of the fleshfly. *Mem. Fac. Sci. Kyushu Univ.,* Ser. E. (Biol.), 4:123–135.

Morita, H., and Shiraishi, A. 1968. Stimulation of the labellar sugar receptor of the fleshfly by mono- and disaccharides. *J. Gen. Physiol.* 52:559–583.

Omand, E., and Dethier, V. G. 1969. An electrophysiological analysis of the action of carbohydrates on the sugar receptor of the blowfly. *Proc. Nat. Acad. Sci. U.S.A.* 62:136–143.

Otter, C. J. den, and van der Starre, M. 1967. Responses of tarsal hairs of the bluebottle, *Calliphora erythrocephala* Meig., to sugar and water. *J. Insect Physiol.* 13:1177–1188.

Pflumm, W. W. 1971. Zur Reizwirksamkeit von Monosacchariden bei der Fliege *Phormia terraenovae. Zeit. vergl. Physiol.* 74:411–426.

Pflumm, W. W. 1972. Molecular structure and stimulating effectiveness of oligosaccharides and glycosides. In *Olfaction and Taste,* IV, D. Schneider, ed., pp. 364–370. Wissenschaftliche Verlagsgesellschaft MBH, Stuttgart.

Rees, C. J. C. 1967. Transmission of receptor potential in dipteran chemoreceptors. *Nature*, 215:301–302.

Rees, C. J. C. 1968. The effect of aqueous solutions of some 1:1 electrolytes on the electrical response of the type 1 ("salt") chemoreceptor cell in the labella of *Phormia*. *J. Insect Physiol.* 14:1331–1364.

Rees, C. J. C. 1970. The primary process of reception in the type 3 ("water") receptor cell of the fly, *Phormia terraenovae*. *Proc. Roy. Soc. London* B 174:469–490.

Robbins, W. E., Thompson, M. S., Yamamoto, R. T., and Shortino, T. J. 1965. Feeding stimulants for the female housefly, *Musca domestica* L. *Science*, 147:628–630.

Shimada, I., Shiraishi, A., Kijima, H., and Morita, H. 1972. Effects of sulphhydryl reagents on the labellar sugar receptor of the fleshfly. *J. Insect Physiol.* 18:1845–1855.

Shimada, I., Shiraishi, A., Kijima, H., and Morita, H. 1974. Separation of two receptor sites in a single labellar sugar receptor of the flesh-fly by treatment with *p*-chloromercuribenzoate. *J. Insect Physiol.* 20:605–621.

Shiraishi, A., and Kuwabara, A. 1970. The effects of amino acids on the labellar hair chemosensory cells of the fly. *J. Gen. Physiol.* 56:768–782.

Shiraishi, A., and Morita, H. 1969. The effects of *p*H on the labellar sugar receptor of the fleshfly. *J. Gen. Physiol.* 53:450–470.

Shiraishi, A., and Tanabe, Y. 1974. The proboscis extension response and tarsal and labellar chemosensory hairs in the blowfly. *J. Comp. Physiol.* 92:161–179.

Steinhardt, R. A., Morita, H., and Hodgson, E. S. 1966. Mode of action of straight chain hydrocarbons on the primary chemoreceptors of the blowfly, *Phormia regina*. *J. Cell. Physiol.* 67:53–62.

Thorsteinson, A. J. 1960. Host selection in phytophagous insects. *Annu. Rev. Ent.* 5:193–218.

Uehara, S., and Morita, H. 1972. The effect of temperature on the labellar chemoreceptors of the blowfly. *J. Gen. Physiol.* 59:213–226.

van der Starre, H. 1970. Tarsal water receptor cell responses in the blowfly *Phormia regina*. *Netherlands J. Zool.* 20:289–290.

van der Starre, H. 1972. Tarsal taste discrimination in the blowfly, *Calliphora vicina* Robineau-Desvoi. *Netherlands J. Zool.* 22:277–282.

van der Wel, H. 1972. Thaumatin, the sweet-tasting protein from *Thaumatococcus Daniellii* Benth. In *Olfaction and Taste*, IV, D. Schneider, ed., pp. 226–233. Wissenschaftliche Verlagsgesellschaft MBH, Stuttgart.

Wiesmann, R. 1960. Zum Nahrungsproblem der freilebenden Stubenfliegen, *Musca domestica*. *Zeit. angew. Zool.* 47:159–181.

Wolbarsht, M. L., and Dethier, V. G. 1958. Electrical activity in the chemoreceptors of the blowfly. I. Responses to chemical and mechanical stimulation. *J. Gen. Physiol.* 42:393–412.

Wolbarsht, M. L., and Hanson, F. E., 1967. Electrical and behavioral responses to amino acid stimulation in the blowfly. In *Olfaction and Taste*, II, T. Hayashi, ed., pp. 749–760. Pergamon Press, Oxford.

7
The Flavor of Things: Codes and Information

The meaning doesn't matter
If it's only idle chatter
Of a transcendental kind.

W. S. Gilbert, *Patience*

To operate efficiently in a diverse and constantly changing environment an animal must be able to detect differences in the external world and in its own internal world and to make appropriate responses. To this end it must not only detect, it must also be able to construct, as it were, in its own central nervous system, where decisions are made, a representation of external events that compares reasonably well with reality. Survival depends upon the animal's achieving a fairly high degree of fidelity to fact. It depends upon detection of those events in the environment that are relevant to survival and the transmission of this information to the central nervous system.

One of the unique features of each of insect's sense cells, in contrast to those of vertebrates, is that they perform the three basic duties of a sensory system without direct assistance from other cells. The chemoreceptive cell detects, that is, is sensitive to, a source of potential energy in the environment; it changes endogenous cellular energy to electrical energy; it generates and transmits a code to the central nervous system. In the gustatory system of vertebrates one cell, a modified epithelial cell, is responsible for detection and transduction to electrical energy, but it does not of itself generate and transmit a code. It is in synaptic contact with neurons, which then generate and transmit messages. A complicating feature of the vertebrate gustatory receptor is its multiple cellular association (Murray, 1969; Murray and Murray, 1970). The receptor is connected to more than one neuron, and more than one neuron makes contact with each receptor. There are, in addition, desmosomes between receptors. Their presence lends credibility to the idea that there may be electrical interaction at these junctions. These various anatomical relationships suggest that there can be complex neural interactions among receptors, and indeed both enhancement and suppression by one receptor of activity in another have been demonstrated electrophysiologically (Beidler, 1969). Because of these influences a message travelling in a single fiber in the vertebrate tongue is an abstraction of events detected by several receptors.

Still another complication is introduced into the vertebrate system by the ephemeral nature of the gustatory receptors. As Beidler and Smallman (1965) have shown, at least in the rat, a taste cell has a life span of only about eleven days; it is then replaced by a new one. The significance of this substitution is that all the complex neural connections must be renewed and, presumably, in some semblance of order.

No single sense organ, however complicated, could possibly assess all the events encountered by an animal. On the other hand, not all events bear on the well-being of an individual. Superfluous stimulation could occupy the sense organs at critical moments or overwhelm the central nervous system with irrelevant information. For these reasons it is just as necessary to be insensitive to some events as it is to be sensitive to others. It is not surprising, therefore, to find "blind spots" in all sensory systems as well as filtering processes in central nervous systems that discard some of the input. This phenomenon of central filtering is well known in vertebrate systems and has been beautifully demonstrated in the acoustic system of moths by Roeder (1966). Certain moths, especially noctuids, possess, hidden beneath the scales on each side of the thorax, an "ear." The crucial neural elements of this very simple ear are two acoustic cells that are so constructed as to detect the ultrasonic hunting cries of moth-eating bats (see Roeder, 1962). Each cell is capable of transmitting a code that contains information about the duration of a bat's cry, the number of pulses in it, the duration of each pulse, and various characteristics of the pulse. This message upon entering the central nervous system is received by a set of interneurons in serial order. Each interneuron in turn abstracts from the information it receives so that by the time the message arrives at higher centers the only portion that remains of the original message transmitted by the acoustic cell is the duration of the bat's cry.

Even allowing for these means of reducing input from the environment, the fact remains that the world is extraordinarily rich in events. To develop a special receptor for each event—a separate receptor for every wavelength of light, every pitch, every shape, every kind of vapor, every different chemical solution—is neither economical nor possible. On the other hand, it is hardly feasible to develop an omnireceptor. Some compromise between total generality and absolute specificity is mandatory. What is the nature of this compromise? Equally important is the evolutionary decision concerning the number of any given kind of receptor with which an animal is to be equipped. How much duplication is desirable? What is the best evolutionary strategy to follow in perfecting sense organs?

As we have seen, the fly has relatively few contact chemoreceptors, roughly 3500 as compared with 9,000 to 450,000 in the tongues of various mammalian species. Some insects have even fewer cells. Caterpillars, for example, have approximately 78 olfactory receptors as compared with the number in one mammalian herbivore, the rabbit, which has 10^8. Although it is undeniably true that animals with large and complex nervous systems are capable of more varied, complex, and flexible behavior than are animals with small nervous systems, the degree of difference between the two seems all out of proportion to the difference in the number of neurons. The idea that possession of few working parts severely limits behavioral capacity is belied by the demonstrably complex activities of animals with small systems. In some instances the patterns of behavior are superficially so imitative of human behavior that comparisons, often invidious, to us, are frequently made—for example (Proverbs 6:6), "Go to the ant, thou sluggard; consider her ways, and be wise." One wonders, therefore, how it is that the so-called simple nervous system is able to accomplish so much, or, an equally reasonable puzzle, why nervous systems with millions of cells do not accomplish more.

From combined behavioral and electrophysiological studies, a wealth of information has emerged that has forced us to revise the initial idea that the fly had two essential chemoreceptive modalities, acceptance and rejection, and also to revise the idea that strict specificities existed. In terms of the neural codes by which information is transmitted to the central nervous system, a much more complex situation prevails, and the gustatory sensory experience of the fly must be much richer than we had imagined (Dethier, 1971).

The wild blowfly is called upon to make appropriate behavioral responses to multitudes of diverse sapid substances encountered in nature—nectar, saps, honeydew; proteinaceous and fatty juices in carrion and feces; alcohols, esters, and other products of fermentation in fruits; pheromones, and a great number of unacceptable or noxious substances encountered routinely in the search for food, mate, ovipositional sites, and resting places. At first it would appear that only a yes or no answer is required by the fly, that its world could be a world of absolutes. This is clearly not the case.

The business of processing information is not intrinsically simple nor is its nature in nervous systems easily understood. Basically there is a set of messages; these are encoded in symbols; the symbols are transmitted to a receiver; the receiver decodes them. In the chemosensory system the messages are chemical stimuli, the symbols are trains of action potentials (spikes) transmitted along axons to the central nervous system where decoding takes place. The problem facing the central nervous system is that of detecting and characterizing the stimulus, given a response. The problem facing the experimenter is to detect and characterize the response, given the stimulus. The central nervous system and the experimenter may not solve their respective problems in the same way, so that the experimenter's way of decoding may or may not represent reality. In any case, the nature of the symbols employed by the neuron imposes considerable limitation on coding capacities. The Morse telegraphic code, for example, uses as symbols dots and dashes from which any number of words can be constructed by combination. The neural code relies solely on one kind of symbol, the spike. Since all spikes in a single axon are the same and are propagated as repetitive all-or-none events, the sole possible variable is the interval between spikes. How rich a treasury of information can such a system transmit?

Theoretically, great versatility of coding is possible. Five candidate codes will be discussed here with the evidence for their existence. They may be classified as: (1) absolute labeled lines; (2) partially labeled lines; (3) across-fiber patterning in silent lines; (4) across-fiber patterning in spontaneously active lines; (5) temporal patterning.

Since much of the normal feeding behavior of the fly can be initiated and controlled by stimulation of a single hair (four chemical sensing elements), the simplest possibility for coding would involve a condition in which each of the four receptors was absolutely specific. This would mean one cell that was absolutely restricted, as regards its sensitivity, to water, one similarly restricted to sugars, and two to salts. As we have seen, such rigid specificity does not in fact exist; however, the postulated condition can be simulated experimentally. An analysis of simulated conditions can reveal the potentialities of the simplest systems for carrying information, and from this base we can elaborate our analyses to encompass actual conditions.

It is essential to remember that there is at the moment no known neural interaction among the chemoreceptors of the fly; each is an uninterrupted insulated line until it enters the central nervous system. If a hair is stimulated with an *optimal* concentration of sugar, only the sugar receptor responds. The axon carrying information to the central nervous system is, in the terminology of Perkel and Bullock (1968), a "labeled line." The central nervous system identifies the source of the message, that is, over which line it is arriving; the message is equated with sugar in the environment; appropriate action is taken. In the absence of modifying or contradictory information from other sources (e.g., information from the stretch receptors of the foregut or body wall (cf. Chapter 8), appropriate action consists of extending the proboscis and feeding. By choosing as a stimulus an optimal concentration of sodium chloride, an experimenter can stimulate another absolutely specific labeled line. In this case the central nervous system identifies the line, equates the message with salt in the environment, and makes the appropriate decision—retraction of the proboscis or inhibition of extension.

With narrowly specific lines of this sort the transmitted information is virtually unambiguous but is severely restricted in content. The only variable that can be accommodated is intensity of stimulus. This is coded in each line as the rate of impulses. How the central nervous system reads rates and changes in rates is not known. It could measure any of several parameters of the code and issue a command that would depend on which parameter was measured. It could average frequency over time, measure the most recent interval, measure rate of change, etc.; see Perkel and Bullock (1968) for a full discussion of candidate codes.

In seeking answers to these questions we must consider two general characteristics of transmission by chemoreceptors—its nonstationary feature and its variability. The first refers to the fact that the pattern of action potentials is not the same from the beginning of stimulation to the end. The line may or may not be silent just before stimulation. A basal rate not exceeding 2.5 impulses per second has been recorded (Evans and Mellon, 1962a; Rees, 1968); however, this could conceivably arise from the stimulation imposed by the side-wall technique of recording, as previously mentioned. Upon stimulation there is a latency period of 1 to 6 msec., followed by an increase in rate of impulse generation from zero (or some low basal level) to a maximum determined by the strength of stimulus (and level of adaptation at time zero). This rate increases rapidly in 20 msec., then falls only slightly less rapidly in the next 25 msec. This is the phasic portion of the response. The new level then declines very slightly and attains a constant rate, the fully adapted state, by time 0.35 sec (Fig. 75) (Rees, 1968). This is the tonic portion of the response. Which portion contains the information critical for feeding? A considered answer is that each phase has some useful information. In Chapter 4 a description was presented of the important role of the interval between the first and second or second and third spike for some phases of feeding behavior. The role of the tonic segment of the response has also been touched upon (Getting, 1971).

The first evidence of the critical nature of the phasic period was derived from a study of discrimination in which electrical activity from a single chemoreceptor was monitored simultaneously with the resulting behavior (Dethier, 1968). In this study a thirsty fly was attached to a wax-tipped stick and an electrode inserted into the lateral

The Flavor of Things: Codes and Information

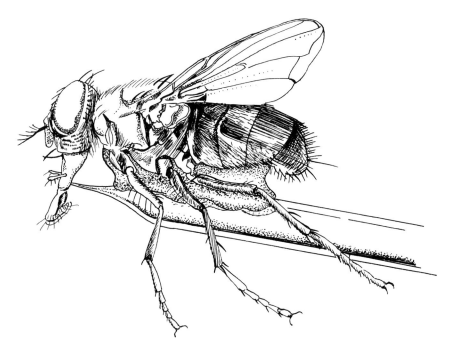

Fig. 86. One method of placing a chronic electrode.

intersegmental membrane of the last two abdominal segments (Fig. 86). A recording electrode in the form of a micropipette containing 0.2 M sodium chloride was placed over one of the "largest" labellar hairs. The record, reproduced in Figure 87, shows that upon contact with the solution one receptor in the hair, the salt receptor, responded by generating 14 impulses in the first 100 msec. Muscle potentials then appeared, and shortly thereafter the fly responded by extending its proboscis. In other words, when 14 impulses had reached the central nervous system from the line labeled "salt," the central nervous system activated the muscles causing extension.

It had always been assumed that a fly that accepts dilute salt solutions does so because of the activity of the water receptor. This assumption was reasonable because as salt concentration is increased the water cell is progressively inhibited until the only activity remaining is that of the salt cells (Evans and Mellon, 1962b). At intermediate concentrations of salt, therefore, both the water cell and the salt cells may be active (Fig. 88b). The point in an ascending series of salt concentrations at which the behavior of a fly changes from acceptance to rejection was believed to be the point at which the frequency of firing of the water cell falls below some critical level and that of the salt cell rises to some critical level. At this point the frequency of the water cell is too low to counteract in the central nervous system the increased firing of the salt cell.

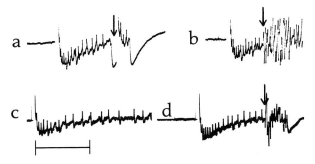

Fig. 87. Electrophysiological responses of receptors in "largest" labellar hair number 5 on intact fly. All records begin just before onset of stimulation. "On" artifact is followed by action potentials from the salt receptor. Large irregular spikes beginning at vertical arrows represent muscle activity of the moving proboscis. (a) Response to 0.2 M NaCl. After 14 sensory spikes muscles contracted, the proboscis extended, and contact with the electrode was broken. (b) Response to 0.5 M NaCl with the proboscis in the extended position. The muscle activity after the salt spikes was followed by proboscis retraction. (c) Response of extended proboscis to 0.2 M NaCl. Salt spikes continued for 5 seconds and no behavioral action occurred. (d) Response of the extended proboscis to 0.5 M NaCl. The muscle activity was followed by retraction. Horizontal bar represents 100 msec. (from Dethier, copyright 1968 by the American Association for the Advancement of Science).

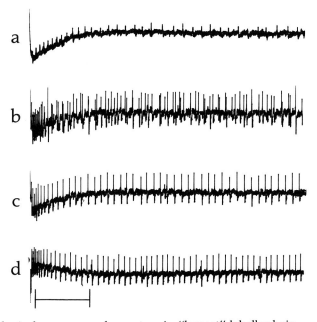

Fig. 88. Electrophysiological responses of receptors in "largest" labellar hair no. 1 on an isolated head. All records begin at the onset of stimulation. (a) Response of water cell to 0.05 M NaCl. (b) Response of water cell and two salt cells to 0.1 M NaCl. (c) Response of classical salt cell (large spike) to 0.2 M NaCl. (d) Response of classical salt cell to 0.5 M NaCl. Solutions in (a)–(c) had been accepted by the intact fly. Solution in (d) was rejected. Horizontal bar represents 100 msec. (from Dethier, copyright 1968 by the American Association for the Advancement of Science).

The simultaneous recording just described, plus a series of sequential experiments in which behavioral responses to stimulation of a single hair were first observed and then electrophysiological recordings obtained from the same hair after the fly was decapitated, agree in showing that flies responded to 0.05 M NaCl and to all concentrations up to and including 0.2 M by extending the proboscis and that the water cell was not firing at all of these acceptable concentrations. Figure 88a shows that only the water cell responded to 0.05 M NaCl. At 0.1 M NaCl, its frequency was much lower and two salt cells were responding (Fig. 88b). At 0.2 and 0.5 M there were no water cell spikes in the first 100 msec of activity, (Fig. 88c, 88d). It is clear, therefore, that acceptance can be mediated not only by the water cell but also by the salt cell. This would agree with observations of preferences of some insects in which salt at low concentrations was preferred to water (cf. Frings, 1946).

Now, let us return to the fly with the chronically implanted electrode, with the proboscis in the extended position; a recording electrode containing 0.5 M NaCl was placed on the hair. Again the salt cell responded. It generated 14 impulses in about 70 msec. The record shows that the frequency of firing was greater than before, and the intervals between spikes shorter. The central nervous system "read" this message and activated the muscles causing withdrawal of the proboscis. One cannot decide at this point whether the frequency was averaged, the intervals measured, the spikes counted, or the rate of change noted. The experiments of Getting (1971) would suggest that the critical parameter was spike interval. In any case, the different rates in a single labeled line can mean different things to the central nervous system. In the example just described, a low rate is interpreted as stimulation by a solution that is acceptable, whereas a high rate indicates an unacceptable solution.

Such precision in decoding on the part of the nervous system implies that the sensory neurons are themselves quite precise. It turns out that this precision is illusory both in terms of response and in terms of behavior of the sensory neurons. The animal does not always respond nor does it always respond correctly. The behavioral response is probabilistic in nature (cf. Dethier, 1974a). There is also a probabilistic character to neuronal excitability and responsiveness. The intervals between impulses discharged by afferent neurons under steady conditions vary considerably even in such extremely reliable preparations as the eye of *Limulus* where, furthermore, the stimulus can be impeccably controlled (see Ratliff, Hartline and Lange, 1968). Not much attention has been paid to variability in chemoreceptive systems although it is much in evidence; see, for example, the ragged response in single fibers of the chorda tympani of the rat in Figure 1 of Sato,

Yamashita, and Ogawa (1969), and Figure 1 of Frank and Pfaffmann (1969). Hodgson and Roeder (1956) and Hodgson (1957) reported that the pattern of impulses from the labellar sugar receptor is irregular, that there is some grouping, and that periods of inactivity are characteristic. Observations of patterns of discharge from labellar salt and water receptors also disclose the existence of considerable variability in interspike interval (Figs. 89-93) (Dethier, 1974a).

With random variability it is to be expected not only that any single message will be inconstant but also that no two responses to the same stimulus will be exactly alike. Fidelity of repetition had not previously been carefully studied in the chemoreceptive system of the fly. Evans and Mellon (1962a) reported that brief tests with stimuli of constant intensity led to complete disappearance of the phasic part of the response to salt after the fourth or fifth presentation and that the effect was not a result of prior stimulation. The tonic phase was reported as "satisfactorily constant." Gillary (1966a), who carried out the most carefully controlled experiments up to that time, reported that "the stimulation procedure was capable of yielding responses within 10% of the mean better than 80% of the time for normally responding receptors, and quite often they were within 5%. In the current investigation a response was considered reproducible if it fell within 10% of the mean of all responses to the same applied stimulus. Experiments yielding nonreproducible responses were considered unreliable and were discarded." His Figure 2 depicts the results of 25 repeated stimulations (every 10 sec.) with 0.40 M and 2.0 M sodium chloride, for a duration of less than 1 sec. It shows a variation of about 10% of the mean. Van der Starre suggested that variability might be a normal feature of tarsal receptors. Rees (1968) reported that for the salt cell in labellar hairs the 93% confidence limits are ±5 impulses in the tonic phase "where the [rate] is constant enough to calculate them." In the unstimulated state the standard deviation (% of the mean) varies from 30.3 to 58.3%. A very detailed study of responses to repeated stimulations by water, salt, and sugar of two types of labellar hairs (side-wall recording, duration of stimulus 2 sec., interstimulus interval 2 min.) revealed the following: (1) there are fluctuations in the response of "largest" and "intermediate" hairs to water, 0.5 M sodium chloride, and 0.5 M sucrose; (2) fluctuations are noncyclic and generally greater than 10% of the mean; (3) fluctuations occur both in the phasic and the tonic sections of the response; (4) the salt receptor is slightly more irregular than the water receptor; (5) the sugar receptor is not characteristically more irregular than the others upon repeated stimulation; (6) the "intermediate" hairs are more irregular than the "largest" hairs.

Several possible causes of the fluctuations in response can be con-

The Flavor of Things: Codes and Information

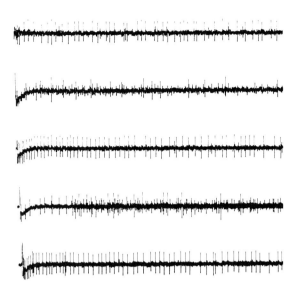

Fig. 89. Response of labellar hair "largest" 2 to repeated stimulation with 0.1 M NaCl. The records (from the top) are from presentations 1, 8, 11, 16, and 20 respectively. Each stimulation lasted 5 sec., of which the first 1.75 sec. are shown. The interstimulation interval was 130 sec. Tip recording was employed (from Dethier, 1974a).

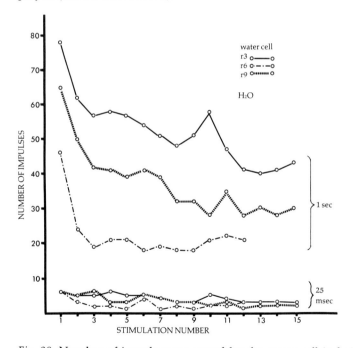

Fig. 90. Number of impulses generated by the water cell in hairs "largest" r3, r6, and r9 in response to repeated stimulation by water. Side-wall recording was employed. Other conditions as in Figure 89. The bottom set of lines represents the impulses in the phasic part of the response (first 25 msec.); the top set, in the first second of the tonic response (from Dethier, 1974a).

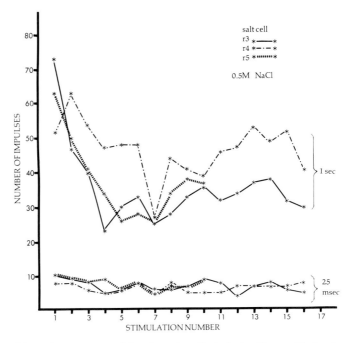

Fig. 91. Number of impulses generated by the salt cells in hairs "largest" r3, r4, and r5 to 16 repeated stimulations with 0.5 M NaCl. All conditions as in Figure 90 (from Dethier, 1974a).

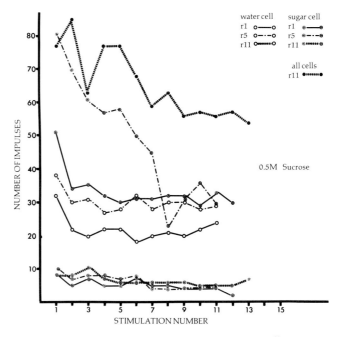

Fig. 92. Number of impulses generated by the water and the sugar cells in hairs "largest" r1, r5, and r11 in response to 0.5 M sucrose. All conditions as in Figure 90 (from Dethier, 1974a).

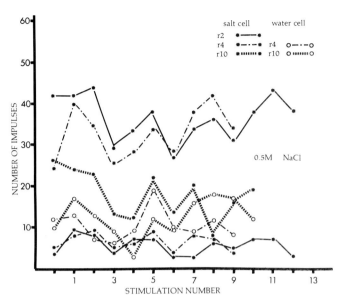

Fig. 93. Number of impulses generated by the water and the salt cells in hairs "intermediate" r2, r4, and r10 in response to 0.5 M NaCl. All conditions as in Figure 90 (from Dethier, 1974a).

sidered. The most obvious is that the variation is an experimental artifact; a second, that changes are extrinsic and occur in the accessory structures of the sensillum; a third, that there are changes in the responsiveness of the receptor itself. The first possibility has not been ruled out beyond all shadow of doubt. The same electrode was used in a repetitive series; the temperature did not vary more than 0.1°C; and changes in concentration resulting from evaporation in the electrode were controlled to the extent recommended by Gillary (1966b). Furthermore, it is significant that responses to water also varied. If indeed the salt and sugar receptors are sensitive to subtle differences in concentration, their very sensitivity could actually be a handicap rather than an asset to the fly in nature because the system would be too "noisy" to be practical.

With respect to the second explanation, two possibilities exist. It could be that the tip of the hair opens and closes. This action has been shown to be a natural feature of some gustatory receptors of the locust *Locusta migratoria* (L) (Blaney, Chapman, and Cook, 1971). Stürckow, Holbert, and Adams (1967) have demonstrated that the tip of labellar hairs of *Phormia* may be single or bifurcate (Fig. 94-96). High humidity and water cause the tip to "open" whereas treatment with such reagents as fixatives cause it to "close." Our electromicrographs show, however, that the pore in which the dendrites lie does not open and close but, rather, that a line of weakness between the terminations of

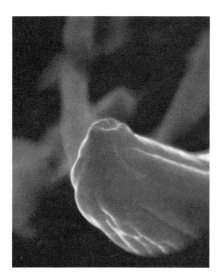 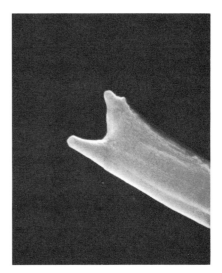

Fig. 94. Picture taken with the scanning electronmicroscope of a labellar chemosensory hair with a "closed" tip (×12,200).

Fig. 95. A labellar hair with an "open" tip (×11,600).

the two lumina allows the hair to "split." The severity of the treatment that is necessary to induce this event argues against its being a natural phenomenon. In any case it is difficult to imagine that successive applications of identical solutions would cause this action.

The other possibility is more likely, namely, that there are changes in the nature or quantity of the extracellular fluid that bathes the tips of the dendrites. There is no doubt that the dendrites are constantly bathed in fluid. Under certain conditions, again not necessarily normal, an excess of this viscous fluid exudes (Fig. 46) (Stürckow, 1967). In a series of experiments in which a stimulating solution was allowed to flow constantly over the tip of the hair, Sturckow observed in two cases a sudden increase in excitation although no change had taken place in the nature of the stimulus. She suggested that a correlation exists between the frequency of action potentials generated by the salt receptor (and the water receptor) and the presence or absence of exudate.

An alternative to these mechanisms as a cause of frequency fluctuation is the possibility that the receptor itself has an irregular responsiveness. Comparisons with other sensory systems, invertebrate and vertebrate alike, make this explanation highly probable. In the visual cell of *Limulus*, the generator potential is due to the superposition of numerous discrete conductive events, triggered by the adsorption of photons. These events cause fluctuations in the membrane potential of the neurons and are the principal source of variations in interspike in-

The Flavor of Things: Codes and Information

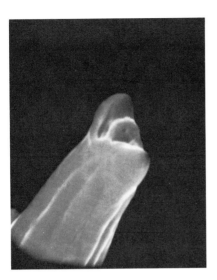

Fig. 96. A labellar hair with an "open" tip (×16,100).

tervals (Yeandle, 1958; Ratliff, Hartline, and Lange, 1968). A comparable situation involving the adsorption of molecules could underlie the variation in chemoreceptors. Some variation in the response of mammalian gustatory receptors to the first few repeated applications of sapid substances, especially postexcitatory depression, has been attributed tentatively to adsorption (Hellekant, 1968).

Inherent "noise" in the receptor must limit its capacity to carry information. Although the variation may be "noise" as far as external events are concerned, it is conceivable that it may carry useful information about the state of the receptor itself (Ratliff et al. 1968). We are still left with the question of whether the "noise" distorts information about the external world or is trivial. One is reminded again of the critical nature of the first or second interspike interval with respect to proboscis extension; however, this very fine temporal discrimination is not incompatible with the existence of variation in the rate of generation of action potentials. If the variability is such that occasionally there are no intervals as short as 20 msec. after the first impulse, then quite simply there is no extension of the proboscis. Indeed, extension of the proboscis does not occur 100% of the time in response to repeated identical stimuli (cf. Dethier, 1974a).

All of the foregoing remarks apply to laboratory situations where behavioral responses can be elicited by stimulation of a single receptor, but that situation is highly artificial. The fly's central nervous system is normally the recipient of information derived from a large population of receptors. If only a few were variable in their responses, their impact on behavior would not be very great; however, most of them seem to be variable. One might argue that variability is tolerated because it is trivial or because it is energetically uneconomical (or even impossible) to build and operate high-fidelity systems and that random variability is smoothed out by interneurons at other levels. In most sensory systems (e.g., the olfactory, visual, and auditory systems of vertebrates and the auditory and visual systems of insects) there are successive levels of convergence that give rise to successive levels of abstraction. As Knight (1972) has pointed out, "unusually precise over-all results arise from the functioning of components which have very modest precision in their individual structure and behavior." Indeed, as he has demonstrated mathematically, a certain lack of precision can be a positive virtue in preventing a population of neurons from falling into step and presenting a false representation of the stimulus (Knight, 1972). Or, to state the case if reverse, if the animal is equipped with noisy units, a large population of such units is required to provide a clear view of the world.

Returning to the simplistic idea of *simulated absolute labeled* lines, what further use of them could the central nervous system make, considering that there are more than one? It could integrate information arriving over two parallel lines that were labeled differently; it could integrate signals with respect to different rates in each line. That some sort of integration can occur was demonstrated by stimulating two different hairs simultaneously, one with salt and one with sugar, and varying the concentrations independently. The experiment is difficult to perform because the nutritional and excitatory states of the fly must be rigorously controlled; however, a weak solution of sugar elicits proboscis extension until a salt solution (0.5 M) is applied to another hair. Then the proboscis is retracted. If, while the salt is still stimulating the hair, the sugar solution on the other hair is replaced by a more concentrated one, extension of the proboscis is reinitiated. Now, when the salt solution is replaced with a still more concentrated one, retraction occurs once again, and so on.

All of the arguments presented thus far have been directed toward the information-carrying capacities of labeled lines derived from receptors assumed to be absolutely specific. We have already seen that they are not. The classical idea of a salt receptor, sugar receptor, etc., imposes a spurious specificity on the receptors. They do in fact exhibit a compromise between specificity and generality. Even when one considers only a single line, it is obvious that by expanding the action spectrum more information can be coded and transmitted, provided that either or both of two conditions are met: different compounds cause different messages to be sent in a single line; several lines with different and overlapping sensitivity spectra combine to carry a common message. Let us consider first exactly what this may mean in terms of a single line.

When, as in the sugar receptor of the fly, there are multiple sites or subunits of a site on the receptor membrane, which are filled preferentially by different sugars, which sites are filled by which sugar makes little difference insofar as response is concerned. As more sites are filled per unit time, the rate of generation of action potentials increases. A receptor with a heterogeneous surface can obviously detect more kinds of compounds, but it transmits the same kind of message regardless of the identity of the compound. Only intensity of stimulation is coded. In short, it should be possible, within limits, to match any concentration of one sugar with another, and the central nervous system would still treat the line as one labeled sugar. An amino acid or glycoside that affected this receptor would be responded to (if this were the only receptor being activated) as though it too were sugar. And a compound that synergized with sugar would in effect merely be increasing the concentration of that sugar.

Evidence has been accumulating, however, that not all stimuli cause a receptor to generate the same message. The classical picture of sensory response, by the salt receptor, for example, is that shown in Figure 88. Here there is an even series of spikes occurring initially at a high rate and then at a decreasing rate, in an orderly fashion; the intervals increase progressively, the increment in each succeeding interval being a measure of adaptation. The beginning of response is unambiguously indicated, and although the cessation of response may not be so certain, the attainment of a steady response is recognizable, within limits. The total impression is one of order. We have already seen, however, that the receptor is not entirely constant. If there is more information contained in the message than that indicated simply by mean firing rate, a statistical description of the response is necessary to reveal it. Spike trains to be analyzed may be considered to be stationary or nonstationary; consult Moore, Perkel, and Segundo (1966) for a more complete discussion of methods of analysis. Stationary situations refer to spontaneously acting units or to the long, so-called completely adapted steady phase of response during stimulation. The nonstationary phase following stimulation can be treated as stationary by dividing it into segments and treating each segment as though it were stationary. The implication is that the sample of spike train is a single example of a particular stochastic point process, stochastic because the neuron has random variability, and point process because each spike is instantaneous and indistinguishable from every other one. In the orderly response to sodium chloride, measures of interspike intervals plotted as a histogram would show the variance about the mean, and the histogram would be the same irrespective of the order in which the intervals were observed in the record and of how they were shuffled. Higher levels of analyses pay attention to the order in which intervals are observed by measuring second-order intervals (the interval from a spike to the second following spike) and higher-order intervals. Such analyses reveal serial dependence among intervals, trends, slow oscillations, and so on. Since these kinds of analyses have not yet been applied to the chemoreceptors of the fly, at the moment we have only the empirical observation that the normal or adequate stimulus for a receptor elicits a spike train with comparatively low variability, whereas certain other compounds elicit a characteristically different response.

A number of years ago Stürckow (1959), working with the taste receptors of the Colorado potato beetle, showed that some alkaloids (tomatin, solanin, demission, chaconin) normally present in the leaves of food plants cause individual receptors to respond with volleys or bursts of impulses. No direct correlation was observed between this bizarre neural activity and behavior. Bursting, or volleying, in response to calcium chloride was recorded also in the oral taste papillae by Dethier and Hanson (1965), and later in the labellar hairs in response to the salts of long-chain fatty acids (Dethier and Hanson, 1968). Volleying in response to high concentrations of alcohols was reported by Hanson (1965) and Steinhardt, Morita, and Hodgson (1966). It was generally believed that volleying was indicative of a pathological condition. More recently, Haskell and Schoonhoven (1969) discovered that volleying is a regular phenomenon in one of the taste receptors (A_3) of the locust *Schistocerca gregaria* responding to the juice of grass, a natural food, and in one of the dome receptors of *Locusta migratoria migratorioides*. In these cases it is hard to dismiss volleying as an artifact.

That volleying and other irregularities in frequency have behavioral significance was amply demonstrated with the tarsal receptors of *Phormia*. A combined electrophysiological and behavioral study revealed some quite unexpected relationships (McCutchan, 1969). When rejection thresholds were measured with HCl and dicarboxylic acids the concentration-response curves derived from the data were of the usual type (Fig. 97). Thirsty flies responded to low concentrations as if these were water, and then, as the concentrations were increased (pH lowered), gave fewer and fewer extensions of the proboscis. Curves describing concentration-response relations were quite different for monocarboxylic acids (Fig. 97). There was the usual extension of the proboscis at low concentrations and progressively fewer responses as concentration was increased; however, a point was soon reached beyond which further increases in concentration elicited an increasingly high percentage of responses. McCutchan noticed two distinct kinds of extension of the proboscis. At low concentrations the proboscis extended rapidly, no segments remaining flexed and the la-

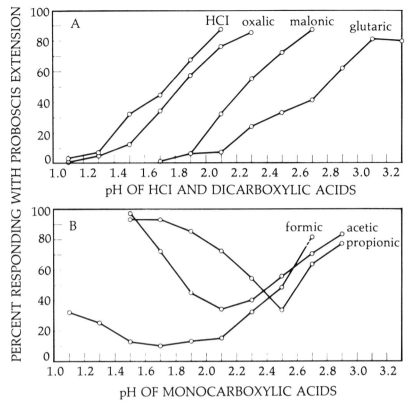

Fig. 97. Behavioral responses of flies to acids. Each point represents testing of at least 100 flies; each fly was used for two tests. All flies were positive to 0.25 M NaCl and negative to 1.0 M NaCl at the beginning of the tests (from McCutchan, 1969).

bellar lobes spread. At high concentrations of the monocarboxylic acids the response latency was long (2–3 sec.), extension was slow and hesitant, not all segments extended so that the proboscis appeared bent, the fly wiped the proboscis on the prothoracic legs, and regurgitation occurred if the fly had recently drunk water.

When the behavioral tests were repeated with water-satiated flies, the expected change occurred in the response to dicarboxylic acids; that is, the high concentrations failed as usual to elicit a response (the low concentrations no longer did because the flies were no longer responsive to water; see Fig. 98). In contrast, the response to monocarboxylic acids at high concentrations continued as usual even though the flies were water-satiated. Clearly this was not a feeding response, but rather a manifestation of avoidance. It was similar to the response to high concentrations of vapors of essential oils.

In the electrophysiological study that followed, complex results emerged. First, different types of hairs and different hairs within a

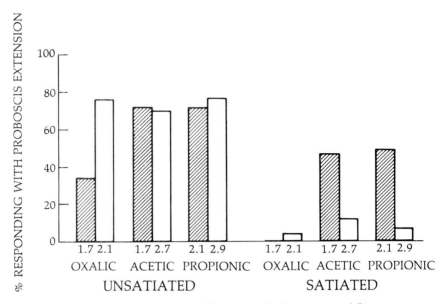

Fig. 98. Responses of water-satiated flies to acids. Responses of flies unsatiated and satiated to water are compared at different pH values of three acids. Results for unsatiated flies were taken from Figure 97. The number of satiated flies used were 26 for oxalic, 49 for acetic, and 55 for propionic acid (from McCutchan, 1969).

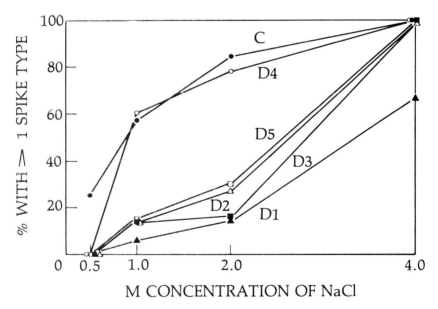

Fig. 99. Number of neurons responding to NaCl. The percentage of hairs of each type that responded with spikes from more than one neuron is shown as a function of NaCl concentration. Responses of more than 45 hairs of each type are included in the graph (from McCutchan, 1969).

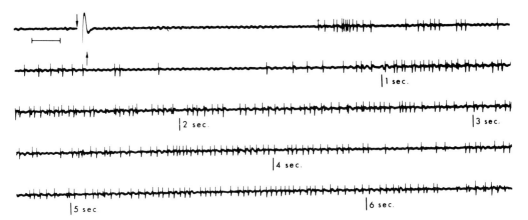

Fig. 100. Prolonged response of a hair to acetic acid. The response of a tarsal D hair to 0.2 M propionic acid recorded with the side-wall technique shows prolonged firing of neurons after removal of the stimulating solution. Arrows indicate application and removal of the stimulus. Time markers give seconds elapsed after removal of the stimulus (from McCutchan, 1969).

type behaved differently (Figs. 99 and 100), not only to acids but to NaCl. No response to water (0.001–0.1 NaCl) was ever recorded, even though it is almost a certainty that a water receptor exists. Nor were A and B hairs tested, because of their very small size and the difficulty of access. C and D hairs responded to NaCl with two and even three spikes. Only one of these showed a clear relation to concentration; this was the salt cell. The remaining two were presumed to be the water cell and the "fifty" cell of Dethier and Hanson. Side-wall recording of acid stimulation revealed many complexities; for example, continued activity following removal of the stimulus, or after-discharge in instances when the stimulus itself evoked no response.

Most of the data were derived from tip-recording. Five different types of response to acids were observed: (1) a small spike fired at a regular frequency; (2) a small spike and a large one (salt?) fired regularly; (3) three spikes appeared, the small one, the salt (?), and another unidentified; (4) two spikes of nearly the same amplitude fired; (5) more than two spikes fired in an irregular fashion and with frequent volleying. In one hair (D_2) a very tiny spike, completely unidentified, appeared regularly. C hairs typically responded as in category 4; D_5, in categories 1, 2, 3, 5; D_4, in category 1.

In general, all hairs responded to decreasing pH of an acid with increased frequency of firing and/or an increase in the number of cells responding. Different acids at the same pH have different stimulating effectiveness. The longer the chain length the more effective the acid. As Chadwick and Dethier (1947) had concluded, the anion clearly

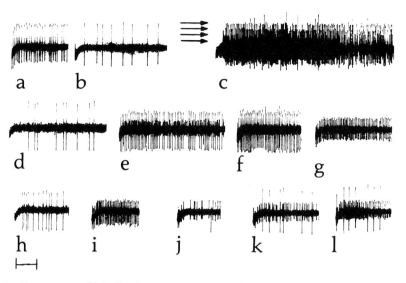

Fig. 101. Responses of labellar hairs to vapors. Sodium chloride used for reference. Horizontal bar represents 0.2 sec. in (a)–(l) and 0.1 in (m)–(s). (a) 0.1 M NaCl on "largest" hair no. 1; (b) vapor of formic acid on same hair; (c) solution of formic acid on same hair which was silent before stimulation, four

plays a contributory role in stimulation; pH is not the sole determinant.

There is a characteristic response of D hairs to acids. It is a small spike whose frequency increases with concentration. It may be the "fifth" cell. Activity by the salt cell is also correlated with the concentration of acid. The blowfly obviously has the sensory capacity of distinguishing acid from salt with the tarsi.

The behavioral response to very low pH values of monocarboxylic acids is correlated with activity from many receptors, irregular firing, and volleying. The behavioral response is not abolished by food or water satiation. The hesitancy of the response, together with the fact that sugar as well as salt and acid fibers fire, suggests an approach-avoidance conflict. This behavior will be discussed in later chapters.

The study of aversive behavior is only just beginning to receive more attention. Observations have been made of electrophysiological and behavioral responses to a wide variety of unrelated compounds, a number of which (formic acid, toluquinone, citronellal, and benzaldehyde) occur naturally in the defense secretions of arthropods. Many of these compounds evoked characteristic aversive behavior: hesitant extension of the proboscis, regurgitation, and cleaning movements involving manipulation of the regurgitate with the legs and rubbing of the mouth parts either on the legs or on the substrate. Neural discharge was frequently irregular and involved several cells (Fig. 101).

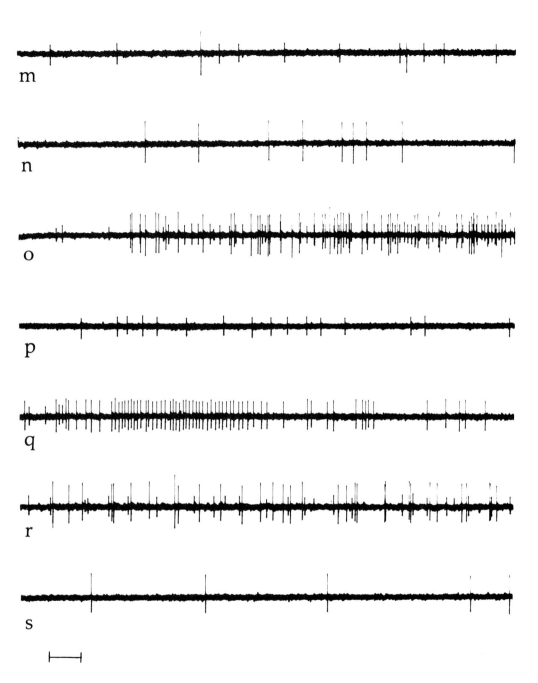

(Fig. 101 Continued)
cells responded; (d) 0.1 M NaCl on "largest" hair no. 5; (e), (f), and (g) responses to 6 M NaCl, vapor of formic acid, and vapor of citronella respectively, baseline silent at onset of stimulation. (h) Active baseline; (i) vapor of benzene; (j) baseline of "largest" hair no. 2; (k) and (l) 0.1 M NaCl and vapor of formic acid respectively; (m) baseline; (n) vapor of formic acid; (o)–(s) vapors of valeric acid, citral, limonene, benzene, and xylene respectively (from Dethier, 1972).

More recently it has been discovered that the labellar and tarsal contact chemoreceptors are sensitive to the vapors of compounds, some of which we would characterize as noxious (e.g., formic acid, hydrochloric acid, etc.) (Dethier, 1972). Not all compounds that are noxious to us evoke the response (e.g., ammonium chloride and allyl alcohol). On the other hand, a number of inoffensive (to us) compounds are effective, among them limonene, citral, benzaldehyde, benzene, xylene, butyraldehyde, etc. The reaction clearly is not caused merely by unphysiologically high or low pH conditions because nonpolar compounds are effective. The reaction is also specific to the extent that different compounds affect different receptors and in different ways. The electrolytes tend to initiate activity in the salt cells. The nonpolar compounds tend to single out other receptors, may inhibit as well as excite, and may cause after-discharges. The responses described are by no means always irreversible.

The various instances of erratic firing and multiple receptor response in the blowfly are all associated with stimuli that are very intense and potentially harmful. One is reminded of the common chemical sense of man. Whether the effects of noxious compounds are reversible or irreversible, whether the irregular firing is a recurring response to potential injury or actual injury that is a prelude to death of the receptor, the important fact remains that the neurons are sending characteristic messages that evoke aversive behavior. From an adaptive point of view the sacrifice of a few receptors is a small price to pay for survival when there are so many receptors available. That a receptor should die complaining, rather than silently, also has considerable adaptive value. In a very real sense this situation is analogous to human pain.

Examples of different kinds of responses to different compounds by a single fiber in chemoreceptive systems are rare. There is one report dealing with vertebrates. Some fibers have been found in the chorda tympani of the rat that give different response patterns to different chemicals, whereas other fibers give relatively constant patterns to all stimuli (Mistretta, 1972). The situation is not strictly analogous to that in the fly, however, because each fiber is served by more than one receptor.

In all the foregoing discussions we have been assuming that each receptor in the fly goes about its business in complete isolation, insulated from its companions, and aloof to the possibility of any cross-talk. In nature, few substances occur in pristine purity. Practically everything encountered by flies is a mixture, frequently extraordinarily complex; consequently more than one receptor is apt to be stimulated at one time. See, for example, the records depicting responses to honey, stinking fish, etc. (Fig. 102). Does traffic in one receptor and its fiber influence what travels in another?

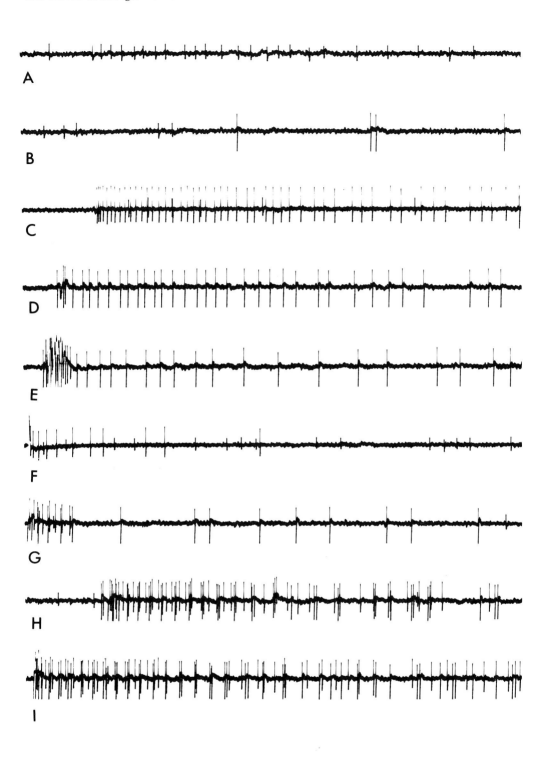

Fig. 102. Responses of labellar hair "largest" no. 2 to (from the top): 0.1 M NaCl, 0.1 M NaCl continued, fish, meat, beer, 0.5 M sucrose, honey, honey repeated, apple. Duration of record 1.75 sec. (from Dethier, 1974b).

A point has been made of the anatomical fact that each receptor sends its axon into the central nervous system without any peripheral synapsing. In the absence of synapses, interaction between receptors is still possible if electrotonic coupling occurs between receptors at any point. In the chemoreceptive sensilla of wire-worm larvae, electronmicrographs have revealed places where the dendrites of adjoining receptors at some points come closer than 250 Å. Here the membranes are clearly modified (Zackaruk, personal communication). In gustatory organs in the buccal cavity of the cockroach *Blabera craniifer,* two kinds of morphological connections have been reported (Moulins, 1968; Moulins and Noirot, 1972). One is a junction between cell bodies where the glial sheath is interrupted. Cisternae of endoplasmic reticulum are associated with this very narrow gap, which in some features resembles gaps where electrotonic coupling between cells has been demonstrated. A second type occurs in the form of septate desmosomes between axons in some instances, and the proximal segments of dendrites in others. It is also claimed that there are efferent fibers with vesicles resembling neurosecretory vesicles in close association with the axons of sensory cells. Thus far none of these structures have been found associated with the chemoreceptors of blowflies. Nevertheless, it has been known for a long time that some kind of peripheral interaction takes place when hairs are stimulated with mixed solutions. All workers have observed an interaction between activity in the salt fiber and in the sugar fiber (Hodgson, 1956, 1957; Morita et al., 1957; Wolbarsht, 1958, 1965; Morita and Takeda, 1959; Morita, 1959; Stürckow, 1959; Barton Browne and Hodgson, 1962; Gillary, 1966a; Rees, 1968). At least four possible modes of interaction could take place: mixing chemicals could cause changes in the physical-chemical properties (e.g., thermodynamic activity coefficient, diffusion coefficient, etc.); action potentials in one neuron could influence activity of action potentials in another; the receptor potential in one neuron could affect the generation of receptor potentials in its neighbors; the different chemicals could be competing for different molecular sites on the dendrite. Since an inverse relationship between the frequencies of the salt receptor and the sugar receptor could not be demonstrated, Hodgson (1956, 1957) concluded that the effects of mixtures could not be due to neurological interaction (cf. also Takeda, 1961). Wolbarsht (1958) agreed and attributed the effects to changes in the physical properties of solutions that result from mixing.

Many kinds of interaction occur in addition to the depression of activity in the salt receptor by sugar, and the converse. Salt and sugar inhibit the water cell (Evans and Mellon, 1962b). The concentration of salt that begins to inhibit the water cell is almost identical to the threshold concentration for the salt cell (Rees, 1968). Salt can enhance the response to sugar (Kuwabara, 1951; Morita, Hidaka, and Shiraishi, 1966). Quinine, calcium chloride, and acetic acid can inhibit sugar and salt receptors by hyperpolarizing them (Morita, 1959). Alcohols and aliphatic amines can inhibit all cells initially and then cause indiscriminate firing (Hanson, 1965; Steinhardt, Morita, and Hodgson, 1966; Hodgson and Steinhardt, 1967). And, as already pointed out, sodium salts of fatty acids can inhibit the fifth cell (Dethier and Hanson, 1968), and some sugars can inhibit or synergize the action of others (Dethier, Evans, and Rhoades, 1956; Morita and Shiraishi, 1968; Omand and Dethier, 1969).

There is still no concrete evidence of electrotonic coupling between the fly's receptors; the interactions reported are attributable either to the changed character of mixed solutions or to membrane effects on single receptors. Therefore, the only hope of increasing the information-carrying capacity of single lines by making them less specific is to allow for changes in patterns of discharge, which have been shown to occur with noxious stimuli.

However, if there is interaction, not at the periphery but in the central nervous system, by which information from many lines is pooled, the capabilities of the sensory system are greatly enhanced. The only requisites are that the receptors be less rigidly specific: they must have different spectra of activity that may overlap, and the central nervous system must be able to integrate the input from all fibers. If these conditions are fulfilled, across-fiber patterning becomes possible and the range of codes becomes enormous. Such a system was proposed for the taste system of the rat by Pfaffmann (1941, 1955) and developed by Erickson (1963, 1967, 1968). These investigators suggested that there is a limited number of receptor types, that each has a moderately broad reaction spectrum, and that these spectra overlap to some extent. In other words, one receptor would respond to more than one compound but with different magnitudes of response; another receptor would respond to some of the same compounds but also to additional ones—also with different magnitudes of response. Any given stimulus applied to the total complement of receptors would therefore evoke a different magnitude of response from each, from zero to maximum. From the point of view of the central nervous system, which is the recipient of this "across-fiber" or combined message, each stimulus would produce a different and characteristic total response profile.

Details of the characteristics of across-fiber patterning in vertebrates are still in an unsettled state. Much of the confusion arises from uncertainty about the specificity of the fibers. Very early, Pfaffmann (1941) had rejected the idea of four specific fibers; and Frank and Pfaffmann (1969) had concluded that the four taste qualities are randomly distributed among fibers. Broad multiple specificities had been found in the cat, rat, rabbit, hamster, and monkey. Probing deeper into the situation in the rat and hamster, Sato, Yamashita, and Ogawa (1969) found that some fibers responded to one stimulus only and others to 2, 3, or 4. Some were also sensitive to cooling, some discharged spontaneously, and some responded more or less easily to some combinations than to others. Sodium-dominant types generally showed both phasic and tonic responses, and sucrose-dominant types showed rhythmic bursts. Generally speaking, even though there is multiple responsiveness by fibers, they can be classed statistically into distinct categories. Frank (1972) finally decided that the sensitivities (in the hamster) are not randomly distributed among fibers. Responsiveness to sucrose is negatively correlated with responsiveness to each of the other three tastes, all of which are positively correlated with each other. Even the sugar fibers are variable because some respond better to sucrose than to fructose (in the squirrel monkey), whereas others behave in an opposite manner (Pfaffmann, 1970).

The stability of across-fiber patterning varies with stimuli (Doetsch, et al., 1969). Lithium, for example, is very stable over 1 sec. but $CaCl_2$ less so. The least stable potion of the response is the phasic (the first 0.1 sec.). It was suggested that the tonic phase carries more information for discrimination than does the phasic portion. Work with the opossum tended to support this idea. Taste discrimination correlated better with across-fiber patterns generated in second 2 than in second 1 (Marshall, 1968). The pattern also changed with the concentration of the stimulus. If the pattern has meaning, then taste should also change with concentration—and it does. Intensity is probably encoded in terms of *over-all* rate in *all neurons*.

How the hypothesis of across-fiber patterning could apply to the fly is shown in the following analysis of responses to some natural foods (Dethier, 1974b). Consider first the total response from a hair during the first second of continuous stimulation by some naturally occurring substances (Fig. 103). The pattern received by the central nervous system from this hair is different in the several cases (Fig. 104). These differences occur because each substance (sucrose and salt excepted) is chemically complex and because each receptor cell of the hair has a characteristic action spectrum rather than a unitary specificity. If the central nervous system is capable of analyzing these patterns, there is contained within them sufficient information to characterize each substance as different.

It has been shown experimentally that stimulation of one hair by water or sugar can elicit a complete behavioral response, acceptance; and similarly that stimulation of one hair by salt or acid can elicit rejection. It has not yet been demonstrated that the patterns generated

The Flavor of Things: Codes and Information 213

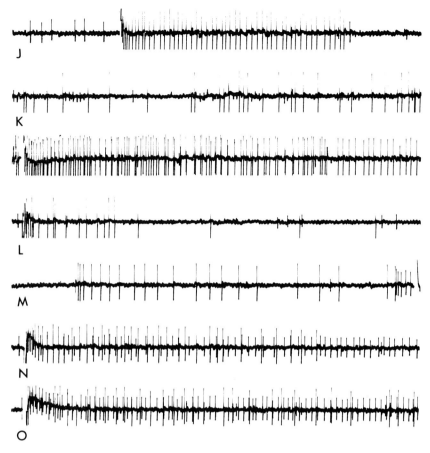

Fig. 103. Responses of labellar hair "largest" no. 6 to (from the top): 0.5 M NaCl, fish, meat, beer, 0.5 M sucrose, honey, and apple (from Dethier, 1974b).

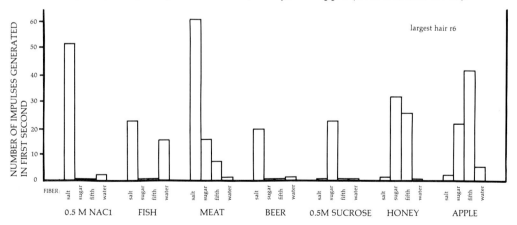

Fig. 104. Patterns of response generated by hair "largest" no. 6 to natural substances. For each substance the bars represent from left to right the number of impulses generated by the salt, sugar, fifth, and water cells respectively (from Dethier, 1974b).

by single hairs are behaviorally significant. The experiments of Dethier (1974b) merely prove that a single hair (i.e., the package of four chemoreceptor cells) can generate different across-fiber patterns, depending upon what substance is applied. What relevance does this fact have in normal circumstances, when the central nervous system is receiving patterns transmitted by approximately one hundred hairs? (It is not known exactly how many hairs are touching a solution during normal feeding.) What is the fate of each individual pattern in the central nervous system? Moreover, does accurate discrimination require that all hairs act in an identical manner?

These questions engender still others to which there are at the moment only partial answers. What advantages accrue from the possession of large receptor fields? What difficulties are raised by interhair variability and by the intrinsic variability of individual receptors? What is the functional unit of the chemosensory system, the sensillum (the hair itself) or the individual receptor cell?

The advantages accruing from a large population of receptors are several: (1) insurance against wear and tear; (2) compensation for progressive loss of receptors with age; (3) provision for stimulus-fractionation (that is, a wide range of sensitivity can be achieved by having different receptors operating most efficiently over different limited ranges, rather than having all receptors operating over the same extremely wide range); (4) opportunity for prolonged sensitivity of the animal as different receptors adapt and disadapt out of phase; (5) compensation for noise in individual receptors; (6) provision for a chemotopographic sensory capacity (that is, adding a two-dimensional character to chemoreception); (7) provision for finer control of motor output and response by allowing many shades of temporal and spatial summation; (8) provision for the generation of many across-fiber patterns.

It might be asked, especially with reference to plant-feeding insects, why, for example, a locust employs several hundred chemoreceptors on its palps for the task of tasting food, whereas caterpillars do equally well with as few as eight. Of the possible answers that suggest themselves, one could relate to developmental requirements. The caterpillar, one ontogenetic stage of a holometabolous insect, is primarily a feeding machine, which serves the purpose of storing energy preparatory to becoming an adult with an entirely different life style. It would not be economical to expend a lot of energy unnecessarily in equipping a caterpillar lavishly with sense organs that will be demolished at metamorphosis. The locust, on the other hand, being hemimetabolous, is building equipment that will constitute part of the adult complement, since the two life stages have essentially the same life style.

Locusts are much more peripatetic than caterpillars (with the possible exception of army worms) and it might also be argued that they require more sophisticated sensory equipment.

Given a large population of receptors, the significance of across-fiber patterns might be impaired by interhair variability. As Table 16 indicates, the patterns generated by different hairs are not identical. Interhair variability results in part from qualitative differences, i.e., differences in the kind of receptor present in individual hairs (cf. Dethier, 1961; McCutchan, 1969); and in part from differences in threshold, in individual hairs, of any given type of receptor (cf. Dethier and Hanson, 1968). The difficulties imposed by interhair variability disappear, however, if one considers the situation in terms of a population of labellar receptor cells rather than a population of hairs. This is permissible because there is no evidence that the units of a single hair are treated as a single unified complex in the central nervous system. A large population of cells with a limited number of characteristic action spectra could generate a labellar sensory pattern from the integration of which an abstraction of the stimulus could be derived.

The question as to what comprises the functional unit of the labellar chemosensory field, the receptor cell or the sensillum with its complement of several physiologically different cells, is not only relevant to the concept of across-fiber patterning but is of considerable theoretical interest as well. Characteristically, the gustatory receptors of insects are assembled together in groups of two or more (usually four to five). Most adult insects as well as the nymphs of hemimetabolous species possess large receptor fields of such multi-innervated sensilla. The situation is somewhat comparable to the field of ommatidia, each consisting of more than one retinal cell, in the compound eye. There the question of which was the functional unit, the retinula cell or the ommatidium, has engendered a wealth of experimental work. In contrast, the analogous question has not stimulated much investigation of the behavioral significance of different functional organizations in chemoreceptive systems. Not every labellar hair in *Phormia* has been thoroughly explored electrophysiologically, but more often than not every one tested contains one cell of each type. When any particular hair fails to respond to any of the compounds employed as stimuli, it cannot be assumed that the relevant receptor is absent, because age and other factors influence the activity and health of receptors. Admittedly, the possibility that different hairs of the labellum have qualitatively different kinds of receptors cannot yet be ruled out absolutely (the sensilla of the tarsi do appear to be qualitatively different). There is more certainty that the corresponding cells of different labellar hairs

Table 16. Responses of Several Labellar Hairs to Natural Foods.

Stimulus	Duration	Number of impulses:				Hair (largest) number
		Salt cell	Sugar cell	5th cell	Water cell	
0.1 M NaCl	25 msec	0	0	0	0	r 2
	1 sec	0	0	0	20	r 2
Fish	25 msec	4	0	0	0	r 2
	1 sec	39	0	0	5	r 2
Meat	25 msec	2	0	0	0	r 2
	1 sec	26	0	0	0	r 2
Beer	25 msec	3	0	0	0	r 2
	1 sec	25	0	0	0	r 2
0.5 M sucrose	25 msec	0	3	0	0	r 2
	1 sec	0	11	0	9	r 2
Honey	25 msec	2	2	0	0	r 2
	1 sec	12	8	0	3	r 2
Honey	25 msec	2	0	0	0	r 2
	1 sec	27	16	20	1	r 2
Apple	25 msec	2	1	1	0	r 2
	1 sec	27	15	17	2	r 2
0.5 M NaCl	25 msec	4	0	0	0	r 6
	1 sec	51	0	1	2	r 6
Fish	25 msec	2	0	0	0	r 6
	1 sec	24	0	0	15	r 6
Meat	25 msec	3	0	0	0	r 6
	1 sec	60	15	7	1	r 6
Beer	25 msec	3	0	0	0	r 6
	1 sec	19	0	0	1	r 6
0.5 M sucrose	25 msec	0	2	0	0	r 6
	1 sec	0	17	0	0	r 6
Honey	25 msec	0	4	0	0	r 6
	1 sec	1	31	25	0	r 6
Apple	25 msec	0	3	2	1	r 6
	1 sec	2	21	41	5	r 6
Water	25 msec	0	0	0	7	r 11
	1 sec	0	0	0	73	r 11
0.5 M NaCl	25 msec	3	0	0	0	r 11
	1 sec	20	0	0	1	r 11
Fish	25 msec	4	0	0	2	r 11
	1 sec	37	0	49	10	r 11
Meat	25 msec	4	0	1	0	r 11
	1 sec	27	0	3	6	r 11
Beer	25 msec	0	0	0	19	r 11
	1 sec	0	0	0	0	r 11
0.5 M sucrose	25 msec	0	2	0	0	r 11
	1 sec	0	12	0	1	r 11

From Dethier, 1974b.

have different sensitivities. The sensitivities of different cells in sensilla apparently do not vary independently of their neighbors; this is not true in some caterpillars, however (Schoonhoven, 1974).

Given the multifarious duties of a large chemoreceptive field characterized by a considerable degree of duplication (but not redundancy), in which way are the ends best served, by making the receptor or the sensillum the functional unit? The question implies two different systems of wiring the receptors into the central nervous system. The proposition that the receptors rather than the sensilla are the units (Dethier and Hanson, 1968; Dethier, 1974b) implies that all of the water receptors (for example) hook up with a common interneuron or group of interneurons at an early stage of central circuitry. The alternative view (van der Starre, 1972; Blaney, 1974) implies that the neurons of each sensillum hook up as a group with an interneuron.

In the first instance a smaller number of interneurons would suffice. All of the advantages of a large receptive field, except a fine chemotopographic sense, would be present. In the second view all of the advantages, including a fine chemotopographic sense, would be possible. With the sensillum as the functional unit it is conceivable that a more subtle analysis of stimulus situations would occur because intimate interaction between receptors (not yet shown to exist peripherally) would be possible at an early synaptic level in the central nervous system. With the receptor as the unit, this degree of interaction is improbable, and the assessment of stimulus situations would be fuzzier. On the other hand, it is conceivable that this kind of circuitry is more parsimonious than the other. In either case, across-fiber patterning can be accomplished.

The ultimate answers will be provided by visual demonstration of the central circuitry and/or by electrophysiological recording from interneurons. A start in the first direction has been made by cobalt back-filling of neurons. The results at the moment are ambiguous. Also, little success has attended efforts to record from interneurons in the central nervous system of *Phormia.*

There are two additional potential complications that must be considered in evaluating the hypothesis of across-fiber patterning. One is the relative contribution of the phasic versus the tonic portion of the receptor response; the other is temporal variability within a receptor. When patterning is assessed solely in terms of events occurring within the phasic portion of the response (± 25 msec.), there is no consistent, distinctive, total pattern. For example, fish, meat, salt, and beer all elicit from 2 to 4 spikes from the salt receptor of largest hair 6. Honey and sucrose elicit 2 to 4 spikes from the sugar receptor. All six of these substances are accepted. The acceptance of the first four is in agreement with the earlier finding that low concentrations of sodium chloride can initiate acceptance (Dethier, 1968). There is not enough information in the phasic element of the response to permit distinguishing among the four. They could be distinguished from sugar and honey but not from one another. On the other hand, apple elicits three spikes from the sugar receptor, two from the "fifth" cell, and one from the water cell in the phasic portion and so preserves the elements of a pattern. The phasic portion of the response may serve merely to set the sign of the response, leaving the task of quality discrimination to the tonic portion.

Intrinsic variability, already discussed, poses a greater obstacle to the hypothesis of patterning. Patterns to be meaningful would have to be stable above the noise level. Preliminary studies seem to indicate that this is so.

Across-fiber patterning has also been proposed as an explanation of how the fly distinguishes between water and sugar solutions applied to the tarsi. McCutchan (1969) had found tarsal sugar receptors but no tarsal water receptors. Van der Starre (1972) concluded that at least two, and possibly more, receptors responded to both solutions but changed their frequency of response differently when solutions were changed gradually from pure water to concentrated sugar. Thus the total pattern of response of all receptors in a single hair for pure water and dilute sugar (that is, sugar below behavioral threshold) differed from that for concentrated sugar. The fly distinguished sugar solutions from water by means of the total pattern.

Across-fiber patterning codes in the gustatory system of the fly operate on essentially silent backgrounds, that is, in the absence of stimulation the afferent fibers do not transmit any action potentials. As a consequence, a stimulus can affect the system in one direction only; it can increase the rate of firing of action potentials. A new dimension would be added to the coding system if the resting state, instead of being silent, maintained some constant level of spontaneous activity. Were this the case, a stimulus could decrease as well as increase frequency. Coupling spontaneous activity with across-fiber patterning would permit a small number of receptors to encode a vast amount of information. The olfactory sense of caterpillars appears to operate in this fashion (Dethier, 1967, 1971; Dethier and Schoonhoven, 1969; Schoonhoven and Dethier, 1966). Each of the sixteen olfactory receptors on the antennae of caterpillars is, in Schneider's (1969) terminology, a "generalist." A generalist has been defined as a receptor that has a broad, constant, but highly individual reaction spectrum. Furthermore, it has three modes of response: maintaining a constant basal rate of firing, increasing the basal rate (excitation), or decreasing the basal rate (inhibition). Some typical response spectra of caterpillar receptors are depicted in Figure 105, which represents four receptors

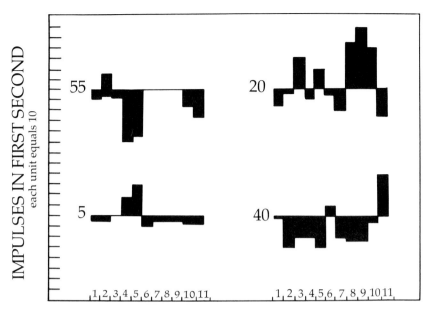

Fig. 105. Activity spectra of each of four olfactory receptors of the tobacco hornworm, *Manduca sexta*, responding to eleven different odors presented successively. Activity is given as the increase or decrease from spontaneous firing level of nerve impulses generated during the first second of stimulation. The spontaneous firing level is given with each block diagram (from Dethier, 1967; calculated from the records of Schoonhoven and Dethier, 1967).

of the tobacco hornworm (*Manduca sexta*) responding to eleven different odors (Dethier, 1967). For example, the receptor with the basal firing rate of 20, represented by the upper right-hand histogram, responds to odors 1, 2, 4, 6, 7, and 11 by decreasing its firing. It responds to odors 3, 5, 8, 9, and 10 by increasing its firing. The receptor with the basal rate of 40 (lower right) responds to odors 6 and 11 with increase and to all others with decrease. Since the response profile of each receptor is different and the caterpillar receives information from all simultaneously, the across-fiber pattern presented to the central nervous system can be rich and varied (cf. Fig. 106).

The ability of the sensory system to code and transmit information could be further enhanced by taking advantage of different latencies, rates of adaptation, and aftereffects. The caterpillar olfactory system has this potentiality (Dethier and Schoonhove, 1969). For example, in the tobacco hornworm there are two receptors that respond to the odor of tomato with a gradual increase of firing, which builds up at about the same rate to about the same maximum, and with similar rates of adaptation (Figs. 107 and 108, third record from the top). These receptors also respond to the odor of geranium with an increase in rate of firing; however, one cell (the smaller spike in Figs. 107 and 108, top

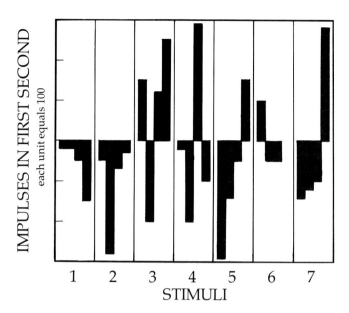

Fig. 106. Profiles of total activity from four olfactory receptors of the tobacco hornworm for each of seven different odors (stimuli) (from Dethier, 1967).

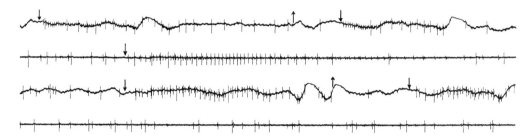

Fig. 107. Extracellular recording from two cells of the lateral sensillum basiconicum on the antenna of *Manduca sexta*. Onset of stimulation indicated by arrows pointing down. Arrows pointing up indicate off. Total duration of each record 2.2 sec. Top line, two consecutive stimulations with geranium (9, 10); second line, linalool (19); third line, two stimulations with tomato (4, 5); bottom line, background activity (15). Numbers in parentheses indicate the position of each stimulation in a series of fifty stimulations (from Dethier and Schoonhoven, 1969).

record) responds with a shorter latency, attains a higher maximum more quickly, and begins to adapt while the other cell (large spike) is still in the rising phase. In response to linalool, the cell giving the smaller spike again anticipates and overshoots the other; but then, as the other begins to adapt, this cell again generates activity rising to a maximum (Figs. 107 and 108, second record from top). A comparable situation is seen in records from another preparation in which two cells responded to limonene and citral (Figs. 109 and 110). Because dif-

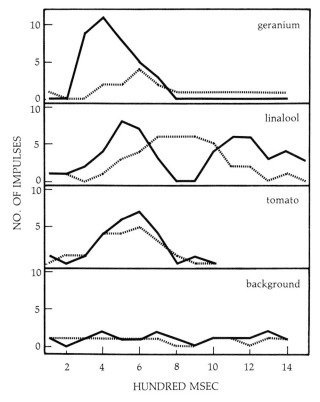

Fig. 108. Frequency of discharge per 100 msec. of the two cells recorded in Figure 107. Solid line, small spike; broken line, large spike (from Dethier and Schoonhoven, 1969).

ferential responses of cells such as these vary from one compound to the next in a consistent manner, it is unlikely that the cells are interacting peripherally. More probably they are totally independent. In other words, one cell does not necessarily inhibit another when some critical rate of firing is attained.

The responses just described establish a temporal patterning extending over a period of approximately one second. The significance of this patterning would depend upon the ability of the central nervous system to read, as it were, a whole sentence rather than taking its cue from the first word.

Finally, some thought should be given to the idea that an animal is not the slave of any one sensory system at a given moment. While one system is generally in command (in the case of feeding, the chemical senses), other systems are still barraging the central nervous system with information. The fly that is feeding is more than likely standing in its food, so that it is receiving chemical information from its tarsi as well as from the labellar hairs. The oral papillae are also contributing

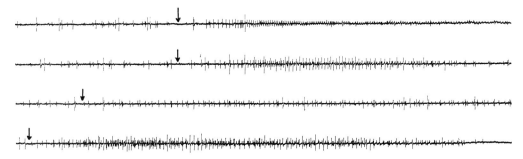

Fig. 109. Extracellular recording from four cells of the lateral sensillum basiconicum on the antenna of *Manduca sexta*. Preparation and electrode position differ from those in Figure 108. Total duration of each record 1.8 sec. Records from top to bottom: limonene (12), citral (11), amyl acetate (9), amyl alcohol (21). Other details as in Figure 107.

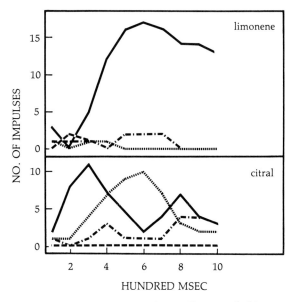

Fig. 110. Frequency of discharge per 100 msec. of the four cells recorded in Figure 109. Largest spike, . . .; smallest, ----; medium, —; second smallest, -.- (from Dethier and Schoonhoven, 1969).

to the input, as is the olfactory system, the tactile sense, vision, and all the rest. The central nervous system obviously can switch off some input, can ignore it, so that the matter of attention becomes an important one. Unfortunately we know little about it. What we do know is considered in Chapter 12.

The one additional system that we do have some information about is that comprising the oral papillae (interpseudotracheal papillae), although this information is scanty because of the very great difficulty in applying electrodes to these sensilla (Dethier and Hanson, 1965). First of all, these receptors are very small, only 10 μ long. A more formidable obstacle is their location *within* the labellum and the hydrophilic nature of that surface, which encourages fluid to flow out of electrodes. To surmount the obstacle of location, blood was forced into the labellum by ruthlessly squeezing the thorax of the fly between thumb and forefinger. This caused the labellar lobes to open. A ligature was then applied to the base of the labellum tightly enough to prevent the flow of blood to the head. The bleeding of electrodes onto the hydrophilic surface was prevented by mixing every stimulating solution with warm agar and forcing this into the electrodes, where it solidified. When recording was finally accomplished, the following facts emerged. Of the four cells known to innervate each papilla, one responded preferentially to salt, one to sugar, and one to mechanical stimulation. An adequate stimulus for the fourth was not discovered, nor was a response to water observed electrophysiologically.

The original idea of two taste modalities, acceptable and unacceptable, has now been shown to be an oversimplification. In the range of unacceptability the fly certainly has the peripheral mechanism for distinguishing between acids and salts. Furthermore, it can clearly distinguish between unacceptable substances and noxious substances. In the realm of acceptability it can distinguish water from other acceptable liquids, which is hardly surprising since maintenance of water balance must be kept separate from maintenance of energy balance. Beyond this, however, as early behavioral studies showed, the fly can distinguish between proteinaceous and nonproteinaceous food. Across-fiber patterning, which may be the basis for this latter discrimination, provides a dimension for even finer discriminations among a variety of foods. Even without the assistance of the olfactory sense the fly can distinguish between liver and brain-heart extract. Perhaps it can distinguish between rotten fish and rotten meat. More behavioral studies are certainly in order. With the assistance of the olfactory sense even more subtle or more extensive discriminations might be possible. And beyond the realm of food lie those odors and/or tastes that assist in courtship and sexual behavior. Flavor may after all be the proper word to apply to the chemosensory activities of the blowfly.

Literature Cited

Barton Browne, L., and Hodgson, E. S. 1962. Electrophysiological studies of arthropod chemoreception. IV. Latency, independence, and specificity of labellar chemoreceptors of the blowfly, *Lucilia*. *J. Cell. Comp. Physiol.* 59:187–202.

Beidler, L. M. 1969. Innervation of rat fungiform papilla. In *Olfaction and Taste,* III, C. Pfaffmann, ed., pp. 352–369. Rockefeller University Press, New York.

Beidler, L. M., and Smallman, R. L. 1965. Renewal of cells within taste buds. *J. Cell. Biol.* 27:263–272.

Blaney, W. M. 1974. Electrophysiological responses of the terminal sensilla on the maxillary palps of *Locusta migratoria* (L.) to some electrolytes and nonelectrolytes. *J. Exp. Biol.* 60:275–293.

Blaney, W. M., Chapman, R. F., and Cook, A. G. 1971. The structure of the terminal sensilla on the maxillary palps of *Locusta migratoria* (L.) and changes associated with moulting. *Zeit. Zellforsch.* 121:48–68.

Chadwick, L. E., and Dethier, V. G. 1947. The relationship between chemical structure and the response of blowflies to tarsal stimulation by aliphatic acids. *J. Gen. Physiol.* 30:255–262.

Dethier, V. G. 1961. Behavioral aspects of protein ingestion by the blowfly *Phormia regina* Meigen. *Biol. Bull.* 121:456–470.

Dethier, V. G. 1967. Feeding and drinking behavior of invertebrates. In *Handbook of Physiology,* Sec. B., *Alimentary Canal.* C. F. Code, ed., pp. 79–96. Amer. Physiol. Soc., Vol. 1, Washington, D.C.

Dethier, V. G. 1968. Chemosensory input and taste discrimination in the blowfly. *Science* 161:389–391.

Dethier, V. G. 1972. Sensitivity of the contact chemoreceptors of the blowfly to vapors. *Proc. Nat. Acad. Sci. U.S.A.* 69:2189–2192.

Dethier, V. G. 1974a. Sensory input and the inconstant fly. In *Experimental Analysis of Insect Behaviour,* L. Barton Browne, ed., pp. 19–31. Springer Verlag, N.Y.

Dethier, V. G. 1974b. The specificity of the labellar chemoreceptors of the blowfly and the response to natural foods. *J. Insect Physiol.* 20:1859–1869.

Dethier, V. G., Evans, D. R., and Rhoades, M. V. 1956. Some factors controlling the ingestion of carbohydrates by the blowfly. *Biol. Bull.* 111:204–222.

Dethier, V. G., and Hanson, F. E. 1965. Taste papillae of the blowfly. *J. Cell. Comp. Physiol.* 65:93–100.

Dethier, V. G., and Hanson, F. E. 1968. Electrophysiological responses of the blowfly to sodium salts of fatty acids. *Proc. Nat. Acad. Sci. U.S.A.* 60:1269–1303.

Dethier, V. G., and Schoonhoven, L. M. 1969. Olfactory coding by leipdopterous larvae. *Ent. Exp. Appl.* 12:535–543.

Doetsch, G. S., Ganchrow, J. J., Nelson, L. M., and Erickson, R. P. 1969. Information processing in the taste system of the rat. In *Olfaction and Taste,* III, C. Pfaffmann, ed., pp. 492–511. Rockefeller University Press, New York.

Erickson, R. P. 1963. Sensory neural patterns and gustation. In *Olfaction and Taste,* I, Y. Zotterman, ed., pp. 205–213. Pergamon Press, Oxford.

Erickson, R. P. 1967. Neural coding of taste quality. In *The Chemical Senses and Nutrition,* M. Kare and O. Maller, eds., pp. 313–327. The Johns Hopkins Press, Baltimore.

Erickson, R. P. 1968. Stimulus coding in topographic and nontopographic afferent modalities: on the significance of the activity of individual sensory neurons. *Psychol. Rev.* 75:447–465.

Evans, D. R., and Mellon, De F. 1962a. Stimulation of a primary taste receptor by salt. *J. Gen. Physiol.* 45:651–661.

Evans, D. R., and Mellon, De F. 1962b. Electrophysiological studies of a water receptor associated with the taste sensilla of the blowfly. *J. Gen. Physiol.* 45:487–500.

Frank, M. 1972. Taste responses of single hamster chorda tympani nerve fibers. In *Olfaction and Taste,* IV, D. Schneider, ed., pp. 287–293. Wissenschaftliche Verlagsgesellschaft, MBH, Stuttgart.

Frank, M., and Pfaffmann, C. 1969. The distribution of taste sensitivities among single taste fibers. In *Olfaction and Taste,* III, C. Pfaffmann, ed., pp. 488–491. Rockefeller University Press, New York.

Frings, H. 1946. Gustatory thresholds for sucrose and electrolytes for the cockroach, *Periplaneta americana* (Linn.). *J. Exp. Zool.* 102:25–50.

Getting, P. A. 1971. The sensory control of motor output in the fly proboscis extension. *Zeit. vergl. Physiol.* 74:103–120.

Gillary, H. L. 1966a. Stimulation of the salt receptor of the blowfly. I. NaCl. *J. Gen. Physiol.* 50:337–350.

Gillary, H. L. 1966b. Stimulation of the salt receptor of the blowfly. II. Temperature. *J. Gen. Physiol.* 50:351–357.

Hanson, F. E. 1965. Electrophysiological studies on chemoreceptors of the blowfly, *Phormia regina* Meigen. Doctoral dissertation, University of Pennsylvania, Philadelphia.

Haskell, P. T., and Schoonhoven, L. M. 1969. The function of certain mouthpart receptors in relation to feeding in *Schistocerca gregaria* and *Locusta migratoria migratorioides. Ent. Exp. Appl.* 12:423–440.

Hellekant, G. 1968. Postexicitatory depression of gustatory receptors. *Acta Physiol. Scand.* 74:1–9.

Hodgson, E. S. 1956. Physiology of the labellar sugar receptors of flies. *Anat. Rec.* 125:555.

Hodgson, E. S. 1957. Electrophysiological studies of arthropod chemoreception. II. Responses of labellar chemoreceptors of the blowfly to stimulation by carbohydrates. *J. Insect. Physiol.* 1:240–247.

Hodgson, E. S., and Roeder, K. D. 1956. Electrophysiological studies of arthropod chemoreception. I. General properties of the labellar chemoreceptors of Diptera. *J. Cell. Comp. Physiol.* 48:51–76.

Hodgson, E. S., and Steinhardt, R. A. 1967. Hydrocarbon inhibition of primary chemoreceptor cells. In *Olfaction and Taste,* II, T. Hayashi, ed., pp. 739–748. Pergamon Press, Oxford.

Knight, B. W. 1972. Dynamics of encoding in a population of neurons. *J. Gen. Physiol.* 59:734–766.

Kuwabara, M. 1951. Effects of inorganic ions on the tarsal chemoreceptor of the butterfly, *Vanessa indica. Zool. Mag.* 60:9 (in Japanese).

McCutchan, M. C. 1969. Behavioral and electrophysiological responses of the blowfly, *Phormia regina* Meigen, to acids. *Zeit. vergl. Physiol.* 65:131–152.

Marshall, D. A. 1968. A comparative study of neural coding in gustation. *Physiol. Behav.* 3:1–15.

Mistretta, C. M. 1972. A quantitative analysis of rat chorda typani fiber discharge patterns. In *Olfaction and Taste,* IV, D. Schneider, ed., pp. 294–300. Wissenschaftliche Verlagsgesellschaft, MBH, Stuttgart.

Moore, G. P., Perkel, D. H., and Segundo, J. P. 1966. Statistical analysis and functional interpretation of neuronal spike data. *Annu. Rev. Physiol.* 28:493–522.

Morita, H. 1959. Initiation of spike potentials in contact chemosensory hairs of insects. III. D.C. stimulation and generator potential of labellar chemoreceptor of *Calliphora*. *J. Cell. Comp. Physiol.* 54:189–204.

Morita, H., Doira, S., Takeda, K., and Kuwabara, M. 1957. Electrical responses of contact chemoreceptor on tarsus of the butterfly, *Vanessa indica*. *Mem. Fac. Sci. Kyushu Univ.* E2:119–139.

Morita, H., Hidaka, T., and Shiraishi, A. 1966. Excitatory and inhibitory effects of salts on the sugar receptor of the fleshfly. *Mem. Fac. Sci. Kyushu Univ.* Ser. E (Biol.), 4:123–135.

Morita, H., and Shiraishi, A. 1968. Stimulation of the labellar sugar receptor of the fleshfly by mono- and disaccharides. *J. Gen. Physiol.* 52:559–583.

Morita, H., and Takeda, K. 1959. Initiation of spike potentials in contact chemosensory hairs of insects. III. The effect of electric current on tarsal chemosensory hairs of Vanessa. *J. Cell. Comp. Physiol.* 54:177–187.

Moulins, M. 1968. Les sensilles de l'organe hypopharyngien de *Blabera craniifer* Burm. (Insecta, Dictyoptera). *J. Ultrastr. Res.* 21:474–513.

Moulins, M., and Noirot, C. 1972. Morphological features bearing on transduction and peripheral integration in insect gustatory organs. In *Olfaction and Taste*, IV, D. Schneider, ed., pp. 49–55. Wissenschaftliche Verlagsgesellschaft, MBH, Stuttgart.

Murray, R. G. 1969. Cell types in rabbit taste buds. In *Olfaction and Taste*, III, C. Pfaffmann, ed., pp. 331–344. Rockefeller University Press, New York.

Murray, R. G., and Murray, A. 1970. The anatomy and ultrastructure of taste endings. In *Taste and Smell in Vertebrates*, G.E.W. Wolstenholme and J. Knight, eds., pp. 3–30. J. A. Churchill, London.

Omand, E., and Dethier, V. G. 1969. An electrophysiological analysis of the action of carbohydrates in the sugar receptor of the blowfly. *Proc. Nat. Acad. Sci. U.S.A.* 62:136–143.

Perkel, D. H., and Bullock, T. H. 1968. Neural coding: *Neurosciences Res. Program Bull.* 6:221–348.

Pfaffmann, C. 1941. Gustatory afferent impulses. *J. Cell. Comp. Physiol.* 17:243–258.

Pffaffmann, C. 1955. Gustatory nerve impulses in rat, cat, and rabbit. *J. Neurophysiol.* 18:429–440.

Pfaffmann, C. 1970. Physiological and behavioral processes of the sense of taste. In *Taste and Smell in Vertebrates*, G.E.W. Wolstenholme and J. Knight, eds., pp. 31–50. J. A. Churchill, London.

Ratliff, F., Hartline, H. K., and Lange, D. 1968. Variability of interspike intervals in optic fibers of Limulus: effect of light and dark adaptation. *Proc. Nat. Acad. Sci. U.S.A.* 60:464–469.

Rees, C. J. C. 1968. The effect of aqueous solutions of some 1:1 electrolytes on the electrical response of the type 1 ("salt") chemoreceptor cell in the labella of *Phormia*. *J. Insect. Physiol.* 14:1331–1364.

Roeder, K. D. 1962. The behaviour of free-flying moths in the presence of artificial ultrasonic pulses. *Animal Behav.* 10:300–304.

Roeder, K. D. 1966. Interneurons of the thoracic nerve cord activated by tympanic nerve fibers in noctuid moths. *J. Insect Physiol.* 12:1227–1244.

Sato, M., Yamashita, S., and Ogawa, H. 1969. Afferent specificity in taste. In

Olfaction and Taste, III, C. Pfaffmann, ed., pp. 470–487. Rockefeller University Press, New York.

Schneider, D. 1969. Insect olfaction: deciphering system for chemical messages. *Science* 163:1031–1037.

Schoonhoven, L. M. 1974. On the variability of chemosensory information. Symposium, *The Host Plant in Relation to Insect Behaviour and Reproduction,* Tihany, Hungary. pp. 27–28.

Schoonhoven, L. M., and Dethier, V. G. 1966. Sensory aspects of host-plant discrimination by lepidopterous larvae. *Archiv. Néerl. Zool.* 16:497–530.

Steinhardt, R. A., Morita, H., and Hodgson, E. S. 1966. Mode of action of straight chain hydrocarbons in the primary chemoreceptors of the blowfly, *Phormia regina. J. Cell. Physiol.* 67:53–62.

Stürckow, B. 1959. Über den Geschmackssinn und der Tastsinn von *Leptinotarsa decemlineata* Say (Chrysomelidae). *Zeit. vergl. Physiol.* 42:255–302.

Stürckow, B. 1967. Occurrence of a viscous substance at the tip of the labellar taste hair of the blowfly. In *Olfaction and Taste,* II, T. Hayashi, ed., pp. 707–720. Pergamon Press, Oxford.

Stürckow, B., Holbert, P. E., and Adams, J. R. 1967. Fine structure of the tip of chemosensory hairs in two blow flies and the stable fly. *Experientia,* 23/9:1–7.

Takeda, K. 1961. The nature of impulses of single tarsal chemoreceptors in the butterfly, *Vanessa indica. J. Cell. Comp. Physiol.* 58:233–244.

van der Starre, H. 1972. Tarsal taste discrimination in the blowfly, Calliphora vicina Robineau-Desvoidy. *Netherlands J. Zool.* 22:227–282.

Wolbarsht, M. L. 1958. Electrical activity in the chemoreceptors of the blowfly. II. Responses to electrical stimulation. *J. Gen. Physiol.* 42:413–428.

Wolbarsht, M. L. 1965. Receptor sites in insect chemoreceptors. *Cold Spring Harbor Symp. Quant. Biol.* 30:281–288.

Yeandle, S. 1958. Evidence of quantized slow potentials in the eye of Limulus. *Amer. J. Opthalmol.* 46:82–87.

8
Avoiding the Temptation of Gluttony

They surfeited with honey and began
To loathe the taste of sweetness, whereof a little
More than a little is by much too much.
Shakespeare, *King Henry IV*, Pt. 1

Few, if any, animals eat all the time, without stopping. If there are exceptions, they are to be found among the filter-feeders inhabiting the sea. Many among this vast assemblage of invertebrates, representing almost all classes, appear to feed continuously, sieving out minute organisms and particulate matter with a bizarre variety of devices: cilia, feet, mouthparts, peristaltic pumps, all working day and night. The nutrients in the sea are so finely dispersed (the total content of finely dispersed or dissolved organic matter in sea water is only 4–5 mg per liter) that the danger of overeating is remote and constant activity is required to maintain a positive energy balance. Even then there are periods of rest or, at the very minimum, oscillations in the rate of feeding. The vast majority of organisms, however, alternate periods of eating with periods of abstinence.

The proportion of time spent eating and the temporal pattern of eating vary all the way from the nearly continuous feeding of filter-feeders to the total abstinence of the adults of giant silkworms and of mayflies whose life-span is only about 24 hours. Some insects eat only once in each intermolt period—for example, the blood-sucking bug *Rhodnius*; some eat once daily; some eat every few minutes. The patterns are genetically fixed in some cases, determined by circumstance in others, limited by physics in others. For example, the frequency with which a predator such as a mantis eats is determined in part by the availability of prey, whereas the frequency with which a leaf-eating caterpillar feeds is not normally limited by the availability of food. Sheer physics may regulate pattern in that some animals will eat until it is physically impossible to cram in any more and a second meal is impossible, until room is made for more food. Among vertebrates are to be found many counterparts of these invertebrate patterns. Grazing animals approximate locusts in their feeding patterns, the python that swallows the donkey and lies physically helpless until digestion occurs resembles *Rhodnius* and a bloated tick, and the bushmen of Australia or the aborigines in the Kalahari Desert resemble other predators in that mealtime is determined by the density of prey. To these patterns we must add one other that is restricted to man; that is, a cultural determinant. Some societies eat three times a day, some two, and some three or four if such social eating as that at coffee break, tea time, or cocktail hour is counted. The nearest approximation to the human cultural pattern is learned feeding by honeybees. Foragers can be trained to come to food sources at any time or times, but this is more truly a regulation of shopping habits than of feeding habits.

Irrespective of the temporal pattern of feeding, there is one feature common to ingestion: it starts and it stops. Implicit in the idea of stopping is the idea that there is a regulation of the quantity ingested. The questions that this simple assertion raises are difficult to answer. Is eating the normal state that must be stopped periodically, or is abstinence the normal state that must be terminated by starting feeding? Or are both states connected in some complex manner which requires that each be controlled? And finally, since the quantity ingested is in many animals affected by antecedent events—the time since the previous meal, the quantity and quality of the previous meal, the amount of energy expended in the interim—how is regulation achieved?

We have seen that the hungry fly is an active fly and that ingestion is initiated by stimulation of contact chemoreceptors. Eventually feeding stops, even though food is still present to stimulate, and locomotory activity is not resumed even if the fly removes itself from contact with the food that has up till now inhibited locomotion. The two phenomena, cessation of locomotion and termination of ingestion, have been the subjects of extensive experimental analyses; and although the picture of these activities is more detailed in the blowfly than in any other organism, it is still incomplete.

The simplest hypothesis to explain the termination of ingestion is that the sensitivity of the chemoreceptors declines as feeding progresses; in other words, that they adapt. In Chapter 4 experiments were described showing that adaptation times to sucrose vary from 1 to 13 sec., depending upon the concentration. Over the lower range, 0.0013–0.05 M, the time is proportional to the logarithm of molar concentration. Above 0.05 M there is no significant change in time with increasing concentration. Continuous stimulation at constant intensity for very long periods of time results in short bursts of activity. For example, a fly with its feet exposed to 0.05 M sucrose will continue to extend its proboscis for a period of 13 seconds on the average. Thereupon, the proboscis is retracted. After a varying period of time, extension recommences; then it again ceases. Such rhythmic activity has been observed over periods of continuous stimulation lasting as long as 3 hours. No correlation between the frequency of activity cycles and concentration was observed. It is possible that this pattern represents periodic alternating disadaption of different receptors. On the other hand, it is equally possible that the response occurs as a consequence of "discomfort"; more specifically, in response to occasional mechanical or proprioceptive input. The adaptation to sugar that did occur had two components, a peripheral one and a central one (Dethier, 1952).

That the receptors themselves do adapt had been demonstrated by behavioral studies (Dethier, 1955). An accurate description of the finer temporal aspects of adaptation was provided by electrophysiological studies (Evans and Mellon, 1962; Gillary, 1966a, 1966b; Rees, 1968). Initially it was believed that adaptation of the salt receptor was completed in 0.25 second; however, further studies (Rees, 1968) showed that 0.35 is a more accurate value. Thus, after reaching peak activity, the receptor adapts very rapidly in the first 0.1 second and then more slowly till time 0.35 second from zero.

Of greater relevance to feeding behavior is the performance of sugar receptors when they are stimulated for prolonged periods of time. Our studies of several hundred records of the responses of largest, large, intermediate, and marginal labellar hairs have revealed that the time course of adaption varies from hair to hair. There is, however, a general pattern applicable to all. As with salt receptors, sugar receptors adapt rapidly during the phasic part of their response and then more slowly during the tonic portion. Beginning with a high frequency of response, sugar receptors reduce the frequency to 50% within 250 to 1000 msec. For any given hair the time required for the sugar receptor to reach the 50% level varies inversely with the concentration of the stimulating solution. In one hair, for example, the 50% level was reached in 250 msec. with 0.01 M sucrose, 650 msec. with 0.1 M sucrose, and 1000 msec. with 1.0 M sucrose. This is the same trend that was observed in behavioral studies. Similar trends have been observed with the interpseudotracheal papillae (Dethier and Hanson, 1965). The sugar receptors in papillae adapt to 50% of maximum frequency in 300–400 msec. The time varies from one papilla to the next. Occasionally a papilla requires 3 sec to reach the 50% level.

A few electrophysiological studies were made of hairs that were stimulated with 0.1 M sucrose continuously for as long as 20 minutes. The phasic portion of adaptation was as already described. The tonic portion of the response adapted to 2–4 spikes per 100 msec. after continuous stimulation for 1 minute and fluctuated over the next 19 minutes from 1–10 spikes per 100 msec. Response ceased completely between 20 and 30 minutes.

Clearly, then, sensory excitation is highest at the start of feeding, declines rapidly during the first few seconds, and finally reaches a low sustained level as feeding progresses. The initial level of excitation and the time of attainment of a basal level are related to the concentration of the stimulus: the higher the concentration, the greater the initial frequency of firing and the longer the time to base level.

It stands to reason that if chemosensory input is required to stimulate and drive ingestion, adaptation of chemoreceptors could reduce the flow of requisite information below a critical level and feeding would stop. That this can happen has been amply demonstrated by stimulating one lobe of the proboscis repeatedly, but not allowing ingestion to occur. After a while there will be no further extensions. If, however, the unstimulated lobe is now presented with sugar, extension resumes.

Critical questions still to be answered, however, are: what level of adaptation is required to terminate feeding? are some receptors disadapting while others are adapting? how much of a reduction is there in total input from the entire population of receptors and what is the time course of this reduction? This last question could be answered by measuring in the labellar nerve the total integrated response of all receptors.

An examination of the detailed pattern of ingestion during a single meal, and measurements of volumes ingested at one drink, show quite clearly that although peripheral adaptation can account in part for temporary cessation of intake it cannot be called upon to explain why a meal is finally terminated and not resumed. For clarification it will be necessary to define meals and drinks. A drink is the volume taken at one uninterrupted extension of the proboscis. A meal is the series of drinks taken before the fly becomes completely unresponsive to food. This distinction between meals and drinks is operationally more useful than the definition of meal given by Dethier and Gelperin (1967) and will be adopted henceforth. Figure 111 shows that a normal fly deprived of food and then exposed to 2 M glucose for 30 minutes will feed for about 58 seconds during the first minute, 20 during the second, 2.5 during the third, 0 during the fourth and fifth, about 10 during the sixth, 0 during the seventh and eighth, 12 during the ninth, and then no more during the next 21 minutes (Dethier and Gelperin, 1967). The actual values vary, depending upon the period of deprivation that the fly has experienced and the concentration of the sugar

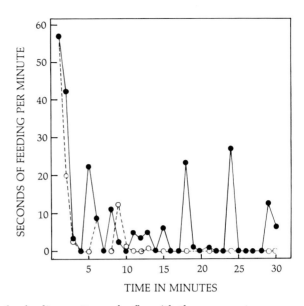

Fig. 111. Comparison of the feeding pattern of a fly with the recurrent nerve cut (solid circles) and a sham-operated (open circles) control during a 30-minute exposure to 2 M glucose (from Dethier and Gelperin, 1967).

presented; however, the broad outline of the pattern is constant, an initial big drink followed by successively smaller ones, and then none at all.

When only the duration of the first drink is considered, it is observed to vary with concentration, the comparison being made with flies all of which were in the same state of deprivation (Table 17) (Dethier, Evans, and Rhoades, 1956). These results are consistent with the idea that sensory adaptation plays a role in terminating the first drink.

Thomson and Holling (1974), employing a smaller number of flies, did not find differences in the volume ingested and the rate of feeding; they concluded, therefore, that sensory adaptation did not regulate the cessation of feeding for concentrations greater than 0.1 M. Considering all available evidence, however, it is premature to deny sensory adaptation any role in the regulation of feeding.

Altering the state of deprivation of the fly changes the duration of the first drink, supporting the idea that other mechanisms are at work. In short, present evidence suggests that sensory adaptation can account in part for the initial temporary cessation of feeding, but cannot account for the termination of a meal nor for changes in the duration of drinks and meals correlated with varying degrees of food deprivation.

Either or both of two mechanisms could be involved: a change in

Table 17. Amounts of Various Sugars Ingested at a Single Feeding.

Sugar	Molar concentration	Number of animals	Mg/fly	Ml/fly ×10³	Duration[a] (sec.)	Rate ml/sec. ×10⁵	Approximate viscosity (centipoises)
Sucrose	2.0	20	8.96	13.0	90	14	—
	1.0	30	4.78	13.9	47	30	—
	0.5	10	1.80	10.5	43	24	—
	0.25	10	0.440	7.05	36	20	—
Glucose	2.0	15	4.92	13.7	61	26	—
	1.0	35	2.27	12.6	44	30	—
	0.5	15	0.820	9.11	38	25	—
Mannose	4.0	15	6.49	9.02	51	18	—
	2.0	15	2.97	8.25	40	21	—
	1.0	10	1.12	6.20	38	16	—
	0.5	10	0.268	2.98	25	12	—
Fucose	1.0	50	0.843	5.14	30	20	—
	0.5	15	0.580	7.08	32	35	—
Lactose	1.0	15	0.903	2.82	18	18	—
Sorbose	3.0	10	1.93	3.58	—	—	—
	2.0	20	1.09	3.04	—	—	—
	1.0	10	0.168	0.934	—	—	—
	0.5	6	0.0481	0.534	—	—	—
Sucrose	1.0	10	4.62	13.5	54	25	2.75
Sucrose 1.0 M in glycerol	2.2	10	5.20	15.2	60	25	4.29
	5.4	10	2.86	8.34	50	17	7.79
	8.7	4	2.20	6.45	55	12	48.5

From Dethier, Evans, and Rhoades, 1956.
[a] Duration times were recorded for fewer flies than were employed in ingestion determinations.

the sensitivity of the receptors consequent upon feeding, or a centrally located change. Although there are many examples of degrees of centrifugal effect on sensory receptors in vertebrates, and demonstrations of efferent pathways to receptors (Livingston, 1959), evidence to support the idea of peripheral change in the receptors of insects has been scanty. A change in the nature of electrophysiological responses of gustatory receptors of the sphinx caterpillar, *Manduca sexta*, as a consequence of a change in diet has been reported by Schoonhoven (1967). A change in the electrical resistance of gustatory receptors of the locust and in the opening and closing of the sensilla has been reported by Bernays and Chapman (1972) and Bernays, Blaney, and Chapman (1972). These changes are associated with changes in the haemolymph and can be induced in insects deprived of food for 4 hours by blood transfusions from newly fed individuals. No changes in the taste receptors of the blowfly as a consequence of feeding have been reported, except in the paper by Omand (1971). On the basis of electrophysiological recording she stated that labellar water, salt, and sugar receptors became more sensitive as the fly was deprived of food, less sensitive after feeding, and again sensitive upon subsequent deprivation. The effect was nonspecific; that is, the nature of the food ingested was immaterial. This provocative report should stimulate more searching investigation into the occurrence of peripheral change.

The search for central mechanisms began with an investigation of behavioral thresholds. Many factors were known to affect thresholds. Nutritional state is by far the most influential (Dethier and Chadwick, 1948). Early work concerning the effect of feeding and starvation on taste thresholds of insects is somewhat contradictory. The change of threshold with starvation is not necessarily the same for all substances nor for each group of receptors on the same insect. Even the various acceptable sugars have been reported to have different sensitivity curves during starvation (Haslinger, 1935). Most investigators have found an increased sensitivity to sugars during a period of starvation [consult Dethier and Chadwick (1948) for a review], but a number of others have reported no change or even a raised threshold. A rise in tarsal threshold to sugar is regularly observed in *Phormia* shortly before death, and it is not improbable that higher thresholds at times of severe starvation signal a moribund condition. Von Frisch (1935) compared the sucrose thresholds of honeybees on a sugar-water diet just adequate to maintain life with that of honeybees that had access to concentrated sugar-water and a hive full of honey. Both groups had a sucrose acceptance threshold between 0.5 and 1.0 M, but the threshold was distinctly lower when natural food was scarce than when it was abundant. Similar observation have been made by others (Kunze, 1927; Núñez, 1966, 1970; Wells and Giacchino, 1968). Elsewhere von

Frisch (1950) has stated that a starvation period of several hours suffices to lower the sugar acceptance threshold to a value of 0.125 to 0.0625 M. The nutritional state of the bees at the start of the period of deprivation was not known. Since the honeybee seemed somewhat anomalous in this regard Evans and Dethier (1957) conducted a few experiments to check the point. After being fed 2 M sucrose to repletion the bees had a sucrose threshold of 2 M or higher. A starvation period of 45–74 hours was required to lower the threshold to 0.125 M. The same decrease could be brought about by an 80-minute flight.

Starvation markedly increases the sensitivity to sugar of the tarsal chemoreceptors of the butterfly *Pyrameis* (Minnich, 1922) and the blowfly *Calliphora* (Minnich, 1929) but does not alter the threshold of the labellar receptors of the latter (Minnich, 1931). The striking feature of the change is its magnitude; the change in the blowfly is by a factor of 10,000,000. Changes by a factor of the order of 100,000 have been reported for other species (Anderson, 1932), and changes by a factor of 10,000 are common (see e.g., Minnich, 1929).

The regular and predictable changes of behavioral threshold with feeding and deprivation, and the obvious dependence of feeding on the state of threshold, suggested that an understanding of the mechanism of change might shed light on the central problem of feeding behavior. The search for a mechanism or mechanisms was begun by investigating what effect was produced on tarsal thresholds for sugar when sugars with different properties were fed to *Phormia* deprived of food (Evans and Dethier, 1957). Four sugars were selected for study: glucose and fucose, which are highly stimulating; mannose, which is poorly stimulating; and lactose, which is nonstimulating. Glucose and mannose support life well; lactose has the same life-supporting effect as, or is perhaps insignificantly better than, water alone; fucose has the same effect as water alone or is perhaps even toxic (Hassett, Dethier, and Gans 1950). The change in threshold with each sugar is diagrammed in Figures 112 and 113. It is clear that changes in threshold with time, after a single feeding to repletion of a given sugar, are different. In each case an elevation occurs after feeding, but the maximum values are not the same and with each sugar there is a characteristic rate of fall as deprivation is prolonged. Both volume and concentration of an ingested sugar influence the level to which the threshold rises (Table 18) (Evans and Barton Browne, 1960).

Three possible mechanisms whereby the ingestion of sugar could bring about an elevation of threshold, and deprivation could produce a lowering, were considered: (1) a peripheral action whereby exposure of the receptors to sugar during feeding brings about a long-lasting refractoriness; (2) response of unknown receptors in some part of the digestive tract to sugars, to distention, or to some other stimulus that

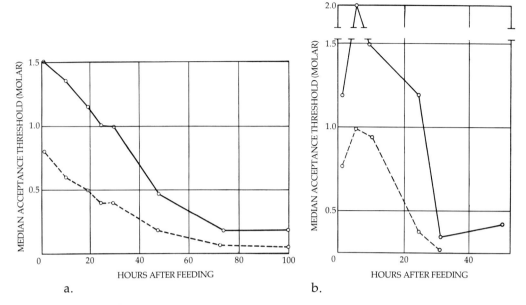

Fig. 112. Change in sugar thresholds following ingestion of (a) 2 M glucose; (b) 1 M fucose. Glucose threshold, —; fucose threshold, ---- (from Evans and Dethier, 1957).

leads directly or indirectly to central inhibition of the probocis response; (3) some internal action of the sugar itself or a metabolic derivative of it that eventually affects the system.

The first possibility was ruled out by the studies of adaptation already described. Sensory adaptation simply does not endure long enough to account for behavioral insensitivity to sugars.

Involvement of some region or regions of the digestive tract itself was next investigated. At this point we must pause in the discussion to look at the anatomy of the digestive system (Fig. 114). The alimentary canal of insects is divided into three major regions:foregut, midgut, and hindgut. Both the fore- and hindgut are of ectodermal origin and are formed embryonically by invaginations of the body wall. As such they are lined with cuticle continuous with that of the external surface of the body. The midgut is of endodermal origin. The foregut extends from the oral cavity to the cardiac valve and consists of the pharynx, oesophagus, and the crop. In many insects the crop is in series with the pharynx and oesophagus and is merely an enlarged section of the foregut. In the blowfly the crop is a separate distensible sac connected to the mainstream of the alimentary canal just anterior to the cardiac valve by a long thin crop duct. The midgut begins at the cardiac (midgut) valve and ends at the malpighian tubules, the principal excretory organ. At this juncture the hindgut begins. It consists of a thin, greatly contorted tube and a rectum.

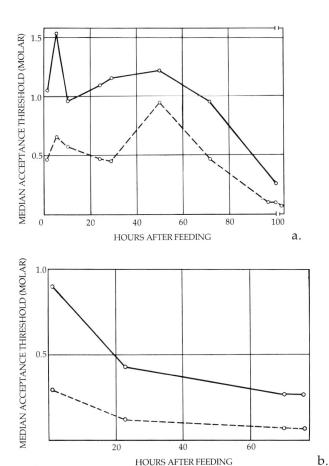

Fig. 113. Change in thresholds following ingestion of (a) 2 M mannose; (b) 1 M lactose. Glucose threshold, ——, fucose threshold, ---- (from Evans and Dethier, 1957).

Table 18. Tarsal Thresholds to Glucose after Ingestion of Fixed Volumes and Concentrations of Glucose

Solution ingested	Mean initial threshold M glucose	n^a 15 minutes after feeding (Mean ± S.E.)	n^a 45 minutes after feeding (Mean ± S.E.)	Significance (P)
10 μl 2 M glucose	0.135	3.5 ± 1.45	3.9 ± 1.24	<0.01
3 μl 2 M glucose	0.175	1.8 ± 2.04	1.7 ± 1.57	<0.01
3 μl 0.1 M glucose	0.238	0.2 ± 1.14	−0.3 ± 1.16	

From Evans and Barton Browne, 1960.

^{a} n is an exponent used as a measure of threshold change. The threshold at 15 minutes and at 45 minutes equals the first threshold × 2^n.

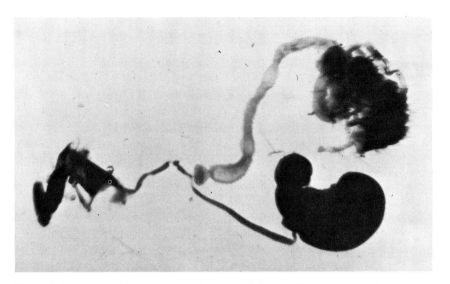

Fig. 114. Complete alimentary canal removed from a fly (courtesy of A. Gelperin).

We have left the fly at the end of Chapter 2 torpid in repletion. How he arrived at this state can be appreciated by making, in a hungry fly, a long dorsal incision that extends from the back of the head to the posterior limit of the abdomen, exposing the alimentary canal throughout its length, and presenting the proboscis of this fly with a sugar solution containing the red dye carmine. Now the entire process of ingestion is clearly visible. The fly eats as though all were right with its world. The cibarial pump sucks fluid into the pharynx, whence it is forced back into the oesophagus. Here waves of peristalsis squeeze the fluid back to the point of bifurcation of the crop and midgut. At first, fluid goes into both the midgut and the crop. When the midgut is full, the cardiac valve closes, the valve at the entrance of the crop duct remains open, and the fluid enters the crop duct where further peristalsis drives it back into the crop itself. When feeding has ceased, the crop valve closes (Knight, 1962; Green, 1964b). The foregut now contains some residual fluid from more anterior regions of the oesophagus and the preoral cavity. Green has pointed out that a full crop exerts considerable pressure on the valve of the crop duct, which is a two-way valve, so that there is some leakage at this point. This situation continues for some time, and during this period the fly regurgitates to a considerable extent. The foregut is almost continuously bathed in ingested fluid.

All this time the crop is roiling and kneading. From now on, as fluid is absorbed into the haemocoele from the midgut it is replaced by transfer from the crop. The crop valve opens; peristalsis in the crop

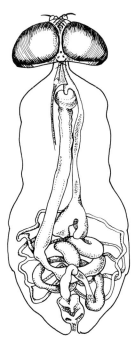

Fig. 115. Alimentary canal of *Phormia regina* in situ, showing normal relationship of parts.

and reverse peristalsis in the duct drive a slug of fluid forcibly into the oesophagus; the crop valve closes; the cardiac valve opens; peristalsis in the oesophagus drives fluid backward into the midgut. As long as the crop contains fluid, for many hours, the transfer is continued. A most astounding feature of this alimentary behavior is that it can be carried out in the absence of the rest of the fly. A full gut, completely dissected from the fly and placed in saline, will continue to transfer sugar from the crop to the midgut on condition that it is maintained under some longitudinal stretch (Knight, 1962).

Of all of the regions of the digestive tract that might possibly be involved in threshold regulation, the crop seemed the most likely source of information. Analogy with human experience suggested that a full crop might signal its condition by pressure, or, conversely, that an empty crop might signal emptiness by excessive motility. These possibilities were checked by performing various operations involving the crop and observing the subsequent feeding behavior (Evans and Dethier, 1957). The technique was described in detail by Dethier and Bodenstein (1958). To ligate or remove the crop the fly was held on the operating table, ventral side up, by two plasticene strips, one crossing the midthorax and the other, the posterior end of the abdomen. The crop lies somewhat to the right side, close to the midline, on the border of the thorax and abdomen (Fig. 115). At this site a small incision was made into the skin of the abdomen and the crop pulled out carefully. A silk loop was slipped over the exposed crop and the duct tied off at its entrance to the crop. The crop was then pushed back through the wound into the abdomen. If the crop was to be removed, a second ligature was tied around the duct and the duct cut between the two constrictions.

The duct was ligated before feeding in some flies and after feeding in others. In the first case, less sugar is ingested because the capacity of the fore- and midgut into which the fluid is directed is less than that of the crop. Nonetheless, the threshold is elevated immediately after feeding and remains elevated long after disadaptation has occurred. The crop itself, therefore, need not contain sugar to cause prolonged elevation of threshold. Almost identical results were obtained when feeding preceded ligation. Fore- and midgut regions contained some solution, even as they did when ligation preceded feeding. In both of these experiments the threshold was elevated and began to fall 4–6 hours after feeding.

Similar experiments with houseflies were reported briefly by Bolwig (1952). Ligation of the crop and removal of an empty crop did not impair the "hunger reaction" (turning toward a drop of sugar solution when it touched the tarsus). Removal of a full crop or removal of the abdomen did not restore the hunger reaction until an hour or two had elapsed. All of these experiments agree in demonstrating that the crop in itself does not regulate the response to sugars.

Again, analogy with human experience might suggest that the level of sugar in the blood changed as ingestion progressed and that this change was instrumental in shutting off feeding. When blood sugar was first investigated, an unknown carbohydrate appeared on chromatograms. It was always present in greater amounts than any other sugar and was present consistently, whereas the occurrence of other carbohydrates depended upon which sugar had been fed. The unknown was a disaccharide, which upon hydrolysis yielded glucose exclusively. Only three disaccharides of glucose linked between the first carbons are possible, and only one, trehalose, was known to occur naturally. At this point in the investigation a report appeared (Wyatt and Kalf, 1956) that described the isolation and chemical identification of trehalose in the blood of the moth *Telea polyphemus*. It is interesting to note that trehalose does not fulfill the definition of "true blood sugar" sought by previous investigators (fermentable reducing sugar), since it is not detected by any of the usual reducing methods. The omnipresence of a low glucose value, quite independent of glucose in the diet, makes it necessary to consider glucose also as an important blood sugar. Since ingested sugars appeared in the blood unchanged in at least small amounts (glucose, fucose, mannose, lactose, and sucrose

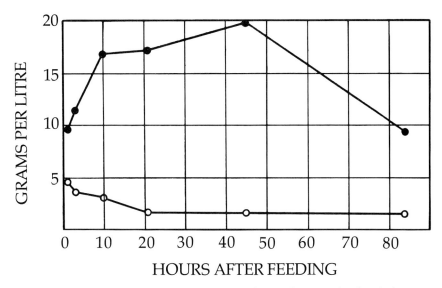

Fig. 116. Blood-sugar levels after ingestion of 2 M glucose. Blood trehalose, ●—●; blood glucose, ○—○ (from Evans and Dethier, 1957).

were tested), the concentration of trehalose, the ingested sugar, and any product of the ingested sugar had to be considered as possibly having an effect on sugar thresholds.

Blood sugar levels were assayed in the following manner (Evans and Dethier, 1957). Forelegs were amputated. A graduated micropipette made from thermometer tubing was applied to the stumps of the legs and 1–4 μl of whole blood withdrawn. The identification and quantitative determination of blood sugars was accomplished by paper chromatographic separation, chromatographic elution, and estimates of carbohydrates in the eluate by standard methods.

The changes in blood sugar levels that occurred with feeding are illustrated in Figures 116–118. In each case the level of dietary sugar in the blood began to climb shortly after feeding and attained high values. At the end of 20–30 hours the blood concentrations of glucose and mannose were regulated to much lower values than fructose, which increased until death (30–40 hours). The fall of blood trehalose after ingestion of fucose illustrates the strict dependence of that sugar on nutritional state. After ingestion of glucose and mannose, blood trehalose bore a simple relation to the curves for the sugar threshold and for crop sugar. As the rate of crop emptying became very slow, blood mannose and glucose did not appear likely as influences on threshold since they assumed nearly constant values, whereas the threshold was progressively decreasing. Trehalose in those cases could have been determining, but the threshold curves of the fucose-fed animals appeared to be unrelated to any of the blood sugars.

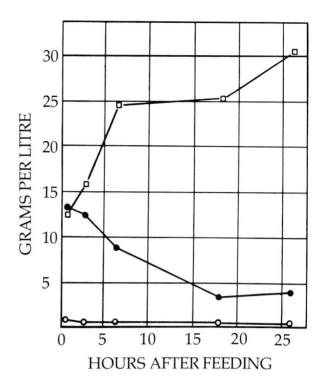

Fig. 117. Blood-sugar levels after ingestion of 1 M fucose. Blood trehalose, ●—●; blood glucose, ○—○; blood fucose, □—□ (from Evans and Dethier, 1957).

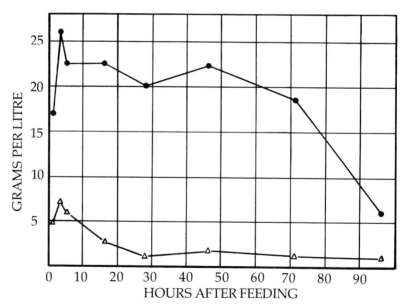

Fig. 118. Blood-sugar levels after ingestion of 2 M mannose. Blood trehalose, ●—●; blood mannose, △—△ (from Evans and Dethier, 1957).

A more direct test of the role of blood sugar in threshold regulation was made by injecting glucose, trehalose, and fucose. Elevations of the levels of these sugars, some to values similar to those achieved after ingestion of the sugars and others many times higher, clearly did not affect sugar thresholds. Tarsal threshold measurements at intervals 5 minutes to 24 hours after injection showed no significant rise. Incidently, blood sugar determinations after injection constitute glucose and trehalose tolerance curves for *Phormia*. This fly can withstand the introduction of phenomenal (1220 µg) quantities of sugar into its blood, when we consider that the total blood volume is probably only 4–6 µl. The concentration of the injected sugar in the blood decreases rapidly, although not necessarily to the original levels. In contrast, blood trehalose maintains a maximum level for many hours after a nutritive sugar has been ingested. The level of glucose under the same circumstances returns to normal after a much shorter period of time. Thus the change in level of blood sugars is not entirely passive although the manner and time course of the regulation differ from those in mammals.

Additional evidence against the involvement of blood sugars in threshold changes was adduced from studies of the effect of flying on threshold. A relationship between taste threshold and the duration of flight of honey bees was described by Evans and Dethier (1957). Hudson (1958) investigated the relationship more thoroughly in *Phormia*. A tethered fly can be induced to fly by yanking its feet off the ground, blowing wind in its face, and providing a moving visual pattern where the ground would normally be. In these studies flies were attached to a roundabout (Fig. 119). After 110 minutes of continuous flight at room temperature (25°C), the tarsal threshold to glucose fell to less than 10% of the initial value. During flight the threshold fell exponentially with time (Fig. 120a) and even after 3 minutes of activity a decrease of 81% of the initial value occurred. At 39°C the decrease occurred more rapidly and at 21°C more slowly than at room temperature. Wing-beat frequency (measured stroboscopically) also increased with temperature. The fact that decrease in wing-beat frequency and the threshold were both accelerated at high temperature suggested that the threshold is in some way related to energy expenditure.

Hudson investigated this possibility by assaying carbohydrate depletion. The changes that occur with starvation are summarized in Figure 121. As expected, the crop emptied rapidly during the first 3 hours after feeding. Correspondingly there occurred a rapid increase in sugar in the midgut, followed by a gradual depletion. The depletion in the gut was reflected by a rise in blood sugar. The rise in blood sugar was maintained until supplies from the crop via the gut could no longer keep pace with withdrawal from the blood, at which point the level of

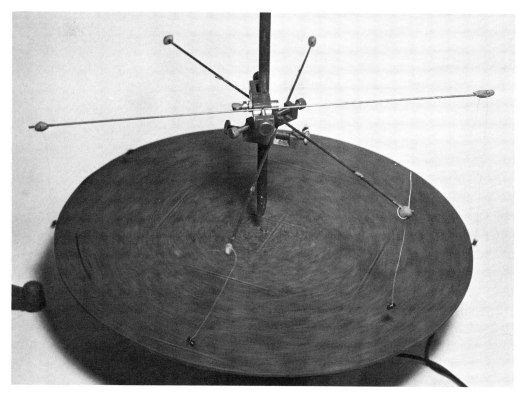

Fig. 119. Roundabout employed by Hudson to fly flies to exhaustion.

blood sugar fell. Glycogen synthesis, which occurred rapidly after feeding (cf. Wigglesworth, 1949), fell to approximately 16.0 g/l, after which glycogen was depleted progressively with starvation. As Hudson pointed out, these results confirm the fact that the threshold at first increases with blood sugar and that only at a critical time, when crop and other sources are unable to maintain the level, does the threshold fall with blood sugar. Blood sugar per se therefore cannot be considered as a direct regulatory factor. The results also show that glycogen must be eliminated as a controlling factor since it is still increasing at a time when the threshold has fallen to 30% of its maximum.

The changes that occurred during flight are summarized in Figure 120b. The most striking difference between starved and flown flies is the rapid disappearance of glycogen in the latter, despite the fact that the crops contained available glucose. The threshold had decreased to 55% of its initial value after 15 minutes. It continued to fall rapidly up to 60 minutes and did not change when flies were allowed to rest 2–3 hours after flying.

On the basis of all of the experiments described thus far it was possible to eliminate crop sugar, blood sugar, and glycogen as factors

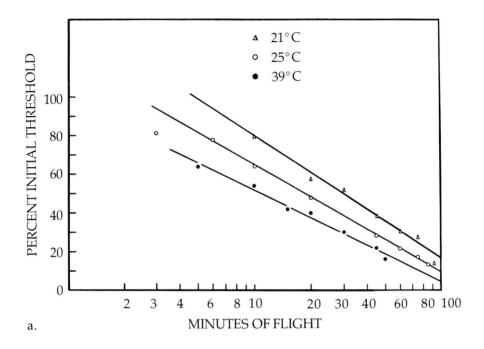

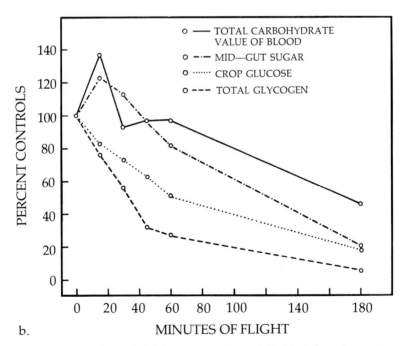

Fig. 120. (a) Relation of threshold decrease to time of flight at three temperatures. (b) The course of carbohydrate depletion during flight (from Hudson, 1958).

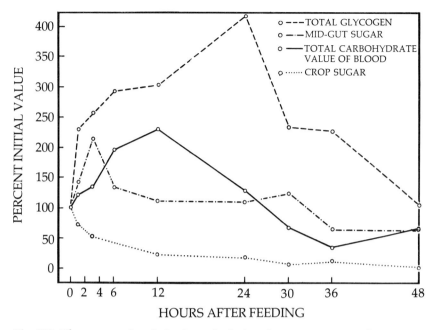

Fig. 121. The course of carbohydrate depletion during starvation (from Hudson, 1958).

immediately regulating threshold (and hence ingestion). The next possible site for consideration was the midgut. Hudson attacked this problem by injecting exhausted flies with a quantity of glucose calculated to maintain flights of definite duration but which, it had been found, did not affect the threshold. At the same time she fed them with a nonutilizable sugar. Since the fed sugar could not be metabolized, its disappearance from the blood could be observed. Elimination of carbohydrate by excretion during flight was monitored by examination of excreta collected on paper chromatograms.

Flies flown to exhaustion had a tarsal threshold of 0.06 M glucose. After being injected with enough glucose to maintain flight for about 40 minutes, they had the same threshold. They were then fed 1.3 mg of fucose, which raised the glucose threshold to 1 M. After this they were flown to exhaustion (38–40 minutes). The threshold remained essentially the same, that is, 1 M. Estimation of crop fucose showed that there was still 95% of the fed amount present. Fucose in the blood was moderately high (15 g/l) and in the midgut high (250–283 μg). The amount of glucose in the blood was negligible. No carbohydrate had been excreted. Obviously the injected sugar had been used for flight, and when it was exhausted flight ceased. Little fucose had been removed from the gut. The threshold was high even after a flight period that would normally result in a decrease to 30% of the initial high value. Thus, energy had been expended to the point of exhaustion, yet the threshold remained high. Had these flies, needing carbohydrate, been offered sugar in the normal fashion, they would not have ingested it.

For comparison some exhausted flies were injected and fed with a utilizable sugar (glucose). Again injection did not raise the threshold but feeding did. After 45 minutes of flight the threshold had decreased by only 2%; that is, it remained high, as in flies fed fucose. In flies fed glucose but not injected the threshold had decreased to 30% in this time. In the injected flies' crops, blood, and midguts, sugar remained high. The injected glucose had been used to maintain flight; from this point on, crop and gut sugar was used and the threshold fell. Here again the threshold remained high while the midgut contained sugar. When the midgut became empty, the threshold fell, but it must be remembered that an empty midgut also means an empty crop. At least under experimental conditions vigorous energy expenditure can occur without a resultant fall in the threshold. Further calculations of total energy expenditure, expressed as total carbohydrate utilization per unit time, failed to reveal any simple relationship between the threshold and energy expenditure (Hudson, 1958).

The search for mechanisms regulating threshold was continued by Dethier and Bodenstein (1958). Although it had been demonstrated that changes in the level of various sugars in the blood of unfed flies had no measurable effect on the threshold of response to sugars, the possibility remained that feeding causes liberation into the blood of an active humoral agent that affects the threshold or, conversely, that starvation causes a humoral change. To test these possibilities blood was transfused from flies that had been fed 2 hours previously to flies that had been starved for 24 hours. When the thresholds of transfused flies were measured immediately after transfusion and at intervals over a subsequent 2-hour period, 80% of the flies had a threshold that did not differ from the previously measured starvation level. There was a possibility that the quantity of blood injected or the effect of dilution with saline might have obscured the action of a hormone, but barring this it was concluded that within a 2-hour period, feeding did not liberate into the blood any humoral substance contributing to a rise in the threshold. This conclusion was strengthened by the observation that an isolated head (with its blood, of course) removed from a fully fed fly with a high threshold will feed.

A more recent search for hormonal intervention was undertaken by means of parabiosis. The technique was developed from that pioneered by Green (1964b). Each member of the presumptive pair was fastened by the wings, ventral side down, to a small paraffin block. On the dorsum of the thorax of each was dropped a small glob of paraffin (Figs. 122, 123). This was then puddled with a warm needle until a

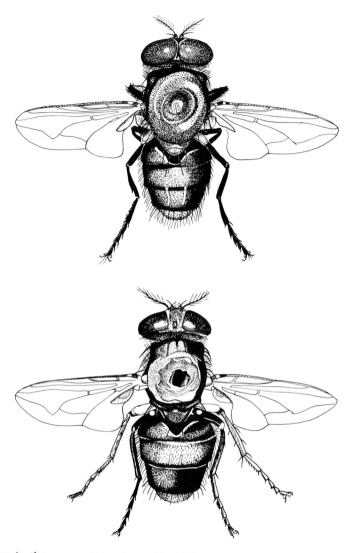

Fig. 122a–c. Steps in the preparation of parabiotic flies.

small crater was formed, at the bottom of which the thoracic cuticle was visible. The floor of the crater with its underlying cuticle was then excised by a circular cut made with a microscalpel. This cut exposed the haemocoele and the dorsal thoracic musculature. A drop of physiological saline was next added until the crater was filled to the brim. Each block was then rotated 90 degrees and the two were moved together so that the flies were back to back and the rims of the two craters were touching. The rims were sealed together with a hot needle. Finally, the flies were freed from their respective gantries and, depending on the use to which they were to be put, attached by their

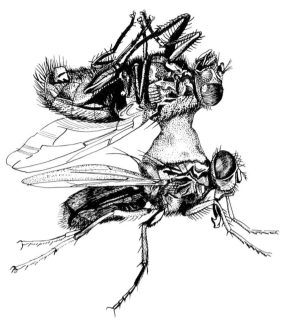

Fig. 122c

Fig. 123. Method of exercising parabiotic flies (from Belzer, 1970).

union to a wax-tipped stick or allowed to perambulate freely. In the latter circumstance they took turns riding piggyback or one had its legs immobilized and so was always carried.

At the time of operation both flies were water-satiated but had been starved for 48 hours. As soon as the operation was completed, one fly was fed to repletion. The pairs were maintained for 3 days, during

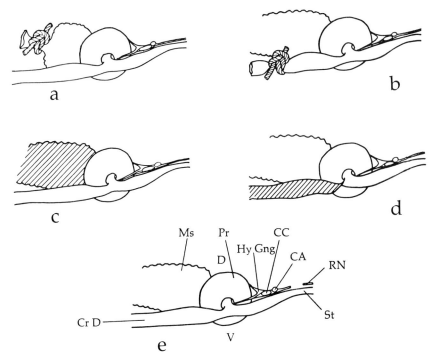

Fig. 124. Diagram of the region of the foregut. (a) Point at which the midgut was ligated. (b) Point at which the crop was ligated. (c) Most anterior advance of fluid injected into the midgut via the anus. (d) Most anterior advance of fluid in the crop duct when the crop valve is closed. (e) Point at which the recurrent nerve was cut. Ms, midgut; Pr, proventriculus; Cr D, crop duct; St, stomodaeum (foregut); Hy Gng, hypocerebral ganglion; CC, corpora cardiaca; CA, corpus allatum; RN, recurrent nerve; D, dorsal; V, ventral (from Dethier and Bodenstein, 1958).

which time the fed member was kept satiated. Tests of tarsal threshold over this period showed a consistently high value for the fed fly and a consistently low value for his unfed partner. It was concluded that neither blood sugar nor humoral agents regulated threshold.

A number of potential regulatory mechanisms had now been eliminated from consideration. The work of Evans and Dethier (1957) had ruled out the crop as the proximate regulatory region of the alimentary canal, but had not provided direct proof of the involvement of any other region. Hudson's (1958) experiments raised suspicions concerning the role of the midgut. Two types of complementary experiments, conducted to supply further information, consisted in eliminating the midgut and hindgut and in loading the midgut and hindgut (Fig. 124).

Five unfed flies, with low thresholds, received ligatures immediately behind the proventriculus (Fig. 124). In order to ligate the midgut, at this point the fly was placed dorsal side up in a depression of wax plate and held in position by two plasticene strips—one across the abdomen and the other across the head. With iridectomy scissors a small wedge of muscle tissue was removed from the prothorax. The cut extended anteriorly almost to the neck, but left intact a small chitin ridge close to the neck. Posteriorly the cut extended to about one-half the length of the thorax. The broadest part of the wedge of tissue was the dorsal surface of the thorax. The sloped cuts on each side met in the midline just above the gut. Removal of this cut wedge exposed the anterior portion of the midgut and showed clearly the junction of the midgut and the proventriculus. A drop of physiological solution was placed in the wound. The excised wedge of tissue was also placed in a drop of physiological solution to keep it from drying. The midgut could now be lifted up with forceps and cut. If ligation was desired, the gut was lifted up with one pair of forceps, while at the same time a fine silk thread, held with a second pair of forceps, was pushed below the gut from one side to the other. The diameter of silk thread used corresponded roughly to that of a human hair. However, silk has the advantage of being more flexible than hair. The two ends of the thread were then manipulated into a sling, which was tightened just behind the proventriculus. After the knot was tied the ends of the thread were cut. If it was desired to cut the midgut, a second ligature had to be applied. The gut was then cut between the two ligatures. Now the excised wedge of tissue was replaced and the wound thus closed. Any excess of blood and physiological solution that was pressed from the wound by replacement of the wedge was soaked up by filter paper. The fly was now ready for testing. Animals thus operated upon are, of course, unable to fly because the various incisions have severed the flight muscles.

The operated flies were then fed to repletion on 1 M sucrose. From then on, thresholds were measured periodically over a 4-hour period. Up to 3 hours after feeding the threshold remained high and only then began to fall. Identical results were obtained if the flies were fed 1 M sucrose before ligation. Furthermore, no differences were observed between the effects of ligation alone and of ligation plus complete section and removal of the midgut. From these experiments it was concluded that the presence of sugar in the mid- and hindgut was not necessary for the maintenance of an elevated threshold.

The complementary experiments consisted in loading the mid- and hindgut. This was accomplished via the anus with an injection apparatus used for work with *Drosophila* (Bodenstein, 1950). To this end the fly was sealed by its wings to the wax table, ventral side up. This injection procedure proved to be rather difficult. The wall of the hindgut close to the anus is exceedingly thin, and the needle often pierced the wall of the gut; this resulted in forcing the injected fluid into the ab-

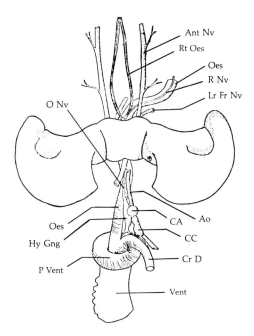

Fig. 125. Dorsal view of the foregut region in relation to the brain and its nerves. Ant Nv, antennal nerve; Rt Oes, retractors of the oesophagus; Oes, oesophagus; R Nv, recurrent nerve; Lr Fr Nv, labrofrontal nerve; Ao, aorta; CA, corpus allatum; CC, corpora cardiaca; Cr D, crop duct; Vent, ventriculus; O Nv, ocellar nerve (stalk); Hy Gng, hypocerebral ganglion; P Vent, proventriculus (from Dethier, 1959).

dominal cavity instead of into the gut. The success of the injection, therefore, was carefully checked in each case by opening the animal after testing. Since a colored fluid (carmine or fuchsin) was used for the injection, it could easily be determined whether or not the gut itself was injected. In successful cases the hind- and midgut were filled up to the proventrical valve, which blocked the solution so that it could not pass further forward into the proventriculus.

In some instances sugar was also injected directly into the midgut through a prepared opening in the thorax, as was employed for ligation of the midgut. Again in 80% of the cases there was no rise in threshold over the starved condition. In a sham operation the threshold remained the same in 76% of the flies. The results of these experiments are in agreement with the results obtained with ligation and sectioning in showing that the presence of sugar in the mid- and hindgut does not cause a rise in the threshold.

If the alimentary canal was in fact involved in the regulation of threshold, the only unexplored region was a small area of foregut extending from the head to and including the proventriculus (Fig. 125). With one exception, attempts to inject sugar into this delicate region

a.

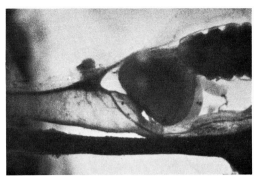
b.

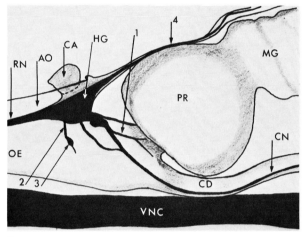
c.

Fig. 126. The stomatogastric nervous system and associated elements behind the brain in *Phormia regina*. (a) Dissection of the neck and thoracic regions showing the relationship of the elements involved. (b) Close-up of the proventricular region showing branches of the recurrent nerve. (c) Camera-lucida diagram of the proventricular region showing recurrent nerve branches in detail. AO, aorta; CD, crop duct; CN, crop duct nerve; HG, complex of hypocerebral ganglion and corpus cardiacum; MG, midgut; OE, oesophagus; PR, proventriculus; RN, recurrent nerve; TG, thoracic ganglion; VNC, ventral nerve cord; 1, branch of recurrent nerve to region of the cardiac valve; 2, branch to oesophagus below hypocerebral ganglion; 3, swelling in branch 2; 4, branch to aorta (from Green, 1964b).

without damaging nerve tissue were unsuccessful. In the single successful case there was a 30-fold rise in threshold. This was a provocative observation. Accordingly an attempt was made to approach a solution by denervating this particular region.

Figure 126 shows that the gut is innervated by the recurrent nerve. This nerve together with its branches and ganglia constitutes the stomatogastric system, the analogue of the vertebrate autonomic system. The recurrent nerve passes from the frontal ganglion posteriorly along the dorsal wall of the oesophagus to the vicinity of the junction of the crop duct. Here it is connected with the hypocerebral ganglion, gives off nerves to the endocrine glands (the corpus allatum and corpora cardiaca) and the aorta, and sends fibers to the oesophagus, the crop, and the proventriculus. Experiments in which the crop had been removed showed that that section of the recurrent nerve was without effect insofar as threshold regulation was concerned (Evans and Dethier, 1957). The experiments involving removal of the midgut demonstrated that that section of the proventricular branch posterior to the cardiac valve was also without effect. One more experiment had to be done. The recurrent nerve had to be cut anterior to the hypocerebral ganglion in the hope of thereby removing any and all innervation to the foregut. This initially difficult and now routine operation was first perfected on *Drosophila* by Bodenstein and then performed successfully on 25 starved *Phormia*.

The procedure for cutting the recurrent nerve was originally performed in the following manner. A shallow depression in the shape of the fly's body (thorax and abdomen) was cut into the wax. The depth of the depression was about two-thirds of the dorsal-ventral width of the thorax. In front of this depression a second smaller one was made to hold the fly's head. Thus a narrow wax bridge separated the two depressions. The fly was placed dorsal side up into the large depression and secured by a strip of plasticene laid across the anterior portion of the thorax. The head of the fly was then carefully manipulated into the second small depression so that the neck stretched across the wax bridge. The head was held forward by a strip of plasticene or a wedged pin (Fig. 127). In this manner the dorsal skin of the neck was brought into clear view. A drop of physiological saline was placed on the neck; it covered part of the thorax and the head. This drop not only kept moist the tissues that were to be exposed but also served as a lens and thus aided considerably in the performance of the next steps. With the drop in place, the skin of the neck was cut along the midline with iridectomy scissors. The cut had to extend posteriorly to the point where the neck meets the thorax. The wound was then widened by pulling the cut skin apart with two pairs of forceps. This exposed the longitudinal neck muscles that extend dorsally close below the skin (Fig. 128). Some of these were severed, but care had to be taken not to destroy too many of the lateral neck muscles. With some of the uppermost dorsal neck muscles cut, the foregut and the two large lateral tracheal trunks came into clear view. Near the prothorax one can also see the anterior border of the proventriculus. Somewhat left and slightly in front of the proventriculus is located the allatum-cardiacum complex. It is rather easily distinguished from the other tissues by the light bluish sheen of the cardiacum. The recurrent nerve, which passes

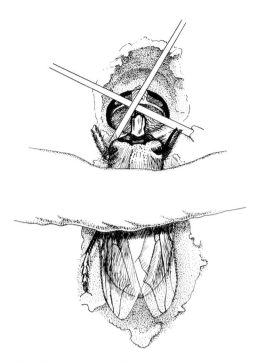

Fig. 127. Position of fly in preparation for operation to section the recurrent nerve.

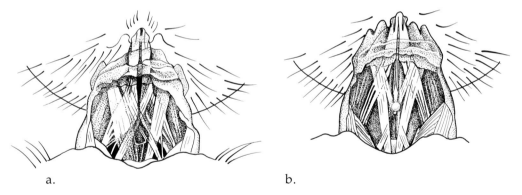

a.   b.

Fig. 128. (a) Appearance of internal organs in neck region after integument is slit. Note the aorta and the longitudinal and oblique muscles. (b) Appearance after lateral tracheal trunks have been teased away from the oesophagus. The corpus allatum and the hypocerebralcardiacum complex are now visible.

from the anterior part of the cardiacum forward, is as yet not visible. The cardiacum adheres laterally to the left tracheal trunk. With the fine tip of the forceps this adhesion can be broken. As soon as this is done the recurrent nerve becomes visible. By holding the nerve close to the cardiacum with forceps one can lift it slightly and cut it (Fig. 129). This finishes the operation. The border of the wound is now pressed

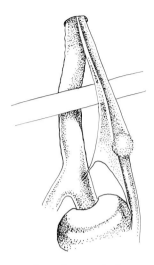

Fig. 129. Enlarged appearance of the recurrent nerve complex when a pin is thrust between the nerve and the oesophagus.

Fig. 130. A hyperphagic blowfly.

together, the physiological solution removed, and the fly freed. After a successful operation the animal behaves normally and, if given food, feeds almost immediately.

Insofar as feeding is concerned the results were decisive and spectacular. Section of the recurrent nerve produced an uncontrolled hyperphagia. Flies began to feed in the normal fashion, and ceased, as was to be expected, when the labellar receptors became adapted; however, as soon as disadaptation occurred vigorous feeding was resumed. The normal prolonged rise in threshold failed to take place. If a hyperphagic fly was fed on 1 M sucrose and disadaptation of the receptors allowed to occur, the fly would still respond to 0.1 M sucrose. As a consequence of the failure of the threshold to rise the flies fed continuously as long as they lived and became enormously bloated. The engorged crop completely filled the abdomen; other organs were squashed into a small compact mass, and the abdominal cuticle was so excessively stretched that the tracheal system could be seen in minute detail and the whole fly was as transparent as a miniature lens (Fig. 130). In these experiments death usually ensued within 24–96 hours. It was concluded that the cause was probably starvation arising from failure of the valves regulating passage of food from the crop to the midgut. Green (1964b) also reported that most of his operated flies died within a relatively short time; however, one lived long enough to be given three feedings. At each feeding the fly became hyperphagic and subsequently was able to pass food to the midgut. Evans and Barton Browne (1960) also reported that hyperphagic flies could pass sugar to the midgut, lost their swollen appearance after having done so, and were still hyperphagic upon subsequent feeding. Since these findings were abundantly confirmed in our laboratory, it is clear that section of the recurrent nerve is not in itself a cause of death.

This observation together with all other available information enabled Dethier and Bodenstein (1958) to construct the following picture of the process of feeding. When an unfed fly encounters food, contact chemoreceptors on the tarsi are stimulated. This sensory input initiates extension of the proboscis. Extension brings the aboral chemosensory hairs of the labellum into contact with the food. Stimulation of these causes the labellar lobes to be opened, so that the oral papillae are brought into contact with the solution. Stimulation of the oral papillae initiates sucking. Feeding is thus initiated and driven by sensory input originating in labellar receptors. At the beginning of feeding the threshold is at a low level. The duration of imbibition depends on the time required for a high level of adaptation of the oral receptors. The sugar that is ingested goes first to the midgut and crop. When the midgut is filled, the cardiac valve is closed and the continued influx is directed into the crop. This sequence insures that food is placed first where it is needed for immediate energy (in the midgut) and after that into storage. Waves of peristalsis passing posteriorly in the crop duct constantly press fluid back into the crop itself. The region of the foregut now contains residual sugar from more anterior regions of the oesophagus and preoral cavity. Green (1964b) subsequently pointed out that a full crop exerts considerable pressure on the valve of the crop duct and that the valve leaks slightly under these conditions. This situation continues for some time, and during this period the fly regurgitates to a considerable extent. As a consequence the foregut is almost continuously bathed in fluid.

When sugar in the midgut is utilized, the crop valve opens momentarily, peristalsis in the duct is reversed, so that a slug of fluid is driven energetically into the foregut region, the crop valve closes, a wave of peristalsis in the foregut drives the fluid toward the proventriculus, where the cardiac valve opens briefly to permit passage into the midgut. As pressure in the crop is decreased there is less leakage around the valve and the foregut contains sugar only at the time when a slug of fluid passes. The point to be emphasized is that the foregut is filled intermittently with discrete slugs of fluid and the frequency of delivery decreases as the crop becomes progressively more empty.

Although disadaptation of the oral receptors has by this time occurred, the threshold of response remains elevated and further feeding is blocked. Dethier and Bodenstein postulated, therefore, that there are receptors somewhere in the region of the foregut that are stimulated every time a slug of fluid passes, that this stimulation causes impulses to ascend the recurrent nerve, and that these impulses inhibit at some central point the sensory input from the mouth (and tarsi), so that the behavioral threshold becomes high. When the recurrent nerve is cut, feeding resumes every time disadaptation of the oral receptors occurs so that an operated fly feeds almost continually in an intermittent fashion.

This hypothesis was criticized by Evans and Barton Browne (1960) on the following grounds: (1) no quantitative criterion of hyperphagia was given; (2) tests were performed with sucrose and there were no published data regarding the effect of ingestion of this sugar on subsequent thresholds; (3) the method of feeding departed from the usual in that the flies were kept in contact with the sucrose solution for long periods or were given repeated opportunities to feed, and since sucrose is among the most stimulating sugars the quantities ingested would be expected to be greater than normal; (4) it was stated that section of the recurrent nerve prevents emptying of the crop; therefore failure of threshold to rise could just as well be explained by the absence of sugar in the region of the foregut that contains the threshold regulating mechanism; (5) the operation may involve damage to a number of other tissues and the observed effects might have been a result of side effects of the surgery.

Criticisms (1), (2), (3), and (5) were easily answered. The criterion of hyperphagia was illustrated in Figures 1 and 2 of Dethier and Bodenstein (1958), and the fact that cutting the nerve does indeed cause abnormal feeding was confirmed by Evans and Barton Browne themselves, by Green (1964b), and subsequently by 210 operations in our laboratory.

With respect to criticism (2), there was no reason to believe that the action of sucrose would differ from that of other sugars, and later tests supported this view. Furthermore, the flies termed hyperphagic after feeding on sucrose were assessed in comparison with sham-operated flies also fed sucrose.

Criticism (3) was irrelevant because control and operated flies were fed exactly the same way, and in any case the *method* of feeding does not influence the quantity ingested in the manner described. Many data relative to this point had been published (Dethier and Rhoades, 1954; Dethier, Evans and Rhoades, 1956; Evans and Dethier, 1957).

Criticism (5) was anticipated by proper control flies that were given sham operations. Furthermore, Green (1964b) repeated these procedures, checked by autopsy to make sure that the recurrent nerve had indeed been cut, and produced hyperphagia in 69% of his animals as compared with 31% reported by Evans and Barton Browne. Green pointed out that "these authors did not report having done autopsies on their test insects to determine whether or not the recurrent nerves had definitely been sectioned, and, in fact they make a point of noting that the 'neatest' operations that they performed did not necessarily produce hyperphagia." Of the 210 operations subsequently performed in our laboratory 91% resulted in hyperphagia.

Criticism (4) was valid, but was answered by the authors themselves and subsequently by Green (1964b) and in our laboratory. In brief, the crop in operated flies can empty.

The conclusion remains that the operation to section the recurrent nerve causes the fly to become hyperphagic; but is the mechanism involved the one proposed by Dethier and Bodenstein? Evans and Barton Browne maintained that threshold regulation is probably not altered by the operation. They noted that tarsal thresholds to glucose were higher for operated and sham-operated flies than for controls (suggesting some nonspecific effect of surgery) and that none of these flies responded to 2 M glucose.

They added: "Tarsal thresholds to sugar solutions can only be obtained after the tarsal chemoreceptors have been adapted to water. Hyperphagic flies often were found to have abnormal response, in that they repeatedly responded to water alone. Because of this it was necessary to retest the response of each fly to water after it had given a positive response to sugar solution so as to verify that the response was, in fact, to sugar. We feel that the abnormal water response might be involved in explaining the discrepancy between our results and those of Dethier and Bodenstein (1958), in that the persistent hypersensitivity to water would tend to mask any elevation of the sugar threshold."

It is indeed true that cutting the recurrent nerve produces polydipsia (Dethier and Evans, 1961). On the other hand, we did find (Dethier and Bodenstein, 1958) that after the water receptors had adapted the operated flies did respond to low concentrations of sugar. We found, furthermore, that hyperphagic flies fed on 0.1 M sucrose continued to attempt feeding each time the receptors became disadapted, and that further ingestion was prevented by the inability of the cibarial pump to overcome the back pressure of the crop. However, when the flies were offered a higher concentration of sucrose, pumping resumed more vigorously than before, more fluid was forced into the crop, and the fly burst. This result suggests that the fly was in fact responding to sugar rather than to water alone.

Two characteristics of the time course of taste threshold after feeding suggested to Evans and Barton Browne that there may be a hormonal link in the chain of events causing threshold elevation. First, there is the slow attainment of peak threshold (but some elevation occurs immediately after feeding). The lag may be 1 to 6 hours, depending upon the sugar eaten (Figs. 112, 113). It is interesting to recall in this connection that the flies operated by Evans and Barton Browne had thresholds close to peak immediately after feeding. Second, there is a long period of threshold elevation after feeding in flies whose crops have been ligated or removed. In these flies the volume ingested after ligation was small and consequently left only a small volume of fluid in the foregut and midgut. It seems unlikely that these small amounts of sugar would remain long in the guts of otherwise starved flies, but the threshold remained elevated for 4 hours or slightly longer. Another point not mentioned by Evans and Barton Browne is that the region of the foregut is bathed by sugar during the process of ingestion, and yet the threshold does not rise during the process.

Some of these points were well taken, and it was obvious that the situation was more complicated than originally imagined. This realization was strengthened by the observation of Núñez (1964) that a related fly, *Lucilia* sp., became hyperphagic when the ventral nerve cord was cut between the brain and the thoracic ganglionic mass. This observation prompted Dethier and Gelperin (1967) to reopen an inquiry into the subject of experimental hyperphagia. First, a criterion of hyperphagia was established. An absolute quantitative criterion is unrealistic because the amount of sugar solution imbibed by a fly

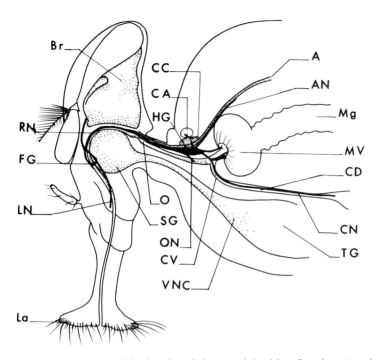

Fig. 131. Diagram of the head and thorax of the blowfly, showing the endocrine complex and the relationships between the nervous system and the alimentary canal. Br. brain; RN, recurrent nerve; FG, frontal ganglion; LN, labrofrontal nerve; La, labellum; CC, corpora cardiaca; CA, corpus allatum; HG, hypocerebral ganglion; O, oesophagus; SG, suboesophageal ganglion; ON, nerve to oesophagus; CV, crop valve; VNC, ventral nerve cord; A, aorta; AN, nerve to aorta; Mg, midgut; MV, midgut valve; CD, crop duct; CN, nerve to crop; TG, thoracic ganglion (from Dethier, 1966).

depends upon varying factors, among them previous nutritional history, degree of activity during the preceding 24 hours, weight of the fly, and ambient temperature and humidity. Within any given sample of flies the standard deviation of the mean intake can vary from 7 to 19% of the mean. Moreover, there is a significant intersample variability from one day to the next. The standard deviation of the sample means may be 13% of the average sample mean. For these reasons, hyperphagia was defined as ingestion of not less than twice the quantity of fluid taken by a control from the same sample.

Two hundred recurrent nerve operations confirmed the original conclusion that hyperphagia followed transection of this nerve. It must be remembered, however, that the nerve was cut posterior to the brain at a point where it also contains fibers of the nervi corporis cardiaci I, which afford neural connection between neurosecretory cells of the pars intercerebralis and the corpora cardiaca and corpus allatum (Fig. 131). These fibers associated with the endocrine glands travel in the recurrent nerve only as far as the posterior edge of the brain, at which point they leave to make independent connections. In the region of the frontal ganglion (anterior to the brain) the recurrent nerve has no associations with the endocrine complex (Fig. 132); therefore a technique was developed for transection of the nerve at this point.

This operation was performed in the following manner. The fly to be operated upon was stood vertically on its tail in a depression in wax molded to its contours. In this position the front (frons and clypeus) of the head lay in a horizontal plane flush with the surface of the wax block. With a microscalpel cuts were made along the two frontal sutures, producing a single V-shaped incision. The apex of the cuticular flap was grasped by forceps and folded ventrally along a natural hinge formed by the epistomal suture. The flap was anchored in this position by a pin thrust across it and into the wax. Internal pressures exerted by opening the flap forced into view the loop of the oesophagus that normally lies between the two furcal prongs when the proboscis is in the retracted position. A certain amount of fat body overlies the oesophagus. When this is removed, the frontal ganglion can be seen lying dorsal to the oesophagus. Also visible are the connectives between the brain and the frontal ganglion and the recurrent nerve extending back into the posterior regions of the body. Observation of these structures is facilitated if a drop of physiological saline is placed in the opening made by the incision. In one variant of the operation the recurrent nerve itself was transected; in another, the connectives to the ganglion were cut; in still another, all three were cut and the ganglion removed. The effect on subsequent events was the same in all cases. After the nerves had been cut the flap of cuticle was released. It fell naturally back into place. In order to make it secure its apex was tucked under the adjoining cuticle. Blood clots sealed the wound.

This operation resulted in hyperphagia indistinguishable from that produced by posterior transection. It is absolutely certain, therefore, that the recurrent nerve is directly concerned with the regulation of feeding and that interrupting its transmission from the foregut to the brain causes hyperphagia.

Next the observation of Núñez with respect to transection of the ventral nerve cord was confirmed.

To sever the nerves connecting the thoracic ganglion and the abdomen the fly was affixed ventral side up on a block of paraffin and all legs secured with strips of modeling clay. The intersegmental membranes on the lateral and anterior edges of the first abdominal sternite were cut. The sternite was folded posteriorly. The field thus exposed contained the abdominal tracheal sacs laterally, with the crop duct and nerve connections between thorax and abdomen situated medially. These nerves were cut, and the cuticular flap was replaced.

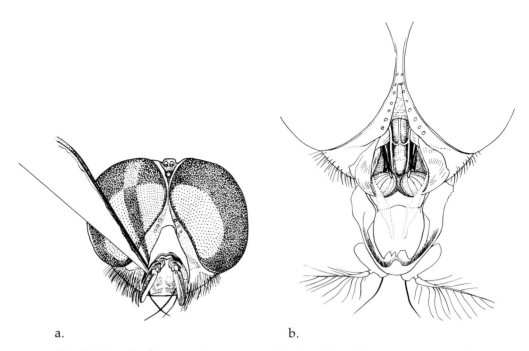

Fig. 132. Steps in the operation to remove the frontal ganglion. (a) Incising and retracting the frontal flap. (b) Appearance of the region of the frontal ganglion and oesophagus when the flap is completely folded back. The large muscles are the retractors of the fulcrum.

The operation is drastic, however. It deprives the brain of all sensory input from the thorax and abdomen (except that transmitted via the stomatogastric system) and silences many potential sources of information relative to feeding. Most particularly, the operation isolates tactile and proprioceptors in the body wall as well as all the tarsal chemoreceptors. Furthermore, none of the activity generated in the ventral ganglionic mass, including thoracic locomotor centers, reaches or is reached by the brain. An operated fly is unable to fly or walk but can stand unsteadily. If a fly maintains a standing posture such that the labellum does not touch the substrate, there is no hyperphagia. The absence of spontaneous proboscis extension argues against the proposition that hyperphagia occurs because some endogenous brain center is released from inhibitory control of the thoracic ganglion.

A refinement of the operation in which only the nerves to the body wall were cut also led to hyperphagia. A comparable condition could be produced by hand-feeding in such a way that sugar was continuously applied to the labellum even when retracted, or by amputating all the legs at the coxotrochanteral joint. The diverse results obtained with all of the various techniques were finally reconciled by examining the temporal pattern of feeding over extended periods.

A normal or sham-operated, unrestrained fly, deprived of food for 24 hours, imbibes from 6.5 to 24.5 μl of 2 M glucose during its first drink. The sham-operated fly whose activity is diagrammed in Figure 111 took a first drink lasting 99 seconds, whereupon the proboscis was retracted. This drink was followed by a series of short drinks over the next 10 minutes. No further drinking occurred over the next 30 minutes. The total duration of drinking within the 30-minute period was 106.0 seconds; the total imbibed, 28.7 μl; the rate of sucking 0.27 μl/second.

The fly lacking an intact recurrent nerve (RN) took an initial drink lasting 116 seconds (Fig. 111). Over the next 30 minutes it took seven drinks of 10 or more seconds' duration. The total duration of drinking in the 30-minute period was 116 seconds; the total volume imbibed, 46.0 μl; the rate of drinking, 0.19μl/second.

The fly with the interrupted ventral nerve cord (VNC) took essentially one drink (Fig. 133), which resulted in bursting after 24 minutes. The total intake just prior to bursting was 42.0 μl. The sham-operated control followed the usual normal pattern of ingestion, resulting in a total intake of 18.1 μl during the 30-minute exposure to 2 M glucose. Flies that had only the nerves to the abdomen cut took an initial drink lasting four times longer than that of the controls. Then they began to walk about. If the ventral nerve cord was then cut between the brain and thoracic ganglionic mass, and the labellum touched to the substrate, feeding recommenced and lasted until bursting.

The patterns of feeding in hand-fed and legless flies are in fact determined by the experimenter since extension of the proboscis occurred only when the labellum was allowed to touch the substrate. When the experimenter attempted to maintain continuous contact, an indication of pattern emerged. The initial drink was normal. The pattern of subsequent bouts resembled that characteristic of the RN fly. The end result was hyperphagia.

When the amount of fluid drunk by each kind of fly was measured at intervals over an 8-hour period, some additional differences were revealed (Fig. 134). At the end of the first 30 minutes hyperphagia had already been established. From 30 minutes to 8 hours all three kinds of flies decreased their intake and reached an asymptote, but the rates of decline differed. The normal fly reached final capacity most rapidly. Lack of further increase resulted from lack of ingestion. The VNC fly had reached nearly maximum capacity within the first 30 minutes. Thereafter there was only a small increase in volume (as was true of the normal fly); however, failure to increase further resulted not from lack of ingestion but from inability to force more fluid into the gut. The RN fly, on the other hand, although clearly hyperphagic at the end of 30 minutes, had not attained the same level of hyperphagia as the VNC fly.

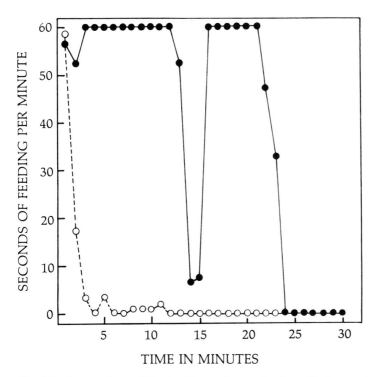

Fig. 133. Comparison of the feeding patterns of a fly with the ventral nerve cord transected and a sham-operated control during a 30-minute exposure to 2 M glucose (from Dethier and Gelperin, 1967).

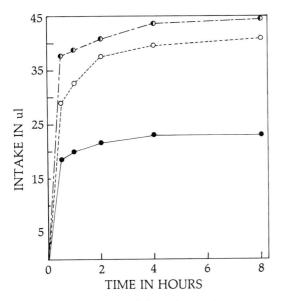

Fig. 134. Comparison of the volume of 1 M fructose ingested by a fly with the recurrent nerve cut (dotted line); with the ventral nerve cord cut (broken line); and a control (solid line) over an 8-hour period (from Dethier and Gelperin, 1967).

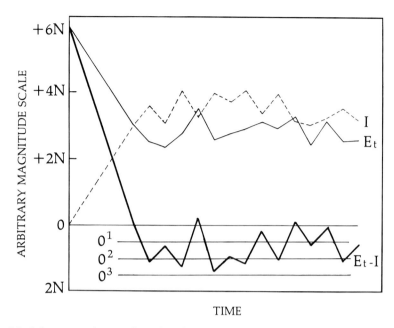

Fig. 135. Model proposed to explain the phenomenon of hyperphagia and the interaction of excitatory and inhibitory factors regulating feeding behavior in the normal fly (from Dethier and Gelperin, 1967). For a full explanation, see text, Chapter 8.

It then continued to increase its weight gradually. In short, as these analyses of feeding patterns illustrate, there is more than one kind of experimentally induced hyperphagia in flies.

The model (Fig. 135) that was proposed to unify all of the data has as its central postulate the idea that feeding occurs as the result of an interplay between excitation and inhibition. Since no evidence has been found that an endogenous neural center drives feeding behavior, the peripheral chemoreceptors on the tarsi and the labellum apparently constitute the sole source of excitatory input that initiates and drives feeding. It must be noted that the excitatory messages from the tarsi and the labellum normally initiate separate phases of behavior. Tarsal excitation initiates extension of the proboscis; labellar excitation initiates spreading of the oral lobes and together with papillar excitation drives sucking. Neural activity that inhibits feeding arises from three sources: (1) stretch receptors in the foregut that send information via the recurrent nerve to monitor the rate and extent of peristalsis in the foregut (I_{fg}); (2) stretch receptors associated with the body wall that send information via the thoracic ganglia and ventral nerve cord to monitor the distension caused by fullness of the crop (I_{bw}); (3) a locomotor center in the thoracic ganglia that produces activity which is inhibitory to feeding (I_{lc}). If total excitation is represented by E and total inhibition by I, then the central postulate of the model states that feeding is absent when $E - I \leq 0$; when $E - I > 0$, feeding is initiated. When the proboscis is retracted, $E = E_{\text{tarsal}}$; when the proboscis is extended $E = E_{\text{labellar}}$; the same inhibition interacts with both E_t and E_l.

Consider a food-deprived fly walking on filter paper saturated with 1.0 M sucrose. At first contact, E_t will be high because the tarsal receptors are completely disadapted and the stimulus is intense; I will be low because the gut is empty. This is the state of events at time 0 in Figure 135. As the fly feeds, its tarsal receptors are adapting, and internal inhibition is increasing as the gut is filled. Feeding is terminated as a result of the accumulation of inhibition and the adaptation of the labellar receptors. The tarsal receptors are now at a steady-state level of adaptation; however, as the fly walks or shifts its position in the sugar field new receptors are stimulated and adapted, thus producing fluctuation in the amount of excitatory tarsal input. Similarly, fluctuations in the inhibitory input are caused by the discontinuous nature of foregut peristalsis and the churning of the crop in the abdomen. The values of both E and I oscillate randomly about their mean values. Consequently the values of $E - I$ also fluctuate, aperiodically becoming greater than zero and initiating a meal. It should be noted that the meal is initiated when $E_t - I$ becomes greater than zero, whereas the duration of the meal is determined by the time that $E_l - I$ is greater than zero. In Figure 135 a proboscis extension is initiated whenever the $E_t - I$ line intersects the zero line with a positive slope, i.e., from underneath.

To determine the effect of cutting the recurrent nerve, the ventral nerve cord, or the posterior connectives, consider the effect of lowering the I curve by a given amount at every point along its length. Lowering the I curve is equivalent graphically to raising the E-I curve; raising the E-I is equivalent graphically to lowering the zero line. Therefore, to determine the effect of removing a source of inhibition, for example, removing I_{fg} by cutting the recurrent nerve, the zero line is lowered by an amount equal to I_{fg} (e.g., to 0^1). The E_t-I line now becomes positive a greater number of times; hence more drinks are taken. This is the experimental observation. If a larger source of inhibition is removed, I_{bw}, for example, then the graphical equivalent is that of lowering the zero line still further, to 0^2, for example. Again the prediction of more drinks per unit time is confirmed. If we assume that E_l is proportional to E_t, then the length of the drink is proportional to the length of time that the E_t-I line is above the zero line; this also increases as sources of inhibition are removed and the zero line is lowered. Finally, if sufficient sources of inhibition are removed, for example, by removing I_{fg} and I_{bw}, continuous feeding results because E-I is continually greater than zero. Cutting the ventral nerve cord produces this effect, predicted graphically by lowering the zero line to 0^3.

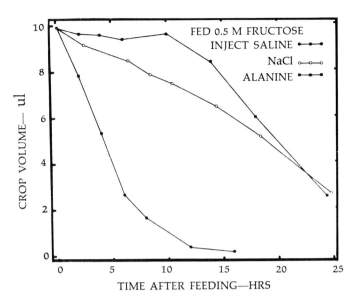

Fig. 136. Effect of increased blood solute concentration on rate of crop emptying (from Gelperin, 1966a).

Recently a question has been raised about changes of labellar threshold associated with feeding and deprivation. Originally no change had been observed in the labellar thresholds of *Calliphora*, although the tarsal threshold did change (Minnich, 1931). Getting and Steinhardt (1972) were unable to detect changes in the labellar sensitivity to sucrose after feeding water-satiated *Phormia* with 10 µl or 15–20 µl of 1 M sucrose. Extension of the proboscis still occurred in response to the application of 0.1 M sucrose to one labellar hair 15 minutes after feeding; however, a more extensive analysis is required before it can definitely be stated that labellar thresholds do not change. On the other hand, the experiments with legless flies do show that the labellum retains its sensitivity long after the tarsi fail. The implication is that the recurrent nerve inhibits tarsal but not labellar input. Motor response (as measured by electrophysiological monitoring of traffic in the extensors and adductors of the haustellum) to labellar stimulation was not altered by starvation or by cutting the recurrent nerve (Getting and Steinhardt, 1972).

Although the crop itself is not *directly* concerned with threshold regulation, its rate of emptying sets the conditions under which the postulated foregut receptor (and the bodywall stretch receptors as well) is stimulated; therefore, experimental alternation of emptying rate should be reflected in the level of acceptance threshold. This prediction was realized by Gelperin (1966b) Injections of salts, sugars, or amino acids into the haemocoele of fed flies were all effective in slowing the emptying of the crop (Figs. 136, 137). They also elevated

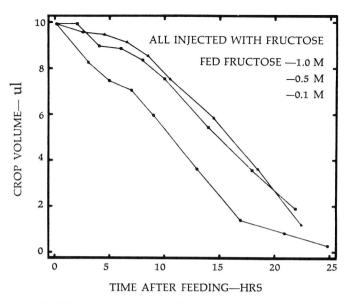

Fig. 137. Effect of injecting a uniform amount and feeding varying concentrations of fructose on rate of crop emptying (from Gelperin, 1966a).

the acceptance threshold (Fig. 138). This result is not in disagreement with the results of injection reported earlier by Dethier and Bodenstein (1958). They had injected sugar into the blood of flies with *empty crops* and did not observe an elevation of the threshold because there was no inhibition from foregut or body-wall receptors under these conditions. Crop emptying was also delayed by feeding high concentrations of large volumes of sugar solutions. As expected, these procedures prolong the elevation of the taste threshold (Figs. 139 and 140).

The rate of crop emptying is obviously an important link in the chain of events controlling feeding. The simplest mechanism controlling emptying of the crop would be the volume of the crop itself. Gelperin (1966a) undertook to test this hypothesis by observing the crop under different experimental conditions. Three methods are routinely employed in studying the volume of the crop. The simplest but least precise is to weigh the entire fly, with the knowledge that the major changes in weight over a short period of time reflect the degree of fullness of the crop. A more direct and exceedingly accurate method is to excise the crop and weigh it. This can be done quickly and without loss of fluid because of the extreme toughness of the crop. By making a small incision in the ventral midline of the first abdominal segment, it is possible to seize the crop duct with a forceps and extract the crop intact. This tough, transparent bag can then be weighed and the volume of fluid calculated. A major improvement in technique which permitted repeated measurements on an intact living fly was introduced by Green (1964b). The rate of crop emptying in individual flies was

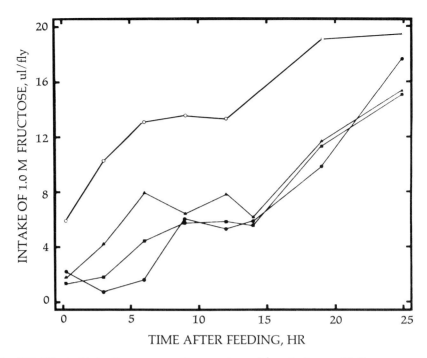

Fig. 138. Effect of injecting sugar, salt, or amino acid on taste sensitivity changes after feeding. All fed 0.5 M fructose. —●—, sorbose injection; —■—, alanine injection; —▲—, NaCl injection; —○—, saline injection (from Gelperin, 1966b).

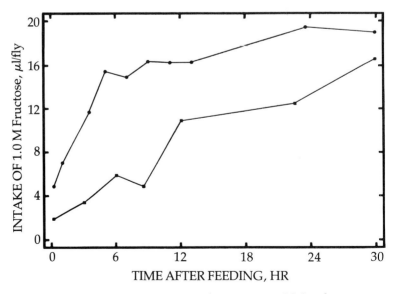

Fig. 139. Effect of sugar concentration ingested on taste sensitivity changes after feeding. —●—, fed 0.2 M fructose; —■—, fed 2.0 M fructose (from Gelperin, 1966b).

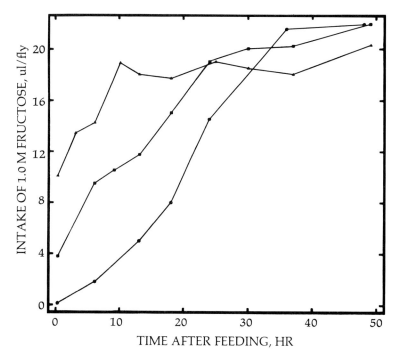

Fig. 140. Effect of meal volume on taste sensitivity changes after feeding. All fed 2 M fructose. —●—, 20 μl meal; —■—, 10 μl meal; —▲—, 5 μl meal (from Gelperin, 1966b).

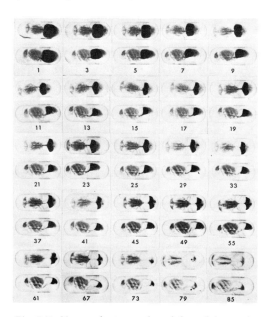

Fig. 141. X-ray photographs of dorsal (upper) and lateral (lower) aspects of a fly fed to repletion on 0.5 M sucrose containing Hypaque sodium and then starved to death. The figures shown represent the time in hours since feeding to repletion (from Green, 1964b).

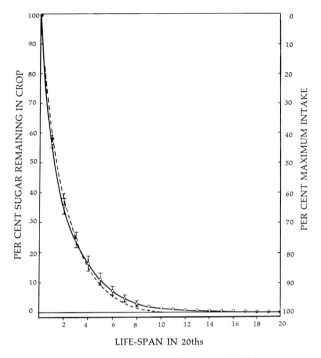

Fig. 142. The rate of crop emptying after feeding to repletion on 0.5 M sucrose containing Hypaque sodium, based on X-ray photographs (solid line, left ordinate) and feeding experiments (broken line, right ordinate) (from Green, 1964b).

determined by repeated measurements at intervals after a meal of sugar mixed with barium sulfate or sodium diatrizoate (Hypaque Sodium) with the aid of X-ray photography (Figs. 141 and 142). Each fly was placed in a tightly fitting gelatine capsule and photographed in dorsal and lateral aspects. The volume of fluid in the crop was calculated from these two areas.

The effect of varying meal size was studied first by feeding flies different amounts of 2.0 M fructose. As Figure 143 shows, the larger the crop volume, the faster the rate of emptying (Gelperin, 1966a). By replotting the data (Fig. 144) one can show that volume is not the only relevant factor. When the 10 μl meal is compared with the last half of the 20-μl meal, it is seen that the 10-μl meal left the crop more rapidly. Similarly, the 5-μl meal left more rapidly than the last quarter of the 20-μl meal. Apparently the prior digestion of sugar retarded the emptying of the crop. By keeping the volume of the meal constant and varying the concentration, Gelperin was able to show that there was an accumulating inhibitory effect of digested sugar on the rate of emptying (Fig. 145).

Several properties of solutions accompany an increase of concentration. Among these are viscosity, nutritive value, stimulating power,

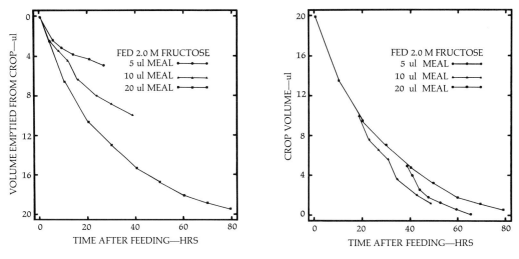

Fig. 143. Effect of varying meal size on rate of crop emptying (from Gelperin, 1966a).

Fig. 144. Effect of varying meal size on rate of crop emptying (from Gelperin, 1966a).

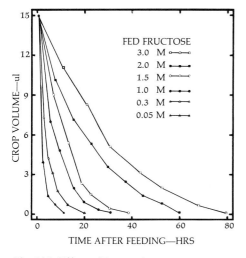

Fig. 145. Effect of increasing sugar concentration on rate of crop emptying (from Gelperin, 1966a).

and osmotic pressure. The irrelevance of viscosity as a factor in crop emptying was proved by adding 4% (wt/vol) methylcellulose, which is nonnutritive and nonstimulating. This increased the viscosity by a factor of 500. The rate of emptying was unchanged (Fig. 146). That nutritive value was not a factor was proven by comparing the rates of emptying of the nonnutritive sugars sorbose and fucose with that of the nutritive sugar fructose (Fig. 147). That stimulating effectiveness

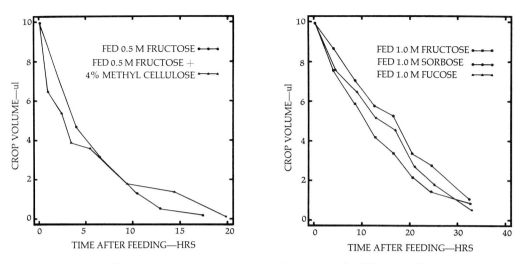

Fig. 146. Effect on rate of crop emptying of increasing by 500 times the viscosity of a fed solution (from Gelperin, 1966a).

Fig. 147. Effect of solutions of high (fructose) and low (sorbose, fucose) nutritional value on the rate of crop emptying (from Gelperin, 1966a).

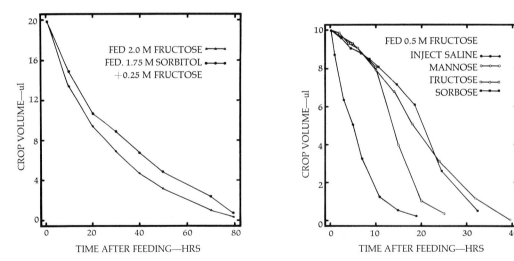

Fig. 148. Rate of crop emptying after ingestion of a strongly stimulating solution (2.0 M fructose) and a weakly stimulating solution (1.75 M sorbitol + 0.25 M fructose) (from Gelperin, 1966a).

Fig. 149. Effect of increased blood sugar on rate of crop emptying (from Gelperin, 1966a).

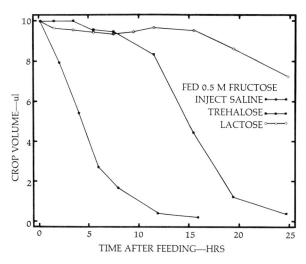

Fig. 150. Effect of increased blood sugar on rate of crop emptying (from Gelperin, 1966a).

was not a factor was proven by comparing various mixtures of stimulating and nonstimulating sugars (Fig. 148). Osmotic pressure is the critical factor. One molar sucrose (a disaccharide) empties at the same rate as 2 M glucose (a monosaccharide); therefore, it is unlikely that the sugar is acting in the crop or foregut because sucrose cannot be hydrolyzed there and would empty at a different rate. This result points to the midgut or the blood as being the site of action following hydrolysis (Gelperin, 1966a).

Ingenious experiments involving, among other things, turning the gut inside out, proved that the transport of sugar from midgut to blood occurs by simple diffusion. Injection of various solutes into the blood showed that increased blood osmotic pressure (not blood sugar per se) retarded the emptying of the crop (Figs. 149–152). Gelperin then varied osmotic pressure of the blood and crop independently, with the results illustrated in Figures 136, 137, 153, and 154. The data indicate that the concentration of solute of the blood is the important variable and that ingested foods exert their effect on the rate of crop emptying by increasing the solute concentration of the blood. Neither neural nor endocrine organs seem to be required for normal action, the mechanism of which remains a mystery.

The physiology of the crop has been studied in a number of other insects, and the results suggest that there is no universal mechanism regulating emptying. A detailed series of papers by Treherne (1957) and Davey and Treherne (1963a, 1963b, 1964) reported on the situation in the cockroach *Periplaneta*. Since the rate of emptying is related in a

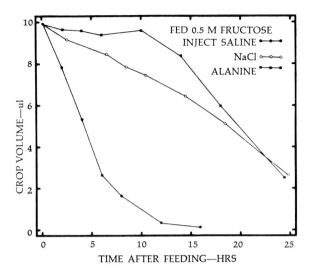

Fig. 151. Effect of increased blood solute concentration on rate of crop emptying (from Gelperin, 1966a).

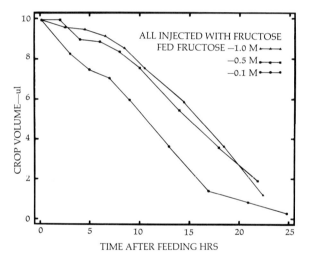

Fig. 152. Effect of injecting a uniform amount and feeding varying concentrations of fructose on rate of crop emptying (from Gelperin, 1966a).

linear fashion to the concentration of ingested sugar, osmotic pressure is an important variable, as in *Phormia*. Changes in viscosity are not critical factors. Ingestion does not cause the volume of the crop to increase, because the inflowing fluid merely displaces air with which the empty crop is filled. Swallowing of air is involved in the emptying processes. The frequency of opening of the proventricular valve seems to be the factor that limits emptying the crop, and it operates in such a way that a constant proportion of a meal is emptied in a given time. The rate of emptying increases as meal size increases.

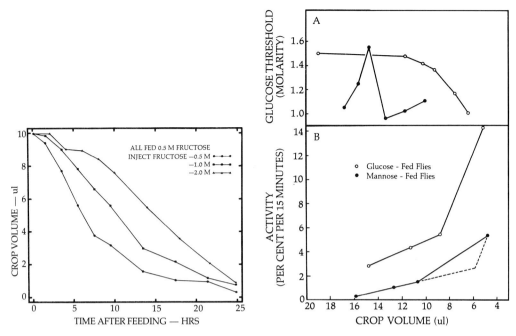

Fig. 153. Effect of feeding a uniform concentration and injecting varying amounts of fructose on rate of crop emptying (from Gelperin, 1966a).

Fig. 154. (a) The relation between crop volume and glucose taste threshold. (b) The relation between crop volume and locomotor activity (from Evans and Barton Browne, 1960).

As in *Phormia*, the valve is innervated by the stomatogastric system. It can operate in the absence of connections between the stomatogastric and the central nervous system; however, separating the proventricular ganglion from the frontal ganglion or cutting the small nerve (N 5) connecting the frontal ganglion to the dorsal part of the pharynx interferes with its action. A receptor in the pharynx, possibly as osmoreceptor, is involved, but since size and viscosity of meals can be compensated for, more than a simple reflex involving this receptor must be invoked.

In another cockroach, *Leucophaea maderae*, a constant proportion of a meal is emptied in a given time, independent of size, and the consistency of the meal is a modifying feature, but the mechanism of neural control is different (Englemann, 1968a, 1968b). Cutting all nerves to the frontal ganglion except the recurrent nerve has no effect. Cutting the recurrent or oesophageal nerve after feeding reduces emptying of the crop by 50%, but the same operation 1 to 12 days before feeding has no effect.

The stomatogastric system is essential for normal functioning of the crop in the locust *Schistocerca*. Extirpation of the frontal ganglion of *Schistocerca* females causes large amounts of food to accumulate in the crop even though the midgut is empty. Furthermore, the normal eating binge that occurs 8 to 10 days after moulting is replaced by a constant low level of intake (Highnam, Hill, and Mordue, 1966). Removal of the frontal ganglion has no effect on the feeding behavior of *Locusta* (Clarke and Langley, 1962).

Knowing something of the factors involved in the emptying of the crop of *Phormia* and the relation between emptying and acceptance threshold made it possible to investigate the characteristics of the postulated foregut receptor. Feeding different kinds of sugars provided this information. When flies were fed 2 M solutions of glucose, mannose, or fructose, all of which move through the gut at the same rate, the flies subsequently showed the same sensitivity to a test sugar. This indicated that the foregut receptor responded identically to different chemical compounds. That the receptor is not an osmoreceptor was proved by comparing the effects of feeding two sugars, which move through the gut at the same rate but are osmotically different. The receptor must be a mechanoreceptor responding to distension of the foregut.

Final proof that a foregut mechanoreceptor does indeed regulate food intake was provided by Gelperin (1967), who found two bipolar neurons in a branch of the recurrent nerve going to the oesophagus. He succeeded in recording electrophysiologically from these when the gut was expanded normally, by spontaneous peristalsis, and when it was expanded artificially, by injecting material through a microsyringe introduced into the foregut via the crop duct. The two cells fire at a sustained and uniform rate when the gut is quiescent but greatly increase their rate whenever expansion occurs. Their response is sensitive to variations in the situation in the foregut. The extent and frequency of foregut peristalsis varies inversely as the rate of crop emptying. Digestion of a dilute sugar solution (0.1 M sucrose), which empties from the crop rapidly, results in *less* output from the foregut stretch receptors than does digestion of a concentrated solution (1.0 M sucrose), which empties from the crop more slowly (McCutchan and Gelperin, 1969).

The abdominal receptors have also been found. By cutting various abdominal nerves selectively and noticing whether or not there was any effect on feeding, Gelperin (1971b) finally localized the effect to the first two pairs of lateral branches arising from the main abdominal nerve (Fig. 155). These branches together with others form a loose basketwork lying over the crop but not innervating it. A histological search revealed 4 to 8 cells here (Fig. 156). When these cells are completely isolated from the rest of the nervous system or when the crop is

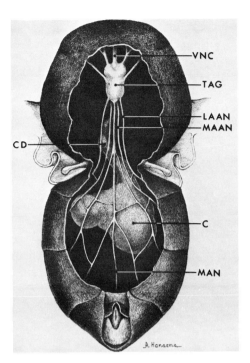
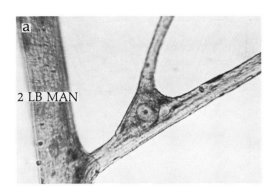
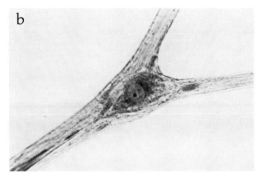

Fig. 155. Semidiagrammatic representation of the abdominal nervous system of *Phormia*. C, crop; CD, crop duct; LAAN, lateral accessory abdominal nerve; MAAN, medial accessory abdominal nerve; MAN, median abdominal nerve; TAG, thoracico-abdominal ganglion; VNC, ventral nerve cord (from Gelperin, 1971b).

Fig. 156. Two cells located in the second lateral branches (2 LB MAN) of the medium abdominal nerve. The cell bodies are 15-20 μ. The axons are on the left. The small nerve arising in the cell body in (a) connects with 1 LBMAN (not shown in the figure) (from Gelperin, 1971b).

quiescent in an intact preparation, the cells fire at a steady basal rate. When the nerve is stretched experimentally or when the crop fills and becomes distended during feeding, the cells increase their average level of output. The more the crop fills, the more the cells respond (Fig. 157). They are also activated by short-term expansion and contraction of the crop; consequently, the in vivo pattern of their activity would be aperiodic fluctuation around an average value determined by the volume of the crop.

Abdominal distension has been implicated in a number of insects as a mechanism for terminating feeding. In *Rhodnius* the pump is stopped mechanically at a critical abdominal pressure. Also there are probably abdominal stretch receptors that shut off feeding. If the ventral nerve cord is transected between the pro- and mesothoracic

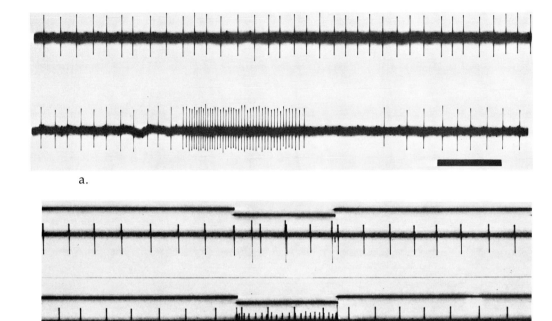

Fig. 157. (a) Activity of cell located in first lateral branch of 1 LBMAN. Top record, ongoing activity in a completely isolated preparation. Bottom record, response to increase in tension of nerve. Time marker = 1 sec. (b) Activity recorded from first lateral branch of the median abdominal nerve (1 LBMAN) with central connection cut and peripheral connection intact. Lower trace in each record signals tactile stimulation at periphery. Stimulus duration, 800 msec. (from Gelperin, 1971b).

ganglia, bugs that normally increase their weight to 330 mg will increase to 550 mg and one week later will still be trying to feed (Maddrell, 1963). The same effect is achieved by cutting the abdominal nerves or making a fistula in the midgut. No particular receptors have thus far been implicated. Each full-sized abdominal segment is equipped with a pair of multiterminal neurons associated with the dorsal intersegmental muscles, and they respond to stretch of these muscles (Van der Kloot, 1960, 1961; Anwyl, 1972); however, their involvement in the control of feeding has not been proven experimentally. Several species of female mosquitoes, insects that like *Rhodnius* take large meals of blood, also are sensitive to abdominal stretch. Transection of the nerve cord anterior to the second abdominal ganglion results in hyperphagia. Intake increases by a factor of four (Gwadz, 1969). The receptors have not been identified.

Thus after a period of sixteen years an almost complete story of the

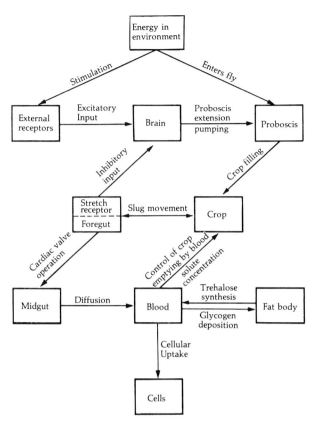

Fig. 158. Homeostatic mechanism for the regulation of energy flow in the blowfly (from Gelperin, 1966b).

control of feeding in the blowfly has been constructed. There are still loose ends, and as more work is carried out revisions can be expected. The main features of the regulation of feeding as now understood are summarized in Figure 158, which shows that there are two feed-back loops involved. One—brain, proboscis, crop, foregut stretch receptor, brain—is fairly well understood although there are still details to be discovered or checked. The other—crop, midgut, blood, crop—still leaves unsolved some of the physiological mechanisms. To explain why there should be two different sets of stretch receptors regulating feeding, Gelperin (1971a) has suggested that they offer an opportunity for range fractionation. The foregut receptors may be more sensitive at low crop volumes. High crop volume may be more accurately signaled by the abdominal neurons when the peristaltic mechanism reaches saturation.

A comparable level of understanding of feeding behavior is gradually being achieved in one other insect, the locust *Locusta migratoria* L. (Bernays and Chapman, 1974a). A comparison with *Phormia* reveals similarities and dissimilarities. Some of the latter reflect species differences and some may reflect the particular approach to the subject that different students have taken. For example, more attention has been paid to hormonal events in *Locusta* than in *Phormia*. Foregut stretch receptors and sensory adaptation play a role in terminating feeding by *Locusta,* but, additionally, hormones released by the corpora cardiaca cause the pores of the chemosensory sensilla on the palps to close, so that there is no more sensory input to promote feeding. A diuretic hormone is also released into the haemocoele and may possibly affect feeding (Bernays and Chapman, 1974b). Finally there is a suggestion that feeding behavior as a whole is driven by some endogenous central nervous activity. For a complete analysis of the regulation of feeding in this insect one should consult the review of Bernays and Chapman (1974a).

The surprising conclusion that at no stage in ingestion does the nutritive value of food regulate intake by *Phormia* raises the interesting question of how this system is adapted to the energy needs of the fly. The initiation of feeding depends on stimulating properties; the termination of feeding depends on mechanoreceptors; the emptying of the crop is regulated by osmotic properties. In the laboratory each one of these can be varied independently of each other and of caloric value. In nature the different properties tend to be correlated. Does the fly possess the capacity to adjust its feeding in response to caloric need in the same sense that mammals do? A series of long-term feeding experiments was undertaken to answer this question (Gelperin and Dethier, 1967). Several measuring techniques were employed: one- and two-choice preference tests over the entire lifetime of flies in individual cages; weighing of flies mounted on wax-tipped sticks immediately before and after a single meal to satiety (intake could be determined to the nearest 0.1 μl); weighing of excised crops (the weight of the crop itself was neglected because it weighed less than 0.1 mg).

The feeding behavior of a fly with respect to sucrose exemplifies the normal pattern of food intake. The daily intake of a male fly given access to 0.1 M sucrose from the beginning of adult life until death is graphed in Figure 159. Sucrose ensures maximum longevity at concentration of 0.1 M or above (Dethier and Rhoades, 1954). For the first day or two intake is low. During this period of adjustment the fly is expanding its wings, hardening its cuticle, and making fluid adjustments; its alimentary canal may contain air and fluids, the former having been swallowed at the time of ecdysis (cf. Cottrell, 1962). Beginning about the second day, the intake of sucrose rises rapidly and reaches a maximum by the second to fourth day. From this time until death some 60 days later, daily intake gradually declines. During this period it fluctuates markedly. In Cottrell's experiment, although

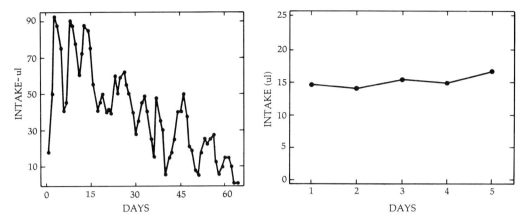

Fig. 159. Daily intake of 0.1 M sucrose by a male blowfly over its entire life span (from Gelperin and Dethier, 1967).

Fig. 160. Quantity of 0.5 M sucrose taken at a single daily meal on each of five consecutive days (from Gelperin and Dethier, 1967).

much of the instability was correlated with changes in the temperature of the laboratory and concomitant changes in the fly's activity, it was not completely abolished by controlling ambient temperature. In the 60-day period a total of 85.5 mg of sucrose, equivalent to 3.3 times the body weight, was ingested.

The causes underlying the gradual decline in intake with age are not known. Green (1964b) found that the activity of flies peaked at about three-fourths of the life span and then declined until death, but was not an age effect. This course does not match the pattern of ingestion. The gradual decline with time could reflect an accumulation of stored reserves. It could also reflect subtle changes occurring with senescence. Over the middle period of life, from day 15 to day 45, the intake is relatively constant despite some fluctuation.

The possibility that a relative constancy does in fact occur was investigated by the more precise method of feeding mounted flies 0.5 M sucrose once daily for 5 days and weighing them to determine daily consumption. The result for a population of 10 flies is shown in Figure 160. This relative constancy is consistent with a hypothesis that a homeostatic mechanism underlies the regulation of taste thresholds. On the other hand, it must be emphasized that the data do not represent ad libitum intake because the flies were fed only once each time. Since the initial termination of a single drink is brought about by sensory adaptation, it is to be expected that the intake will be constant as long as there is no inhibitory feedback from residues from previous meals. The constancy observed here suggests that the fly completely utilized each meal and that each feeding was regulated by adaptation.

One would not expect to see here the gradual decline in daily intake noted in the previous experiment because this experiment represented only a short segment of the total life span.

Although one cannot rule out the possibility that this constancy reflects constancy of sensory adaptation to a given concentration, he can rule out the hypothesis that the fly is an opportunist that merely fills itself to capacity every time the occasion presents itself. That this is not the case is shown by the changes introduced into the feeding pattern by different concentrations of sugar. Compare, for example, the daily ad libitum intakes of 0.0001, 0.001, 0.1 and 1.0 M sucrose. Very little 0.0001 M sucrose is ingested, and the fly survives only 2.5 days. Because of its low power of stimulation this concentration is ineffective in inducing prolonged drinking. Adaptation sets in quickly. A 0.001 M solution generates the same pattern as 0.1 M, but the intake is consistently higher. With 1.0 M sucrose the volume consumed daily is lower than with 0.1 M for every day except the first. This suggests that the fly has accumulated a reserve. Conversion of volume intake to intake by weight shows that although the fly decreases its intake by volume as concentration increases in the range 0.001–1.0 M, the weight of ingested sugar increases. In short, the adjustment of volume consumed does not result in maintenance of a constant weight intake.

The adjustment of intake at different concentrations is even more strikingly demonstrated by alternating the presentation of concentrations (Fig. 161). The outcome is basically the same regardless of which concentration is presented first. When a fly on a diet of 0.1 M sucrose is offered 1.0 M in its place, the daily intake falls. After the fly has been on 1.0 M for several days and is then presented with 0.1 M, the intake increases. The average daily intake of 11 flies whose diet was alternated over a 20-day period with an equal number of days on each concentration was 53 μl of 0.1 M and 38 μl of 1.0 M.

The same relation between intake and concentration was observed when single daily drinks were determined by weighing. Twenty flies were fed either 2.0, 1.5, 1.0 or 0.5 fructose once daily for 5 days. The average daily volume intake for the various concentrations was 18.2 μl of 0.5 M, 14.9 μl of 1.0 M, 13.1 μl of 1.5 M and 11.1 μl of 2.0 M. Again, although volume intake decreased with increasing concentration, the weight of sugar increased. It appears that daily sugar intake at one concentration is maintained at a relatively constant level (e.g., Fig. 160) but that daily intake across concentrations is only crudely regulated. Nonetheless, some sort of regulation is apparent.

Other insects are also known to adjust their meals. Females of *Hierodula crassa*, a praying mantis, eat about 34 house flies each day and males about 5, even when the supply is unlimited (Holling, 1966).

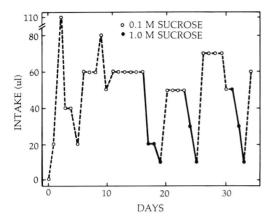

Fig. 161. Daily intake of 0.1 M and 1.0 M sucrose in alternation (from Gelperin and Dethier, 1967).

If larger flies are presented, fewer are consumed (Gelperin, 1971a). Twelve *Phormia* are equivalent to 4 *Calliphora* (for *Paratenodera sinesis*). When foods are diluted, many insects make adjustments that appear to be compensatory. Caterpillars of the hawkmoth *Celerio euphorbiae* increase their consumption of artificial diet as this is progressively diluted. If the consumption of standard diet is valued at 100%, consumption of diets that were diluted to 85, 70, and 50% of the standard was respectively 96, 83, and 65% (House, 1965). A similar response to dilution was observed with German cockroaches (Gordon, 1968). None of these experiments prove that caloric value is at the basis of regulation.

In all of the long-term feeding experiments with *Phormia* described thus far, the stimulating value and the nutritive value of the sugars varied in the same direction as the variation in concentration. The experiments did not serve to separate these two factors. To this end studies were made of long-term ingestion of sorbitol and fucose. Sorbitol is a nutritious compound; LD_{50} on 0.1 M is 15 days (Hassett, Dethier, and Gans, 1950). It is nonstimulating to tarsal and labellar hairs in this concentration range although stimulating to oral papillae (Dethier and Hanson, 1965). Fucose is a highly stimulating non-nutritious sugar; LD_{50} on 0.1 M is 2.5 days (Hassett, Dethier, and Gans, 1950). Use of these two compounds thus presented an opportunity to assess independently the role of stimulating value and nutritive value.

To determine a base line, preliminary experiments were conducted with each compound alone. In a comparison of a one-choice situation, where the only solution offered was 1.0 M sorbitol, and a two-choice situation, where the choice lay between 1.0 M sorbitol and water, the flies ingested as much sorbitol alone as the total intake of sorbitol and water. This was expected because sorbitol is so poorly stimulating that the flies were responding essentially to water and the choice was random. They lived longer in the one-choice situation because they actually got more sorbitol; there was no water bottle to compete.

With 1.0 M fucose as the only fluid available, the flies died within 2.5 days even though they imbibed 80 μl of fluid. Fucose is not only nonnutritive, it is also apparently slightly toxic. In a choice situation with 0.1 M fucose and water, fucose was definitely preferred. During the first 2.5 to 4 days the flies took more fucose than they would have taken sorbitol, a fact demonstrating the superior stimulating power of fucose and the very great importance of stimulating power as a factor influencing intake.

When the choice lay between 1.0 M sorbitol and 1.0 M fucose, fucose was the preferred solution. The average life-span was approximately the same (13+ days) as that when the choice lay between 1.0 M sorbitol and water (15 days), and the volume of sorbitol taken was the same in both cases. In other words, sorbitol was being treated as though it were water. Because the flies took more fucose than water, the total fluid intake in the sorbitol-fucose choice was greater than the total fluid intake in the sorbitol-water choice.

When the only solution offered was a mixture of 1.0 M sorbitol and 0.1 M fucose, the LD_{50} was the same as when the two solutions were offered separately in a two-choice situation. The total intake was about the same as of 1.0 M sorbitol alone, more than of sorbitol versus water, and less than of sorbitol versus fucose.

A striking reversal in preference can be brought about by adulterating fucose with sodium chloride. Thus in a choice situation involving 1.0 M sorbitol versus 0.1 M fucose plus 1.0 M NaCl, fucose was no longer the preferred solution. The total intake here was less than of sorbitol alone, sorbitol versus water, or sorbitol versus fucose; nevertheless, enough sorbitol was ingested to keep the LD_{50} at 21 days.

All of the foregoing experiments show how strongly intake is controlled by stimulating effect. The flies preferred the more stimulating compound (fucose) even though it is nonnutritional and possibly even toxic. They still imbibed enough sorbitol, however, to stay alive. When fucose was rendered unpalatable by the addition of sodium chloride, the preference shifted from fucose to sorbitol, but survival did not increase because sorbitol is too nonstimulating to cause it to be eaten in sufficient quantity.

Comparable experiments were conducted with additional sugars to test the importance of stimulating power more generally. Arabinose is another pentose that is highly stimulating but low in nutritive value (LD_{50} is 2 days). It appears to be slightly toxic to *Phormia*. Both 0.1 M and 1.0 M D-arabinose are preferred to water in a choice situation.

When 0.1 M D-arabinose was paired with 1.0 M sorbitol, the preference for D-arabinose was not marked; the preference was clear when the concentration of D-arabinose was increased to 1.0 M. Because the flies got more sorbitol in the former two-choice situation than in the latter, they lived longer in the first case. They also lived longer on a mixture of 0.1 M D-arabinose plus 1.0 M sorbitol than on 0.1 M D-arabinose versus 1.0 M sorbitol because the arabinose in the mixture was sufficiently stimulating to induce greater intake, hence more sorbitol. They lived longer on sorbitol alone than on sorbitol plus 0.1 M D-arabinose probably because of the toxicity of arabinose. They lived an even shorter period of time on sorbitol plus 1.0 M D-arabinose. Again intake was determined by stimulating power independently of nutritive value.

These results suggested that the intake of various mixtures of sugars should be a function of the total stimulating power and that there should be no adjustment when the nutritive value is diluted by the addition of a stimulating nonnutritive sugar. The following experiments confirmed this expectation. The daily intake by each of 19 flies was measured over a 16-day period. The single solution offered was a mixture of 0.1 M sucrose and 0.9 M sorbitol. The average total intake for 16 days was 2,250 μl. A mixture of 0.1 M sucrose plus 0.1 M galactose presented under the same circumstances was consumed to the extent of 3,390 μl. The first mixture is not very stimulating but is highly nutritious, whereas the second mixture is more stimulating but less nutritious. The flies consumed more of the stimulating mixture than of the highly nutritious mixture, yet the increased intake did not compensate for the deficiency in food value. Detailed data are presented in Table 19.

Although all of the foregoing experiments negate the direct involvement of nutritive value as a factor controlling intake, they do point to the existence of some sort of regulation. Earlier we have discussed factors controlling the rate of crop emptying. When crop volume is measured daily in flies maintained on different regimens, it is clear that there is a steady-state regulation. From a population of 60 flies permitted ad libitum access to a sugar solution on filter paper in a petri dish, 10 were removed each day for 6 days, their crops excised and weighed. It was clear that different volumes were maintained for sucrose, fructose, and mannose.

The negative feedback relationship between blood osmotic pressure and the rate of food transport (Fig. 158) theoretically could form the basis of a homeostatic mechanism in which blood osmotic pressure, food transport through the foregut, taste threshold, and hence food intake are maintained at a relatively constant level. The nutritive value of the sugar plays no role in this mechanism nor in determining daily in-

Table 19. Fluid Intake, Preference, and Life Span of Flies Offered Various Carbohydrate Solutions.

Solution	No. flies	Average life-span (days)	Spread (days)	Preference	Total 15-day intake (μl)
1.0 M sorbitol	5	24	20–28	—	480
1.0 M sorbitol versus water	3	15	13–16	none	400
1.0 M sorbitol versus 0.1 M D-arabinose	6	8	3–10	slight arabinose	—
1.0 M sorbitol + 0.1 M D-arabinose	7	19	7–21	—	410
1.0 M sorbitol versus 1.0 M D-arabinose	9	7	2–13	arabinose	—
1.0 M sorbitol + 1.0 M D-arabinose	10	4	2–8	—	—
0.1 M D-arabinose versus water	4	4	2–5	arabinose	—
1.0 M D-arabinose versus water	4	2	—	arabinose	—
1.0 M sorbitol versus 0.1 M fucose	6	13	5–15	fucose	560
1.0 M sorbitol + 0.1 M fucose	6	14	9–18	—	460
1.0 M fucose versus water	6	3	1–4	—	80
0.1 M fucose versus water	7	4	1–4	fucose	180
0.1 M sucrose + 1.0 M NaCl versus water	6	4	3–4	water	—
0.1 M fucose + 1.0 M NaCl versus 1.0 M sorbitol	2	15	14–15	sorbitol	200

From Gelperin and Dethier, 1967.

take. The important parameters are stimulating power and concentration. The first determines the magnitude of excitation via the external chemoreceptors; the second, the rate at which solutions move through the gut and hence the magnitude of internal inhibition. The resultant of these two opposing inputs determines daily intake.

The proposed mechanism accounts for most of the observed feeding behavior of the fly (Dethier, 1969; Gelperin, 1972). Dilute solutions are ingested in greater quantity (not at a single drink) than concentrated solutions because the dilute solutions empty from the crop rapidly, causing threshold to fall rapidly. Excessive activity that creates great energy demands—for example, flight—causes the crop to empty even more rapidly. Flies do not prefer a nutritious solution over a nonnutritious one because foregut and body wall receptors respond only to mechanical stimulation and crop emptying is regulated only by osmotic pressure (of the blood).

The sugar preference-aversion curves characterizing the fly's behavior can also be understood in terms of this scheme. It will be recalled that the volume of sugar imbibed, for example, sucrose, increases steadily from 10^{-8} M to 0.01 M and then decreases. In the range 0.01–0.1 M the rate of crop emptying is at its maximum and hence the inhibitory effect internally is at a minimum. Above 0.1 M the crop empties at a slower rate; therefore, inhibition is greater. From 10^{-8} M to 0.01 M stimulating effectiveness is increasing while internal inhibitory effect is more or less constant; hence intake steadily increases. Above 0.01 M stimulating power continues to increase but inhibition probably increases more quickly. The net result is a decline in daily intake. At higher concentrations the additional impending effect of viscosity also begins to be influential.

The most abundant source of carbohydrates in nature is nectar. It accounts for over 90% of the available sugar. The principal sugars found in nectar are maltose, sucrose, fructose, and glucose (Beutler, 1953), all of which are both stimulating and highly nutritious to the fly. Hence, even though the mechanisms regulating the ingestion of food are not dependent upon caloric value for their operation, the fly in the wild is assured of an adequate caloric diet.

For only one other organism in the animal kingdom do data of comparable completeness exist relating to the regulation of feeding. This is the white rat. A comparison of fly and rat as to the mechanisms each has evolved to control food intake is instructive. Both animals maintain a relatively constant intake when tested for several days at a single level of nutrient density. In the face of dilution both animals increase their intake, the rat maintaining constancy more precisely than the fly. The rat is not usually tested over a range of nutrient dilutions covering four orders of magnitude, as has been done with the fly. If the fly's behavior is observed over a narrow range of dilutions, the precision of regulation seems greater. Before experimental dissection, then, the precision of the behavioral regulation of food intake by the rat does not appear greatly different from that by the fly.

In contrast to the fly, the rat appears to measure some characteristic of its food that is closely related to metabolic value. According to the glucostatic theory (Mayer, 1955) receptors in the hypothalamus monitor the level of utilization of glucose and stimulate or suppress feeding accordingly. The thermostatic theory (Brobeck, 1960) postulates that higher animals eat to keep warm. The extra heat released in the assimilation of food, called "the specific dynamic action," is sensed by the hypothalamus and results in the observed regulation of

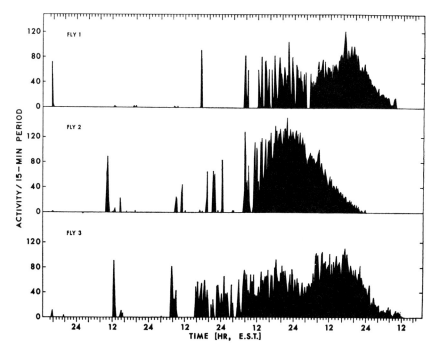

Fig. 162. Locomotor activity patterns of flies exposed to constant light from emergence to starvation death. Fifteen-minute activity intervals have been used to demonstrate intermittent periods of activity early in the deprivation schedule (from Green, 1964a).

ingestion. The ability of food to contribute to either blood glucose or bodily heat is closely related to metabolic value.

There remains one final major aspect of behavior that fluctuates as an animal alternates between states of food deprivation and satiation. It is general activity and, more specifically, the shift from appetitive behavior to quiescence following consummatory behavior—in this instance, eating. Simple observation shows that a satiated fly is less active than a deprived one. Experimental analyses have shed some light on the nature of the mechanisms involved but have not yet yielded as complete a picture as that we have for the regulation of feeding. Green (1964a), after having established base lines for activity under constant conditions, investigated the relation between locomotion and feeding (Green, 1964b). First, by measuring activity at intervals of 15 minutes (Fig. 162) he found that activity occurred in bursts punctuated by inactivity. As food deprivation increased, activity increased. This could mean that the fly moved faster, more often, or both. By comparing the speed of movement with time and the percent of time inactive (Figs. 163 and 164) he was able to show that the speed remained almost constant but the percent of time spent in activity changed. As deprivation increased the rest periods became less frequent and

Avoiding the Temptation of Gluttony

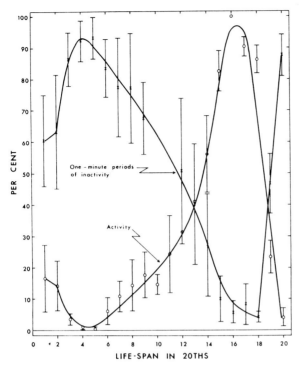

Fig. 163. The effect of one-minute periods of inactivity on the activity patterns of flies fed to repletion on 0.5 M sucrose containing Hypaque sodium and exposed to constant light (from Green, 1964a).

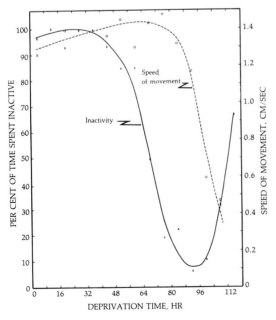

Fig. 164. Changes in the speed of movement and proportion of time spent in locomotor inactivity with deprivation time in adults of *Phormia*. Plotted points are mean for three flies (from Green, 1964a).

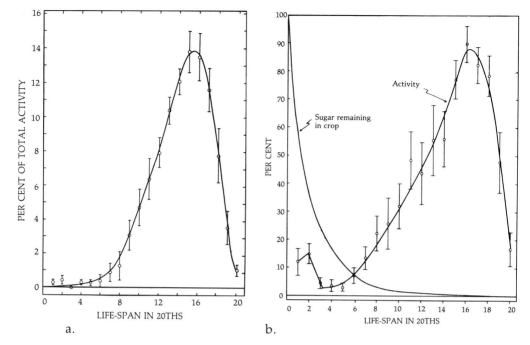

Fig. 165. (a) The relationship between activity and deprivation time (based on life span) in flies exposed to constant light from emergence to starvation death. Plotted points represent the means and standard errors for twelve individual flies. Mean life-span unit equals 5.5 ± 0.0880 hours. (b) The relationship between locomotor activity and the rate of crop emptying in flies fed to repletion on 0.5 M sucrose containing Hypaque sodium and exposed to constant light (from Green, 1964a).

shorter. The tsetse fly *Glossina morsitans morsitans* West behaves the same way, and the short bursts correspond to short field flights (Brady, 1972).

The relation between activity and deprivation is best seen by plotting activity against life-span rather than absolute time (Fig. 165). Life-span was determined by subdividing the length of life into 20 equal parts and converting the activity observed in these parts to a percent of total life activity. Viewed in this manner, the first third of life from emergence to starvation is characterized by little activity. Activity peaks at about three-fourths life-span and then decreases until death. Apparently there is no age effect because new, that is, young, flies, compared with flies that were 7 days old and kept alive by feeding, exhibited the same pattern of activity. Increases in activity, therefore, are clearly related to food deprivation.

That the relationship is a subtle one is suggested by the observation that a virgin female deprived of all food will satiate herself on carbohydrate when it is presented, but will then continue her activity until

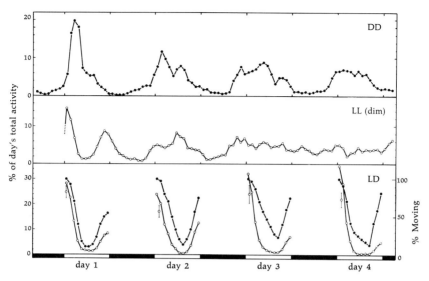

Fig. 166. Spontaneous locomotor activity of teneral male *Glossina morsitans* under constant conditions and different lighting regimes. Upper figure: mean activity of 19 flies in constant darkness with individual records synchronized. Middle figure: mean activity of 18 flies in constant light. Lower figure: mean activity of 10 flies in LD 12:12; open circles (also LL and DD curves) are the mean hourly amounts expressed as a percentage of each 24 hours of total activity (left-hand ordinate); closed circles (right-hand ordinate) are the proportion of flies moving in any one hour; all curves are smoothed by three-point sliding means; isolated open circles are the true mean of the second hour of each day (±S.E.). Abscissa shows the real light cycle of LD animals and the subjective light cycle of DD and LL animals during four days' isolation in actographs (from Brady, 1972).

her "need" for protein is satisfied. The phenomenon is also illustrated by a comparison of the patterns of activity of starving tsetse flies in different physiological states (Fig. 166). The course of increasing activity is the same with males and virgin females, but teneral (newly emerged, unfed) males and pregnant females display radically different patterns (Brady, 1972). This observation is particularly interesting in view of the point made earlier that an animal may cease appetitive behavior with respect to one goal, only to replace it with appetitive behavior associated with another goal.

Feeding inhibits activity. The rate at which activity is resumed decreases as the concentration of fed sugar is increased (Fig. 167). The following factors were eliminated by Evans and Barton Browne (1960) as modifiers of locomotor activity: increase in weight after a meal, nutritional state, concentration of sugar in the blood, and level of potassium in the blood. Green (1964b) then eliminated age, stretch receptors in the abdomen, crop, and crop duct, limitations on oxygen reaching the thorax and legs from abdominal air sacs as a consequence of crop

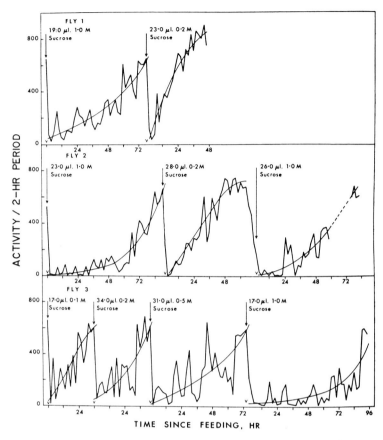

Fig. 167. The effects of feeding to repletion on different concentrations of sucrose on the locomotor activity of flies exposed to constant light (from Green, 1964a).

enlargement, and constant stimulation of oral receptors during regurgitation.

Although sugar and potassium levels in the blood are not involved in the means whereby feeding affects locomotion by blowflies, some blood constituent does seem to be. Green performed the ingenious experiment in which two starved flies were placed in parabiosis, one was then fed, the two were later separated, and each was placed in an actograph to monitor activity. Unexpectedly, both partners exhibited the arrest in activity characteristic of the state of satiety. The fly, like other insects, probably has a locomotor center that is under the control of an inhibitory center. Removal of the suboesophageal ganglion is known to result in akinesis, suggesting the presence here of an excitatory center. The animal can still walk briefly in a co-ordinated fashion if strongly stimulated; hence, the thoracic center by itself can pattern walking. Removal of the supraoesophageal ganglion promotes continuous locomotion. Green has suggested that spontaneous locomotor activity (and hence activity in the centers) is affected by a hormonal factor derived from the neurosecretory cells of the brain or the corpus cardiacum, or both, and that these change with the feeding state of the animal. It has been demonstrated that there are histological changes in the corpora allata associated with ingestion of carbohydrates, but these changes have not yet been correlated with locomotion (Strangways-Dixon, 1961).

With respect to the tsetse fly, it has been assumed that changes in locomotory activity of starving flies result from a monitoring of falling reserves, but no good correlation was found between spontaneous activity level and fat, water, or blood meal reserves (Brady, 1972). Consistently high correlations were found between activity and abdominal weight and total body weight. It has therefore been suggested that these flies modulate their behavior by some complex integration including the measurement of body weight or volume (Brady, 1973, 1974).

Thus, although the immediate mechanisms influencing locomotion appear to be different from those regulating threshold in *Phormia* and other insects studied, the two obviously work in harmony, ensuring that the fly is active and in a receptive state when in metabolic need, and quiet and indifferent to food when satiated.

Literature Cited

Anderson, A. L. 1932. The sensitivity of the legs of common butterflies to sugars. *J. Exp. Zool.* 63:235–259.

Anwyl, R. 1972. The structure and properties of an abdominal stretch receptor in *Rhodnius prolixus*. *J. Insect Physiol.* 18:2143–2153.

Bernays, E. A., Blaney, W. M., and Chapman, R. F. 1972. Changes in chemoreceptor sensilla in the maxillary palps of *Locusta migratoria* in relation to feeding. *J. Exp. Biol.* 57:745–753.

Bernays, E. A., and Chapman, R. F. 1972. The control of changes in the peripheral sensilla associated with feeding in *Locusta migratoria* (L.). *J. Exp. Biol.* 57:755–763.

Bernays, E. A., and Chapman, R. F. 1974a. The regulation of food intake by acridids. In *Experimental Analysis of Insect Behaviour,* L. Barton Browne, ed., pp. 48–59. Springer-Verlag, Berlin.

Bernays, E. A., and Chapman, R. F. 1974b. Changes in haemolymph osmotic pressure in *Locusta migratoria* larvae in relation to feeding. *J. Ent.* (A) 48:149–155.

Beutler, G. 1953. Nectar. *Bee World,* 34:106–116.

Bodenstein, D. 1950. The postembryonic development of *Drosophila*. In *Biology of Drosophilia*, M. Demerec, ed., pp. 275–367. John Wiley and Son, New York.

Bolwig, N. 1952. The hunger-reaction of flies (*Musca*) and the function of their stomatogastric system. *Nature* 169:197–198.

Brady, J. 1972. Spontaneous circadial components of tsetse fly activity. *J. Insect Physiol.* 18:471–484.

Brady, J. 1973. The physiology and behaviour of starving tsetse flies. *Trans. Roy. Soc. Trop. Med. Hyg.* 67:297.

Brady, J. 1974. The physiology of "hunger" in tsetse flies. *Trans. Roy. Soc. Trop. Med. Hyg.* 68:159.

Brobeck, J. R. 1960. Food and temperature. *Recent Prog. Hormone Res.* 16:439–466.

Clarke, U. K., and Langley, P. 1962. Factors concerned in the initiation of growth and moulting in *Locusta migratoria* L. *Nature,* 194:160–162.

Cottrell, C. B. 1962. The imaginal ecdysis of blowflies. *Trans. Roy. Ent. Soc. London* 114:317–333.

Davey, K. G., and Treherne, J. E. 1963a. Studies on crop function in the cockroach (*Periplaneta americana* L.), I. The mechanism of crop emptying. *J. Exp. Biol.* 40:763–773.

Davey, K. G., and Treherne, J. E. 1963b. Studies on crop function in the cockroach (*Periplaneta americana* L.). II. The nervous control of crop emptying. *J. Exp. Biol.* 40:775–780.

Davey, K. G., and Treherne, J. E. 1964. Studies on crop function in the cockroach (*Periplaneta americana* L.). III. Pressure changes during feeding and crop emptying. *J. Exp. Biol.* 41:513–524.

Dethier, V. G. 1952. Adaptation to chemical stimulation of the tarsal receptors of the blowfly. *Biol. Bull.* 103:178–189.

Dethier, V. G. 1955. The physiology and histology of the contact chemoreceptors of the blowfly. *Quart. Rev. Biol.* 30:348–371.

Dethier, V. G. 1959. The nerves and muscles of the proboscis of the blowfly *Phormia regina* Meigen in relation to feeding responses. In *Studies in Invertebrate Morphology,* Smithson. Misc. Coll 137:1138–1145.

Dethier, V. G. 1966. Insects and the concept of motivation. In *Nebraska Symposium on Motivation, 1966,* D. Levine, ed., pp. 105–136. University of Nebraska Press, Lincoln.

Dethier, V. G., and Bodenstein, D. 1958. Hunger in the blowfly. *Zeit. f. Tierpsychologie* 15:129–140.

Dethier, V. G., and Chadwick, L. E. 1948. Chemoreception in insects. *Physiol. Rev.* 28:220–254.

Dethier, V. G., and Evans, D. R. 1961. Physiological control of water ingestion in the blowfly. *Biol. Bull.* 121:108–116.

Dethier, V. G., Evans, D. R., and Rhoades, M. V. 1956. Some factors controlling the ingestion of carbohydrates by the blowfly. *Biol. Bull.* 111:204–222.

Dethier, V. G., and Gelperin, A. 1967. Hyperphagia in the blowfly. *J. Exp. Biol.* 47:191–200.

Dethier, V. G., and Hanson, F. E. 1965. Taste papillae of the blowfly. *J. Cell. Comp. Physiol.* 65:93–100.

Dethier, V. G., and Rhoades, M. V. 1954. Sugar preference aversion functions for the blowfly. *J. Exp. Zool.* 126:177–204.

Englemann, F. 1968a. Endocrine control of reproduction in insects. *Annu. Rev. Ent.* 13:1–26.

Englemann, F. 1968b. Feeding and crop emptying in the cockroach *Leucophaea maderae. J. Insect Physiol.* 14:1525–1531.

Evans, D. R., and Barton Browne, L. 1960. Physiology of hunger in the blowfly. *Amer. Midland Nat.* 64:282–300.

Evans, D. R., and Dethier, V. G. 1957. The regulation of taste thresholds for sugars in the blowfly. *J. Insect Physiol.* 1:3–17.

Evans, D. R., and Mellon, De F. 1962. Electrophysiological studies of a water receptor associated with the taste sensilla of the blowfly. *J. Gen. Physiol.* 45:487–500.

von Frisch, K. 1935. Über den Geschmackssinn der Biene. *Zeit. vergl. Physiol.* 21:1–156.

von Frisch, K. 1950. *Bees: Their Vision, Chemical Senses, and Language.* Cornell University Press, Ithaca, N.Y.

Gelperin, A. 1966a. Investigations of a foregut receptor essential to taste threshold regulation in the blowfly. *J. Insect Physiol.* 12:829–841.

Gelperin, A. 1966b. Control of crop emptying in the blowfly. *J. Insect. Physiol.* 12:331–345.

Gelperin, A. 1967. Stretch receptors in the foregut of the blowfly. *Science,* 157:208–210.

Gelperin, A. 1971a. Regulation of feeding. *Annu. Rev. Ent.* 16:365–378.

Gelperin, A. 1971b. Abdominal sensory neurons providing negative feedback to the feeding behavior of the blowfly. *Zeit. vergl. Physiol.* 72:17–31.

Gelperin, A. 1972. Neural control systems underlying insect feeding behavior. *Amer. Zool.* 12:489–496.

Gelperin, A., and Dethier, V. G. 1967. Long-term regulation of sugar intake by the blowfly. *Physiol. Zool.* 40:218–228.

Getting, P. A., and Steinhardt, R. A. 1972. The interaction of external and internal receptors on the feeding behaviour of the blowfly *Phormia regina*. *J. Insect Physiol.* 18:1673–1681.

Gillary, H. L. 1966a. Stimulation of the salt receptor of the blowfly. I. NaCl. *J. Gen. Physiol.* 50:337–350.

Gillary, H. L. 1966b. Stimulation of the salt receptor of the blowfly. II. Temperature. *J. Gen. Physiol.* 50:351–357.

Gordon, H. T. 1968. Intake rates of various solid carbohydrates by male German cockroaches. *J. Insect Physiol.* 14:41–52.

Green, G. W. 1964a. The control of spontaneous locomotor activity in *Phormia regina* Meigen. I. Locomotor activity patterns of intact flies. *J. Insect Physiol.* 10:711–726.

Green, G. W. 1964b. The control of spontaneous locomotor activity patterns in *Phormia regina* Meigen. II. Experiments to determine the mechanism involved. *J. Insect Physiol.* 10:727–752.

Gwadz, R. W. 1969. Regulation of blood meal size in the mosquito. *J. Insect Physiol.* 15:2039–2044.

Haslinger, F. 1935. Über den Geschmackssinn von *Calliphora erythrocephala* Meigen und über die Verwertung von Zuckern und Zuckeralkoholen durch diese Fliege. *Zeit. vergl. Physiol.* 22:614–639.

Hassett, C. C., Dethier, V. G., and Gans, J. 1950. A comparison of nutritive values and taste thresholds of carbohydrates for the blowfly. *Biol. Bull.* 99:446–453.

Highnam, K. C., Hill, L., and Mordue, W. 1966. The endocrine system and oocyte growth in *Schistocerca* in relation to starvation and frontalganglionectomy. *J. Insect Physiol.* 12:977–994.

Holling, C. S. 1966. The functional response of invertebrate predators to prey density. *Mem. Ent. Soc. Canada,* No. 48.

House, H. L. 1965. Effects of low levels of the nutrient content of a food and of nutrient imbalance on the feeding and the nutrition of a phytophagous larva, *Celerio euphorbae*. *Canad. Ent.* 97:62–68.

Hudson, A. 1958. The effect of flight on the taste threshold and carbohydrate utilization of *Phormia regina* Meigen. *J. Insect Physiol.* 1:293–304.

Knight, M. R. 1962. Rhythmic activities of the alimentary canal of the black blowfly *Phormia regina. Ann. Ent. Soc. Amer.* 55:380–382.

Künze, G. 1927. Einige Versuche über den Geschmackssin der Honigbiene. *Zool. Jahrb. Physiol.* 44:287–314.

Livingston, R. B. 1959. Central control of receptors and sensory transmission systems. In *Handbook of Physiology,* Section I, Vol. 1, H. W. Magoun, ed., pp. 741–760. *Amer. Physiol. Soc.* Washington, D.C.

McCutchan, M. C., and Gelperin, A. 1969. Unpublished observations. Cited in Gelperin, 1971a.

Maddrell, S. H. P. 1963. Control of ingestion in *Rhodnius prolixus. Nature,* 198:210.

Mayer, J. 1955. Regulation of energy intake and body weight. *Ann. N.Y. Acad. Sci.* 63:15–43.

Minnich, D. E. 1922. A quantitative study of tarsal sensitivity to solutions of saccharose in the red admiral butterfly, *Pyrameis atalanta* L. *J. Exp. Zool.* 36:445–457.

Minnich, D. E. 1929. The chemical sensitivity of the legs of the blowfly, *Calliphora vomitoria* Linn., to various sugars. *Zeit. vergl. Physiol.* 11:1–55.

Minnich, D. E. 1931. The sensitivity of the oral lobes of the proboscis of the blowfly *Calliphora vomitoria* Linn. to various sugars. *J. Exp. Zool.* 60:121–139.

Núñez, J. A. 1964. Trinktriebregelung bei Insekten. *Naturwiss.* 17:419.

Núñez, J. A. 1966. Quantitative Beziehungen zwischen den Eigenschaften von Futterquellen und dem Verhalten der Sammelbienen. *Zeit. vergl. Physiol.* 53:142–164.

Núñez, J. A. 1970. The relationship between sugar flow and foraging and recruiting behaviour of honey bees (*Apis mellifera* L.). *Animal Behaviour.* 18:527–538.

Omand, E. 1971. A peripheral sensory basis for behavioral regulation. *Comp. Biochem. Physiol.* A38:265–278.

Rees, C. J. C. 1968. The effect of aqueous solutions of some 1:1 electrolytes on the electrical response of the type 1 ("salt") chemoreceptor cell in the labella of *Phormia. J. Insect Physiol.* 14:1331–1364.

Schoonhoven, L. M. 1967. Loss of hostplant specificity by *Manduca sexta* after rearing on an artificial diet. *Ent. Exp. Appl.* 10:270–272.

Strangways-Dixon, J. 1961. The relation between nutrition, hormones and reproduction in the blowfly *Calliphora erythrocephala* (Meigen). I. *J. Exp. Biol.* 38:225–235.

Thomson, A. J., and Holling, C. S. 1974. Experimental component analysis of blowfly feeding behavior. *J. Insect Physiol.* 20:1533–1563.

Treherne, J. E. 1957. Glucose absorption in the cockroach. *J. Exp. Biol.* 34:478–485.

Van der Kloot, W. G. 1960. Neurosecretion in insects. *Annu. Rev. Ent.* 5:35–52.

Van der Kloot, W. G. 1961. Insect metamorphosis and its endocrine control. *Amer. Zool.* 1:3–9.

Wells, P. H., and Giacchino, J. 1968. Relationship between the volume and the sugar concentration of loads carried by honey bees. *J. Apic. Res.* 7:177–182.

Wigglesworth, V. G. 1949. The utilization of reserve substances in Drosophila during flight. *J. Exp. Biol.* 26:150–163.

Wyatt, G. R., and Kalf, G. F. 1956. Trehalose in insects. *Proceedings, Fed. Amer. Soc. Exp. Biol.* 15:388.

9
Food For the Next Generation: Specific Hungers

One generation passeth away, and another generation cometh.

Ecclesiastes 1:4

A human being grows and develops continuously until he dies. Beginning with birth he increases in weight, stature, and morphological complexity until the end of adolescence. From that time on he grows and develops in the sense that cells are constantly replacing those that are dying—notably blood cells, cells in the taste buds, gonadal cells, uterine cells, and epidermal cells. Until death he is still capable of regenerating tissue to heal wounds and knit broken bones. Until death he still has to cut his hair and trim his nails. For the replacement of cells, the regeneration of tissue, and other purposes he relies upon a balanced diet, the balancing being proper proportions of carbohydrate, fat, protein, vitamins, salts, and trace elements. The fly, in startling contrast, does not grow in size, weight, or complexity once it has emerged from the pupal stage. Wound healing involves only clotting; there is no associated tissue regeneration. There is continual amino acid turnover as proteins in the body are dismantled and reassembled, as saliva, intestinal enzymes, and others are expended and replaced, as the structural volume and crytochrome content of flight muscle sarcomeres increase (Rockstein, 1956; Levenbook and Williams, 1956; Herold, 1965; Brosemer, Vogell and Bucher, 1963; Lennie and Birt, 1967; Balboni, 1968), and as biosynthesis of mitochondrial proteins goes on in the fat body (Larsen, 1970; Keeley, 1972); but little that is new is added to the fly. The attrition of time and wear and tear leaves its marks so that the fly becomes ever more tattered and woebegone as it ages. The constructive work that it accomplishes is contributed by the gonads. For this work an unrelieved diet of carbohydrate will not suffice. The female, especially when eggs must be produced, packed with yolk, and enveloped in protective membranes, has insufficient parental protein for completion of the task. An exogenous source of protein must be found.

Not all adult insects must supplement their diet with protein in order to produce eggs (Blake, 1961; Davies and Peterson, 1956; Robbins and Shortino, 1962). The giant silk moths that lack functional mouthparts have already been mentioned. Among mosquitoes are to be found all degrees of need for protein. Within a single species, *Culex pipiens,* for example, there are autogenous strains, that is, those that do not require a blood meal in order to develop eggs and those that do. In nature this case is not strictly one of requiring or not requiring exogenous protein because the nectar comprising the usual food does contain some nitrogenous compounds. On the other hand, in the laboratory autogenous mosquitoes restricted to a diet of pure carbohydrate are able to develop eggs by tapping protein sources in their own bodies. Of the insects that require protein, some need it only to trigger the hormonal secretion required for egg development because there are enough protein reserves to suffice for at least one clutch of eggs (Braken and Nair, 1967; Johansson, 1955, 1958; Larsen and Bodenstein, 1959). Other anautogenous insects need the protein not only to initiate hormonal secretion but also to supply raw material for synthesis of yolk. The blowfly falls into this category.

The intriguing aspect of protein ingestion by many insects is its restriction to time of need. In flies this manifests itself first in the orientation stages of appetitive behavior and, after food is found, in the feeding preferences exhibited. Many saprophilic flies orient positively to the odors of such compounds as skatol, which are associated with decaying proteinaceous material, only during periods of egg development (Pospíšil, 1958). This is true even when the flies have already fed to repletion on carbohydrate. At other times in the reproductive cycle flies satiated with carbohydrate are unresponsive to the odors.

In choice situations, as when food is presented in "two-bottle" experiments, the flies again show a preference for protein, correlated with particular stages in ovarian development (Strangways-Dixon, 1959, 1961a, 1961b; Dethier, 1961; Belzer, 1970). Since flies are not able to survive longer than 4 days at the most on a diet of protein alone, it was necessary in these experiments to provide a source of carbohydrate. Strangways-Dixon in his studies with *Calliphora* mixed the two materials and measured volumes ingested under different circumstances on a comparative basis. In the studies with *Phormia* the solutions were not mixed; absolute measures of ingestion were made. All three studies clearly showed a relation between the ingestion of protein and the sex and reproductive condition of the fly. Males, whether mated or not, gradually increased their consumption of protein from the time of emergence from the puparium until the 4th to 8th day of life (Fig. 168). Thereafter the intake reached a low level that was maintained until death. The pattern was similar for virgin females (Fig.

Food For the Next Generation: Specific Hungers

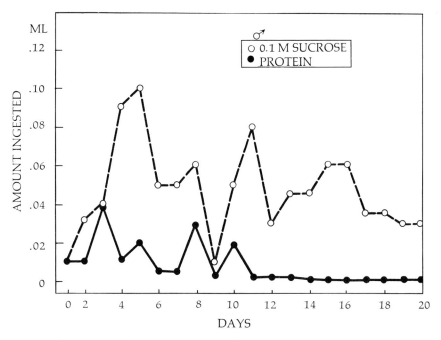

Fig. 168. Daily intake of 0.1 M sucrose and brain-heart extract by a male blowfly in a two-choice situation (from Dethier, 1961).

169), but the volume of protein consumed was greater. Mated females showed a similar pattern during the first 4 to 8 days, but laid a clutch of eggs sometime between the 10th and 15th day and within 24 hours again increased their protein consumption (Fig. 170). Following each successive oviposition there was a renewed bout of eating of protein. The quantity consumed daily by flies that had not laid eggs was always greater than that consumed by flies at any other period after they had laid their first clutch of eggs. This pattern of ingestion was similar to that observed with *Calliphora*. The maximum increase in ingestion of protein coincided with the times during which eggs were growing. Increased ingestion of protein is accomplished, not by more drinks, but by longer drinks (Belzer, 1970).

During these cycles of ingestion of protein there is cyclic consumption of carbohydrate. Strangways-Dixon reported that the ingestion of carbohydrate is practically constant except during periods of yolk depposition, when it is high. This is in general true except for the first day, when there is an initial rapid rise in intake by both males and females, and a slow decline occurs with age. These two features were not noted with *Calliphora* because those flies were on an ad libitum diet of sugar for 6 days preceding experimental measurement (thus the initial use was missed). The slow decline was not evident because the experiments lasted only 14 days.

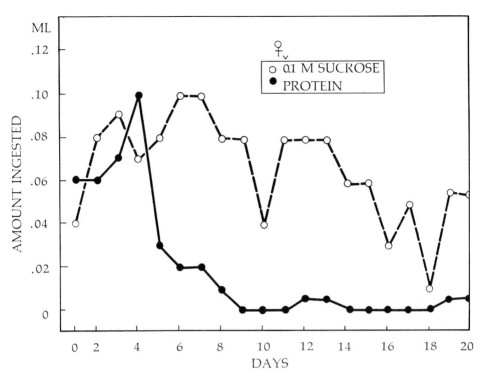

Fig. 169. Daily intake of 0.1 M sucrose and brain-heart extract by a virgin female blowfly in a two-choice situation (from Dethier, 1961).

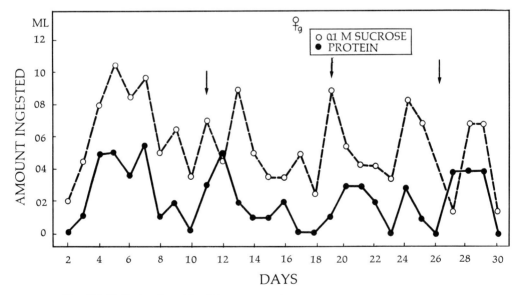

Fig. 170. Daily intake of 0.1 M sucrose and brain-heart extract by a gravid female blowfly in a two-choice situation. Arrows indicate days on which eggs were laid (from Dethier, 1961).

Strangways-Dixon concluded that the cyclic nature of consumption of carbohydrate is an independent phenomenon and is regulated by a specific mechanism. He thought that the increase in ingestion of sugar during periods of yolk deposition is dependent upon the presence of the corpus allatum, and he reported a consistently low intake of carbohydrate after allatectomy. This did not occur with *Phormia* nor was there any increase after ovariectomy. The fact that the corpus allatum increases in size during early stages of egg development and decreases in volume during the deposition of yolk (Strangways-Dixon, 1959) does not necessarily support the hypothesis that the activity of this gland directly influences the consumption of carbohydrate. The multifarious influences of the corpus allatum constitute another reason for caution in assigning to it a direct regulatory role in the ingestion of carbohydrate. If the consumption of carbohydrate does indeed rise after ovariectomy, and fall after allatectomy, it does follow, as Strangways-Dixon asserts, that the change cannot result simply from the failure of egg development per se. On the other hand, it is not impossible that the effect of the corpus allatum on ingestion, if indeed there is one, could be derived from the influence that this gland has on respiration, general activity, and so on. Dethier (1969) argued that it is a side effect of cyclic ingestion of protein. If, as stated, the total intake of fluids by *Calliphora* is constant, any decrease in the ingestion of protein is bound to be compensated for in a choice situation by an increased ingestion of carbohydrate, and vice versa. This interpretation is supported by data obtained by Belzer.

The cyclic nature of ingestion of protein and especially its timing in relation to need is one example of a widespread behavioral phenomenon comprehended under the term "specific hunger." Specific hunger refers to a restricted diet selection, usually temporary, related to a particular deficit. In one respect specific hungers are merely special cases of the general phenomenon of eating when deprived; however, the precision of the discrimination and selection in relation to very particular needs is spectacular and points to the existence of intricate metabolic influences on neural control of behavior. Among mammals, for example, a hunger for carbohydrate develops when violent and prolonged exercise causes a drop in the level of blood sugar. After extreme sweating that causes excessive loss of salt an acute hunger for salt develops (Denton, 1967). Calcium deficiency creates a demand for calcium. Thiamin deficiency creates a demand for thiamin (Rozin, 1967). Pregnancy creates other specific demands (Richter and Barelare, 1938). How these behavioral changes come about is still a great mystery. Adding to the mystery is the knowledge that some demands are anticipated before they come into being. One recalls to mind the gluttonous behavior of hibernating animals that gorge to excess in the absence of any immediate metabolic need. Cases like this suggest that some event or imbalance antecedent to the ultimate need triggers the change in behavior that eventually satisfies the ultimate need.

Ingestion of protein by blowflies fully exemplifies a specific hunger because it relates quantitatively as well as qualitatively to a deprivation. If flies have free access both to protein and to carbohydrate at all times from the day of emergence, carbohydrate is nearly always taken in greater volume than protein. If, however, the flies are denied access to protein and maintained on a minimal (0.001 M) carbohydrate diet for the first 5 days of adult life, the subsequent pattern of carbohydrate and protein intake is quite different. In the case of males the intake of protein remains very low, as before, but consumption of sugar is very high the first 2 days (Fig. 171) (some males in Belzer's experiments behaved as did females). Virgin females by contrast consume a great volume of protein on the first 2 days; the consumption usually exceeds that of carbohydrate (Fig. 172). If females are denied protein for 10 days, the subsequent preference is even more marked (Belzer found no additional increase after 2.5 days). After 20 days of deprivation of protein a preference for it still exists but is no more pronounced than it was on the 10th day. It is also significant that the first drink of protein by protein-deprived flies is correspondingly longer as the period of protein deprivation lengthens (Belzer, 1970).

Observing the behavior of females deprived of protein one obtains the impression that there is a distinct "hunger" for protein. Their reaction is particularly interesting because protein by itself is inadequate for survival. Thus during these periods female flies eschew a nutritionally adequate diet (carbohydrate) in favor of one that meets their reproductive needs. From an evolutionary point of view, reproduction of the species takes precedence over survival of the individual. For these many reasons the study of specific hungers is an especially intriguing one.

The occurrence of a specific hunger implies that an animal can differentiate between the particular food and other food, that it prefers that food over others at that time, and that unusual events trigger and terminate its appetitive behavior. How these changes in behavior are brought about is poorly understood.

An alternative hypothesis to the implication that the particular required food is recognized for itself has been proposed by Rozin (1967) to explain how thiamin-deprived rats in the laboratory select a remedial diet containing thiamin. According to this hypothesis the deficient rats select any *novel* food that is available. When in the course of this selection they happen to eat food containing thiamin, they experience a feeling of well-being some time later and come to associate well-being with the particular novel food most recently eaten.

Food For the Next Generation: Specific Hungers

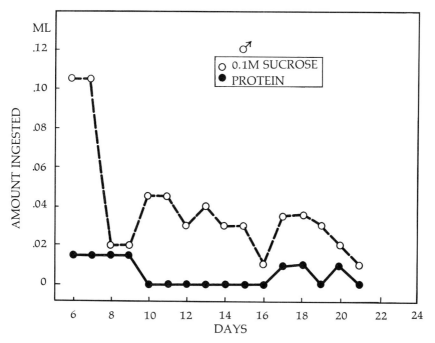

Fig. 171. Daily intake of 0.1 M sucrose and brain-heart extract by a male blowfly in a two choice situation when it has been deprived of protein for five days and maintained on a low carbohydrate (0.0001 M sucrose) diet (from Dethier, 1961).

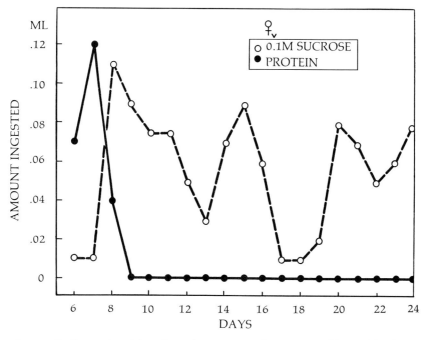

Fig. 172. Daily intake of female under same conditions as in Figure 171 (from Dethier, 1961).

Thus, if a food supplemented with thiamin also contains an identifying flavor, as for example, peppermint, the rats come to associate health with peppermint. Peppermint may be thought of as a token stimulus. This kind of association necessitates an ability to learn. Learning ability of this sort is known to occur in *Phormia*.

Irrespective of whether learning is involved or not, a capacity for sensory discrimination between proteinaceous and nonproteinaceous food is a sine qua non. As indicated earlier, *Phormia* does not possess specific protein receptors, but probably distinguishes protein from carbohydrate (and also from salt solutions lacking protein) by the total gustatory sensory pattern. The fly is quite able to make the choice on the basis of gustatory information alone. To demonstrate this, flies from which the olfactory receptors had been removed were placed on a sheet of nonabsorbent hydrophobic paper, and three concentric rings of solution were drawn around them with a camel's-hair brush. In one set of tests the rings from center to outside were in the order water, protein, sucrose; in another set, water, sucrose, protein. Three kinds of flies were tested: 5-day-old males starved 24 hours and previously maintained an 0.1 M sucrose; 5-day-old virgin females with a similar history; 5-day-old virgin females starved 24 hours and previously maintained on protein solution.

All flies upon encountering water stopped and drank to repletion; thereupon the ring of water became a barrier. Each time a leg touched the water the fly turned away from the ring, whereas it had previously turned toward it. This change in response was in itself interesting because it showed that a solution that was initially acceptable had truly become a deterrent. In tests in which the water had been absorbed into the paper, thus presenting no raised surface, flies, after sucking to repletion, merely walked across the damp surface instead of being repelled. The difference in behavior in the two cases may be explained if we assume that the original ring of water stimulated two sets of receptors, the water receptors and the mechanoreceptors (the effect of surface tension), the former mediating acceptance, the latter rejection. In the thirsty fly the acceptable stimulus overrides the unacceptable one. This balancing of antagonistic stimuli acting on tarsi, and the change in effectiveness with change in internal state have already been discussed.

Having drunk water, the flies turned away from it. Upon each new encounter they avoided it. After a few minutes, however, their behavior changed. They now waded through and continued until they reached the next ring of solution. When 5-day-old females that had been maintained since emergence on a protein-free diet encountered the ring of protein first, they drank the solution avidly, then turned away from it, and then followed one of three patterns upon encountering 0.1 M sucrose: (1) drank some 0.1 M sucrose, then ignored it, but would drink 1.0 M sucrose when it was presented; (2) ignored 0.1 M sucrose, drank 1.0 M; (3) ignored all sucrose. If they encountered the ring of sucrose first, they fed fully on sugar, and thereafter drank protein while repeatedly ignoring sucrose. If, however, after drinking protein they were offered 1.0 M sucrose, they invariably drank it.

Evans and Barton Browne (1960) reported, on the one hand, that flies that had fed to repletion on whole liver subsequently had glucose thresholds as high as though they had fed on 2 M glucose: on the other hand, they stated that flies "which had been fully adapted to liver" will respond to 1 M sucrose. The reverse is not true. There appears to be a contradiction here unless "fully adapted" means peripheral (sensory) adaptation instead of satiation. In any case, it is difficult to assess the meaning of these data since no mention is made of the sex, age, or reproductive state of the flies.

That flies can detect the difference between proteins and carbohydrates before ingesting them, and in the absence of olfactory information, is firmly established.

Additional supporting evidence for the critical role played by gustatory stimulation was provided by preference studies in which protein was paired with different concentrations of carbohydrate (Dethier, 1961). When the concentration of available sugar was high (1 M), the intake of protein was greatly reduced (Fig. 173); when the concentration was low (0.0001–0.001 M), the intake of protein was enhanced (Fig. 174). It is clear, therefore, that the relation between the ingestion of protein and that of carbohydrate is not rigid. There is a strict dependence on the stimulating effect of the two substances on chemoreceptors. Changing the kind of protein also alters the volumes drunk. Homogenized liver is preferred to a 10% solution of Difco brain-heart infusion and both are preferred to crystalline bovine haemoglobin; yet each suffices for egg production, although haemoglobin is only marginally satisfactory.

Since the behavioral response to protein and carbohydrate varies concurrently with events in the reproductive cycle, it must be inferred that the sensory contribution varies. The two most likely alternatives as to the neural level at which changes occur are a peripheral level in the sense organs themselves or some intermediate level in the central nervous system. Either the relative sensory thresholds to protein and carbohydrate change periodically, or some changes occur in the central nervous system where the incoming sensory information is possessed. Little can be stated with certainty about the truth of the first alternative. By analogy with other sensory modalities in the fly, it is unlikely that changes occur here. In mammals intensive investigations have failed to uncover changes in the patterns of impulses in fibers of the

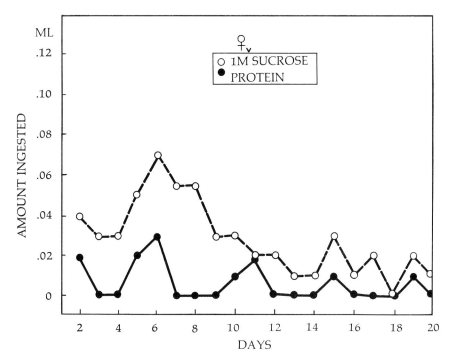

Fig. 173. Daily intake of 1 M sucrose and brain-heart extract by a virgin female blowfly in a two-choice situation (from Dethier, 1966).

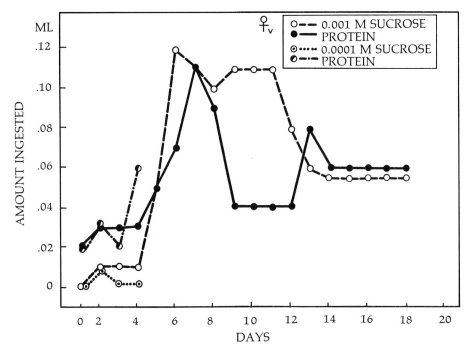

Fig. 174. Daily intake of dilute sucrose and brain-heart extract by a virgin female blowfly (from Dethier, 1961).

chorda tympani associated with a deficit of salt (Pfaffmann and Bare, 1950). Recent electronmicroscopical studies of the fine structure of mammalian taste buds hint at reafference; however, this system, if it in fact exists, has not been implicated in sensory behavior in deficit states. There is considerable evidence that the condition of the internal milieu may in fact influence taste buds, as indicated by the responsiveness of gustatory receptors to materials (e.g., carbohydrates)injected into the blood (Bradley, 1972). In insects there are some suggestive observations that hormones in the haemocoele may affect labellar receptors.

The alternative, that changes in behavior associated with deficits operate via mechanisms that alter the central nervous system's integration of sensory information arriving there, is more attractive. In any case, the behavioral changes that occur at different times in the reproductive cycle in relation to the ingestion of protein and carbohydrate can occur only if some internal variable becomes linked with the nervous system. The protein-deprived fly and the the protein-satiated fly are metabolically different in many respects. Marked differences are to be found in the ovaries, fat body, blood, endocrine, and neurosecretory systems (Thomsen and Thomsen, 1974). Respiratory rates are also different (Calabrese and Stoffolano, 1974). Any one or combination of these changes could affect the events in the nervous system that regulate specific feeding behavior. At the same time, some thought must be given to the fact that there may be protein demands of a nonreproductive nature. By refining the techniques of measuring ingestion Belzer (1970) was able to detect a biphasic character in the first feeding cycle. Most virgin females showed a slight decline in the ingestion of protein on the 3rd day after emergence. This was succeeded by a sharp rise from days 3 to 5 and then a marked decline (Fig. 175). This second peak, the "yolk" peak, was associated with vitellogenesis. The weight of protein consumed during this period, 2 to 2.5 mg, matched the weight of protein that Orr (1964b) had found in a pair of ovaries in a gravid female. Some variation exists in the relative magnitude of the two peaks: the first may be larger, equal to, or smaller than the second. In some few flies it is nonexistent. The first peak coincides in time with tanning, the breakdown of the larval fat body and the development of the adult fat body.

The biphasic nature of the protein intake, together with the fact that males show an initial protein preference, suggests that not all protein feeding is associated with egg development. Nonetheless, the major metabolic changes associated with maximum ingestion of protein are reproductive. It is necessary at this point, therefore, to describe the many complex changes that accompany and/or regulate changes in the reproductive system.

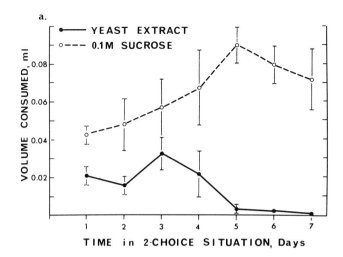
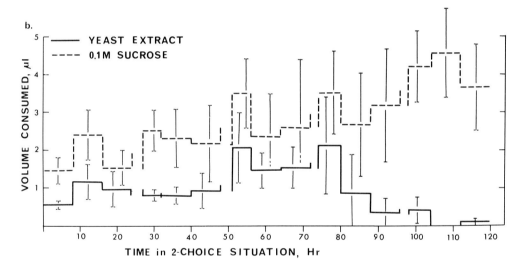

Fig. 175. Ingestion of 10% yeast extract versus 0.1 M sucrose by 4- to 6-hour-old females (newly emerged) in a two-choice situation. Each point is the average of seven flies. Vertical lines depict standard deviations. The decrease from day 1 to day 2 and the increase from day 2 to day 3 are statistically significant. (a) Data presented on a daily basis; (b) calculated hourly consumption for the first 120 hours (from Belzer, 1970).

The primary parts of the female reproductive system consist of paired ovaries, each of which is connected by a lateral oviduct to a common oviduct that opens posteriorly to the exterior at the gonopore. Each ovary is a bundle of tapering tubes, the ovarioles. Anteriorly they taper to a small filament and the filaments of the bundle are joined together terminally to form a suspensory ligament by which the ovary is attached to surrounding tissue. Each ovariole consists of a terminal filament, an egg tube, and a pedicel which is a duct connecting the ovariole to the lateral oviduct.

In the young ovary the egg tube consists of groups of germ cells, and the whole is called a germarium. They rapidly develop into oögonia and these into oöcytes, usually accompanied by nurse cells and follicle cells. As development proceeds two zones can be distinguished in the egg tube: the germarium proper and the zone of growth (the vitellarium). As the oöcytes multiply, continuously form the oögonia, and grow, the egg tube extends beyond the germarium to accommodate them. This extension, the vitellarium, becomes distended into a series of egg chambers or follicles. The first and developmentally most advanced egg lies at the posterior end of the tube. Other less advanced eggs are progressively smaller as they lie nearer the germarium. The extension of the tube is accomplished by rapid growth of follicular cells. The last, so-called "mature," egg is encased in a chorion but has not yet undergone maturation. Properly speaking it is still an oöcyte. It does not undergo its maturation divisions until after it has been inseminated. This event occurs just as the egg is about to leave the oviduct. Here it passes the opening of the spermathecal duct where the sperm was stored following copulation. The growth stage of the follicles is that period when yolk is accumulated, follicular cells proliferate, and a chorion is secreted. The accumulation of yolk, however, is the principal activity.

In a fly deprived of protein, the ovary contains a fully formed first egg chamber, but this remains in the germarium for the first 24 hours. It then descends from the germarium and thereafter remains at that stage. The second egg chamber appears in the ovariole on the 5th day. By the 12th day some degenerative changes appear in the nutritive cells. The fat body is completely imaginal by the 5th day.

In a fly that has ingested protein the first egg chamber is completely formed at age 6 hours and is in the germarium. In 24-hour-old flies the egg chamber is almost completely separated from the germarium and within the next 24 hours the second egg chamber descends. When the fly is 72 hours old, yolk appears in the oöcyte of the first egg chamber, and the fat body changes from a pupal to an imaginal one. A fully mature egg is present in the ovariole at age 120 hours. If copulation occurs, this egg is laid within the next 24–48 hours, and the second egg chamber develops rapidly. Unmated flies retain their eggs in the ovaries without any resorption, and the second egg chamber never develops.

The control of yolk formation and egg maturation by the corpus allatum, first demonstrated by Wigglesworth (1936) in *Rhodnius,* has since been confirmed in many other insects; for reviews consult Wigglesworth (1954), Bodenstein (1953), and Johansson (1958). Development does not take place in the absence of the corpus allatum. The medial neurosecretory cells are also involved. In *Calliphora* yolk deposition fails to take place in their absence (Thomsen, 1948, 1952). They in turn are probably controlled by the corpus allatum; see Lea and Thomsen (1962), and Harlow (1956) (*P. terraenovae*). Even in the presence of these organs development is arrested if no proteinaceous food is eaten. The ovaries never reach the stage of vitellogenesis, and the descent of the second egg chamber from the germarium is long delayed. Besides supplying raw material for vitellogenesis, proteins activate the neurosecretory system. In a long series of experiments involving allatectomy, by the method of Thomsen (1942); ovariectomy, by the method of Strangways-Dixon (1961b); blood analysis, by the method of Evans and Dethier 1957, and analyses of other tissues, Orr (1964a, 1964b) proved that the presence of the corpus allatum was required for a specific period of time if eggs were to develop and that activation of the corpus allatum required the ingestion of a specific amount of protein. In flies deprived of protein the concentration in the blood was less than 20 $\mu g/\mu l$ (Fig. 176). Three days after a protein meal the level was 35 $\mu g/\mu l$. After 6 days it was 60–70 $\mu g/\mu l$, i.e., 7% protein in the blood. From the 6th day on, the level dropped because the ovaries were withdrawing protein. The fact that the first increase in titer did not occur until the 3rd day after feeding indicates that protein was not immediately available after a meal (in contrast to carbohydrate). The level remained high after the 6th day in males and in allatectomized females because no demand was being made upon it.

The corpus allatum also controls metabolism of the fat body. Allatectomy results in hypertrophy of the fat body whereas ovariectomy does not. The fat body, which combines the function of the liver and adipose tissue of mammals, requires protein for its development and also requires hormone from the corpus allatum, which itself needs protein.

The relationship between neurosecretory cells and ovaries found in *Calliphora erythrocephala* was also found in *Sarcophaga bullata* (Wilkens, 1968), and it was suggested that vitellogenesis is controlled directly by a gonadotropic hormone from the neurosecretory cells and that the hormone from the corpus allatum affects only the specific metabolism of proteins. An attempt to clarify some of these relationships was made by Bennettová-Řežábová (1972), who approached the problem by ovariectomy, ovarian implantations, parabiosis, and selective denervation.

Parabiotic flies were prepared by the method of Green (1964) and Dethier, Solomon, and Turner (1965), except that no physiological solutions or antibiotics were administered (Figs. 122 and 177). The parabionts were maintained on a sucrose solution for 24 hours; the

Food For the Next Generation: Specific Hungers

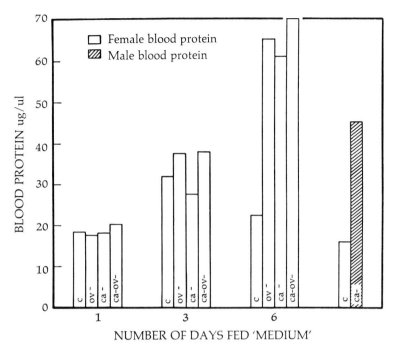

Fig. 176. The blood protein concentration of female flies after 1, 3, and 6 days of feeding a complex medium at the rate of 20 µl/fly thrice daily: C, control; CA⁻, allalectomized; OV⁻, ovariectomized; CA⁻OV⁻, allalectomized and ovariectomized (from Orr, 1964b).

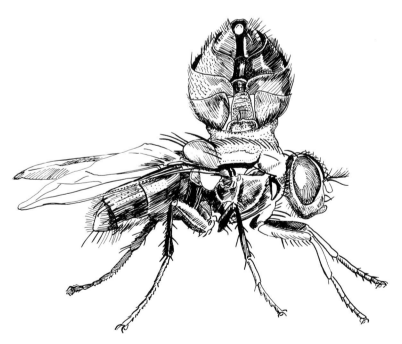

Fig. 177. A female blowfly in parabiosis with an isolated abdomen.

proboscis of one partner was then sealed with paraffin and the other partner was supplied with a solution of sucrose plus yeast. In other series, in order to eliminate the possible influence of the unfed fly's neurosecretory system, fly-abdomen parabiosis was employed. The parabionts were prepared in the usual manner by connecting the isolated abdomen of one fly with a circular opening in the thorax of the recipient fly. All parabionts were 6 to 12 hours old (Fig. 177).

One purpose of parabiosis was to determine whether protein actually had to be acquired per os in order to be effective, or could be equally effective if transferred from the gut without previous stimulation of the taste receptors. When the parabionts were both females the deposition of yolk began in the fed partner on the 5th day. By the 9th day a fully developed egg was present in the ovariole. The unfed partner developed up to the point of depositing yolk but no further. The development of these ovaries went further than in flies fed solely on sugar but even after 10 days had not caught up with that in the fly that had fed. No pathological changes occurred. The fat body of the unfed fly changed from the pupa to the imaginal one on the 3rd day, on schedule. When the fed partner was a male, development in the female was similar to that in the female-female parabiosis. Parabiosis of the female (fed) and an isolated abdomen from an unfed female followed a similar pattern. In the ovaries of the abdomen a second egg chamber appeared on the 6th day, and there were occasionally traces of yolk in the first chamber. Parabiosis of 3-day-old bilaterally ovariectomized females with 3-day-old sugar-fed females or with 3-day-old bilaterally autotransplanted females led to a development of ovaries similar to that in the female-female paraboisis.

In none of the experiments with parabiotic flies did the ovaries of the unfed fly develop beyond the previtellogenetic stage; however, development proceeded much further than in sugar-fed flies. A free exchange of haemolymph clearly took place. The haemolymph of females with developing eggs (or of males) was simply unable to induce vitellogenesis in unfed females.

The failure of ovaries to develop in a nonfeeding female that is attached to a protein-feeding male could be attributed to the lack of some factor that is normally present in the female or to some inhibitory factor in the male. The failure of ovaries to develop in a nonfeeding female that is attached to a normal female could be attributed to inhibition by an oöstatic hormone produced by the developing ovaries of the feeding partner (cf. Adams, Hintz, and Pomonis, 1968). The failure of ovaries to develop completely in a 3-day-old sugar-fed female joined to a 3-day-old bilaterally ovariectomized or bilaterally autotransplanted female is difficult to understand. The ovariectomized partner produces no oöstatic hormone and, having fed on protein, possesses activated neurosecretory cells, corpus allatum, and fat body. After 3 days the fat body presumably has transferred protein to the haemolymph. Usable protein and hormones should therefore be available to the nonfeeding partner because there has unquestionably been a free transfer of haemolymph from one fly to the other. Bennettová-Řežábová suggested that the failure was due to the inability of this female to activate here neuroendocrine system even with contributions from her partner. This implies that certain quantitative criteria must be met.

The search for the impediment to vitellogenesis was continued by means of several different series of transplantations, namely: transplantation of a single additional ovary; unilateral autotransplantation; bilateral ovariectomy and unilateral or bilateral autotransplantation in 24-hour-old flies fed protein; implantation of ovaries from 24-hour-old flies into bilaterally ovariectomized flies that were two days old and fed protein; implantation of fat body and ovaries from protein-fed females into protein-fed or sugar-fed males. Consistently the presence of an intact developing ovary inhibited development in transplants. Surplus ovaries transplanted into young females with intact ovaries not only failed to develop but often degenerated. In unilateral autotransplantation the intact ovary developed normally but the transplant did not. Ovaries placed in males also failed to develop, but they did not degenerate.

Different results were obtained when both ovaries of a young female were removed and then replaced in the same female. The first egg chambers began irregularly filling with yolk. Some attained nearly the size of a fully developed egg; others degenerated in early vitellogenesis. There was also a surplus production of eggs; that is, vitellogenesis occurred simultaneously in two or three egg chambers but never reached completion. Vitellogenesis followed the same course in bilaterally ovariectomized flies into which one or two young ovaries had been placed. It can be concluded that development is not possible in the male body, although no degenerative changes take place, and an oöstatic hormone in the female arrests development and causes degeneration of the transplants, and that, in the absence of developing ovaries and neural connections, implanted ovaries can develop but do so with abnormal vitellogenesis and faulty coordination in individual egg chambers. The formation of egg chambers in the germarium is also disturbed. In view of these results one would have expected development with abnormal vitellogenesis in a normal nonfeeding female put in parabiosis with an ovariectomized protein-fed female.

It is clear that an ovary must have neural connections in order to develop normally. This conclusion was checked by a series of experiments involving denervation.

The principal innervation of the abdomen of *Diptera* is provided by the abdominal nerve cord and its four paired branches (Fig. 178) (Lowne, 1890–1895; Hewitt, 1914; Power, 1948; Bennettová-Řežábová, 1972). The first three branches send fibers to the intersegmental dorsal and ventral muscles and trachea of the lateral area of the abdomen. All four branches send fibers into the area of the fat body where individual fibers are attached to the sheath of the fat body. The last pair of branches innervate the reproductive system. Each trunk of this terminal pair subdivides. The lateral subdivision innervates the fat body, the spiracles and muscles in the posterior area of the abdomen, and the base of the ovipositor. The median subdivision innervates in the following order from anterior to posterior: lateral oviduct, common oviduct, accessory glands, receptacula seminis, vagina, base of ovipositor, rectum. There is no direct innervation to the ovary; however, one of the branches to the lateral oviduct terminates in the immediate vicinity of the ovarian calyx.

Contained among the abdominal nerves are many neural cell bodies. They are especially numerous in the fine branches either at junctions or along unbranched stretches of nerve (Gelperin, 1971). They are 20-30 μ in diameter and are often paired. Approximately 8 to 10 are situated on each nerve. Those in the reproductive neural system have been carefully mapped by Bennettová-Řežábová, from whose work this description is taken.

The branch to the lateral oviduct has three cells in the terminal divisions that are firmly attached to the lateral oviduct close to the ovary. Branches to the common oviduct contain 10 cells, also firmly attached to the oviduct. In the vaginal region of the nerve running along the oviduct there are 9 to 10 cells present in groups and 7 to 10 cells in fibers on the vaginal wall. Among branching fibers at the base of the ovipositor there are 8 to 9 cells. Between the principal reproductive nerve and the accessory gland there is a group of three cells, 30–50 μ in diameter, connected to the main reproductive nerve by two short fine nerves (Fig. 179).

Some of these cells are the stretch receptors that respond to distension of the abdominal wall (Gelperin, 1971); others may quite possibly be neurosecretory cells. Neurosecretory cells have been found in widely scattered parts of the bodies of other insects. Among these are neurohaemal areas of the medial nervous system of phasmids (Raabe, 1965; Raabe and Ramade, 1967), peripheral neurosecretory cells in the neurohaemal tissue of *Carausius morosus*, neurosecretory cells of the median nerves of larvae of *Phormia regina* (Finlayson and Osborn, 1968), neurosecretory tissue associated with the rectal papillae of *Calliphora erythrocephala* (Gupta and Berridge, 1968), and others; for a review, consult Maddrell (1970).

Food For the Next Generation: Specific Hungers

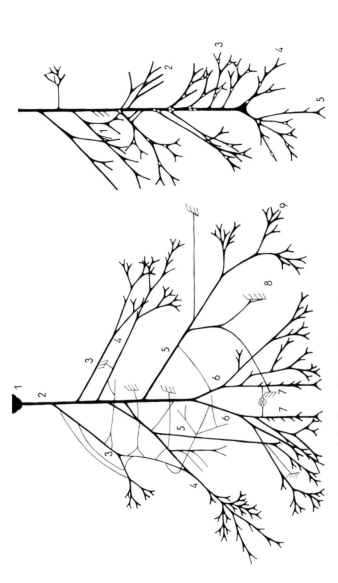

Fig. 178. (a) Diagrammatic representation of the branching of the median abdominal nerve of a femal blowfly. Some interconnections between branches and muscles are indicated (8). 1, thoracico-abdominal ganglion; 2, median abdominal nerve; 3, first pair of abdominal nerve branches; 4, second pair; 5, third pair; 6, fourth pair; 7, reproductive nerve; 9, terminations of nerves in the fat body. (b) A detail of the branching of the reproductive nerve and the position of the nerve cells in the area of the reproductive organs. 1, branches terminating in the lateral oviduct; 2, branches to the common oviduct; 3, branches to the vagina; 4, branches to the base of the ovipositor; 5, branches to the rectum (from Bennettová-Řežábová, 1972).

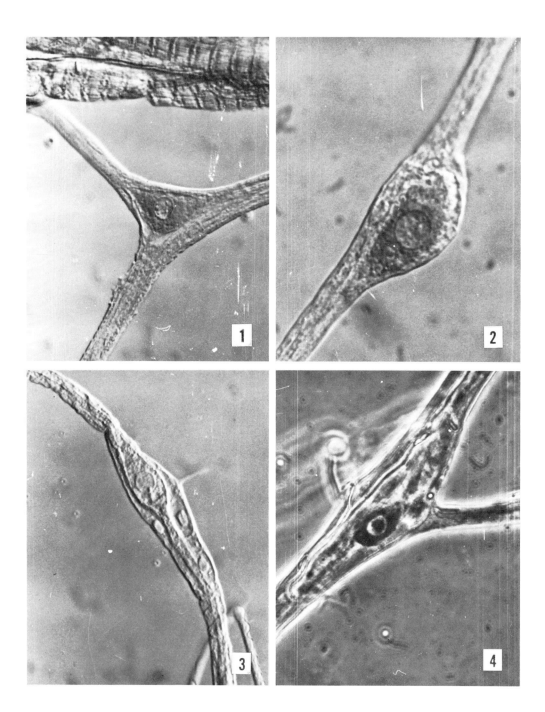

Fig. 179. (a) Nerve cells associated with the abdominal nerve stained with methylene blue and photographed with interference contrast. 1, from peripheral region of median abdominal nerve (×1,100); 2, from area of reproductive system (×1,700); 3, from the same area (×1,100); 4, from area of fat body (×1,100).

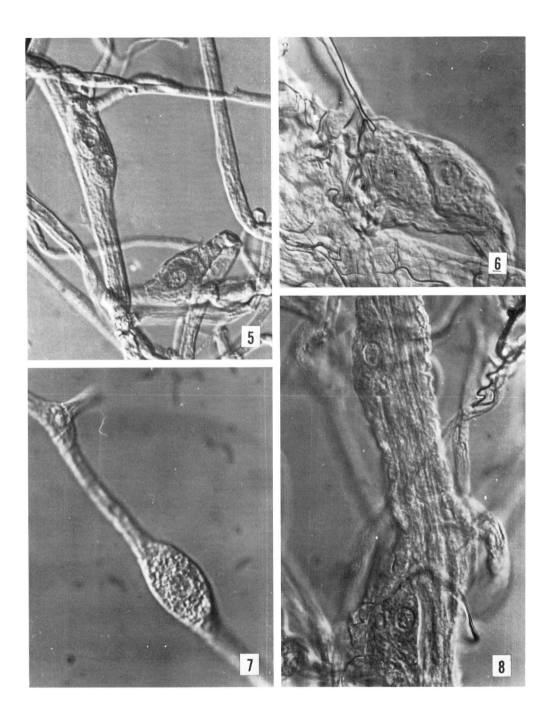

(b) 5 and 8, from area of vagina (×1,000); 6, detail of cluster of two cells from the reproductive nerve branches of the area of the accessory gland (×1,700); 7, a cell from the fine branches of the reproductive nerve. Note the conspicuous granulation of the cytoplasm (from Bennettová-Řežábová, 1972).

The reproductive system of *Phormia* is therefore liberally supplied with neural cells. When the ventral nerve cord is transected, neither neural nor hormonal information is able to pass to or from the brain. Under these conditions the fat body shows marked structural changes, becoming almost translucent; there is an arrest of ovarian development, and the germarium no longer functions normally. Bennettová-Řežábová suggested that ovarian development in *Phormia* is controlled directly by neurosecretory cells in the brain, either by a hormone dispersed along the ventral nerve cord or, if the cells in the neural supply to the reproductive area are neurosecretory, by a neural message to them from the brain.

These, then, are the complicated and only partially understood relations that exist among the reproductive system, the endocrine system, the nervous system, and protein metabolism in the gut, fat body, and blood. Consequently, experimental analyses of the behavioral aspects of ingestion are exceedingly difficult. Removal of any one of the organs whose functions are so closely interlocked interferes with many components of the system. The two most attractive hypotheses to explain the specific hunger for protein are (1) that the neural mechanisms that mediate ingestive behavior are under hormonal control, and (2) that they are influenced by a protein deficit in the blood or elsewhere. The first hypothesis is favored by Strangways-Dixon (1961b) and the second by Dethier (1961) and Belzer (1970).

The female fly emerges from the puparium with an appetite and preference for protein. At this time the corpus allatum and possibly the median neurosecretory cells are inactive (insofar as reproduction is concerned). The fat body is pupal. The titer of protein in the blood is low. Two possibilities for endocrine control can be imagined, namely, that hormones increase sensitivity to proteins (that is, that they cause the central nervous system to act upon information received from sense organs stimulated by protein); or that they decrease sensitivity. For the first alternative to be true the corpus allatum and median neurosecretory cells would have to be active *before* the protein is ingested, which is manifestly impossible since protein is required to activate them. Furthermore, were they active, removal of the corpus allatum or the median neurosecretory cells should abolish the initial peak of ingestion of protein (if neurosecretory cells elsewhere in the body are assumed not to be involved). With *Phormia,* allatectomy did not greatly alter the pattern of initial ingestion of protein. No gross difference appeared between operated and sham-operated flies (Figs. 180). The first peak of the initial protein cycle was unaltered; the second ("yolk") peak was reduced (Belzer, 1970).

It might be argued that there was enough residual hormone to prevent a change; however, this is unlikely since there was not enough

Food For the Next Generation: Specific Hungers 321

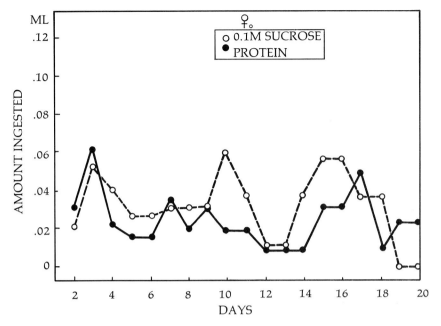

Fig. 180. Daily intake of 0.1 M sucrose and brain-heart extract by a virgin female blowfly from which the ovaries have been removed (from Dethier, 1961).

to permit egg development. Furthermore, allatectomized flies could be held for 8 days on carbohydrate and still show a preference for protein. Strangways-Dixon reported that protein intake was high after extirpation of the corpus allatum (although it appears lower than usual in his graph). The conclusion that preferential ingestion of protein (at least the initial rise) is not stimulated directly by hormones from the corpus allatum seems justified. Belzer concluded not only that the corpus allatum hormone has no *direct* influence on protein ingestion, but also that the protein hunger ("yolk" peak) is not dependent on *growing follicles* in any *specific* way. Several experiments support this last contention: first, the yolk peak in normal females anticipates vitellogenesis; second, sugar-fed females that have been deprived of protein for some time show a maximum consumption of protein as soon as it is made available *even though eggs have not begun their development;* third, sterilized females (by allatectomy or by feeding "apholate," an alkylating agent that presumably affects the synthesis of DNA, RNA, and/or protein) still exhibit a yolk peak (Fig. 181).

The effect of removal of the median neurosecretory cells in the brain is not unequivocal. Extirpation of these cells in sugar-fed females of *Phormia* that were 2.5 days old or older did not reduce the protein hunger (Dethier, 1961). Belzer has pointed out that at this age protein intake would have been maximum, so that no effect should have been

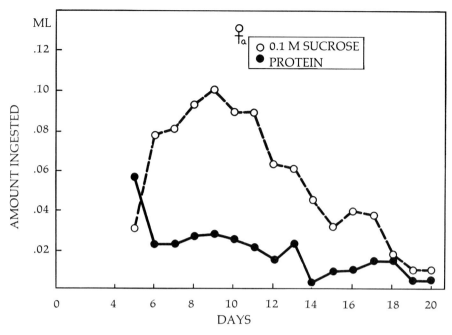

Fig. 181. Daily intake of 0.1 M sucrose and brain-heart extract by a virgin female blowfly from which the corpus allatum has been removed (from Dethier, 1961).

expected. In *Calliphora*, removal of these cells within 24 hours of emergence resulted in failure of females to ingest protein except for a brief initial period. As Strangways-Dixon pointed out, however, his result is not in agreement with the observation of Thomsen and Møller (1959) that females lacking these neuroendocrine cells continued to feed on meat. The data from *Phormia* are in closer in agreement with the last mention results. It is difficult, therefore, to agree with the conclusion that the ingestion of protein is under the *direct* control of the median neurosecretory cells. This does not mean that these cells do not control the enzymatic breakdown of proteins once they are ingested, although Engelmann and Wilkens (1969) have pointed out that there is no evidence for neurosecretory involvement in the synthesis of digestive enzymes in *Sarcophaga*. They maintain, furthermore, that there is no unequivocal information for any insect species that hormones directly control such synthesis.

The other side of the coin is that feeding on protein is the normal situation and that hormones inhibit it. Were this true flies from which the endocrine glands had been removed should continue to ingest protein, if not in preference to carbohydrate, at least in appreciable quantity. Clearly they do not.

The situation with respect to subsequent peaks of protein ingestion is even more difficult to investigate experimentally because removal of any of the glands or of the ovaries themselves does not interfere maximally with the first peak but prevents the development of eggs and abolishes subsequent peaks. The condition of the operated flies depends upon which operation has been performed and when. If any of the operations are performed before the fly is first presented with protein, there are no hormones, the titer of protein in the blood is low, and the fat body is pupal. In a choice situation, protein is preferred. If allactectomy is performed *after* the first protein meal, there are no hormones, the titer of protein in the blood is high, the fat body hypertrophies (that is, becomes filled with lipids), and ovarian development is arrested. If ovariectomy is performed instead, the corpus allatum is active, the fat body does not hypertrophy, the titer of proteins in the blood increases. The titer of proteins in the blood also increases after treatment with aspholate (Painter et al., 1972). In none of the cases just mentioned is there a new peak of ingestion of protein. If ingestion were stimulated by hormones, one would expect protein intake to continue after ovariectomy because presumably the corpus allatum is still active. If ingestion were inhibited by hormones, one would expect it to continue after allectomy.

In the normal fly the condition of the endocrine glands immediately following the first oviposition is not known. If they retain their initial activity and are responsible for stimulating the ingestion of protein, we are left with the question of why ingestion declined before the eggs were laid. Two possibilities suggest themselves: either the intake of protein is switched off by pressure of the developing eggs in the abdomen or in the ovaries themselves, or there is a positive stimulus from somewhere for preferential ingestion of carbohydrate. The operations described in the preceding paragraphs argue against both of these interpretations, but we shall look more closely into the possible role of abdominal pressure again. Finally, if the endocrine glands are inactive in the interovipositional period, a hormonal stimulus to ingestion of either protein or carbohydrate cannot be invoked.

The deficit hypothesis overcomes some of these dilemmas, but it too encounters some difficulties. One of the problems is to identify the deficit. It is assumed that flies of each sex emerge from the pupal state with an accumulated protein deficit. This could stem from requirements that had to be met for tanning the new adult cuticle (Belzer, 1970), or could be related to active biosynthesis of mitochondrial proteins in flight muscle, fat body, and elsewhere. In response to this deficit both sexes are sensitized to protein; that is, the central nervous system accepts and acts upon neural information derived from stimulation by proteins. Accordingly males as well as females show an initial preference. Greenberg (1959) stated that there was no difference in the consumption of protein by male and virgin female houseflies; however, his conclusion was based upon a comparison of mean daily intake of several days, a measurement that tends to minimize the difference occurring shortly after emergence. After this initial difference the intake of each sex is essentially the same. In the newly emerged female the deficit is presumed to be larger because of a start in ovarian development. Males also require proteins for the development of the male accessory glands (Stoffolano, personal communication); however, the glands probably require less protein than do developing ovaries. Although the oocyte is not incorporating protein, nurse cells and other follicular cells are synthetically active. They are capable of very rapid synthesis of macromolecules from amino acids in the blood (Telfer, 1965), and active amino acid turnover is taking place at this time. As protein is ingested in response to this initial deficit, the corpus allatum and median neurosecretory cells are activated, the fat body begins to synthesize special proteins, thus increasing the deficit, and after about 3 days these are released into the blood, whence they are removed by the oöcytes for the formation of yolk. The hypothesis therefore envisions a positive feedback system in which an initial deficit stimulates ingestion of protein, which in turn causes the demand to be accelerated until finally, when the eggs are fully formed, the cycle is broken. If the female is mated, the eggs are laid, a new deficit is initiated, and the cycle is repeated.

According to Orr's (1964b) timetable, a period of 3 days must elapse after the ingestion of protein before the fat body can synthesize protein and release it into the blood. The ovaries now begin to remove it. The peak period of protein ingestion is between the 4th and 6th day.

One great difficulty with the deficit hypothesis is its inability to explain why in sugar-fed virgin females the protein "hunger" increases with time as deprivation is continued up to 6 days. The longer the period of deprivation, the greater the amount of protein consumed when it is finally presented. How does the deficit increase, since there is no continued ovarian development?

As Belzer has pointed out, it is not follicular growth per se that induces protein hunger, but is probably a depletion of some raw material, resulting from increased protein synthesis at that time. If the corpus allatum controls protein synthesis in *Phormia* as it does in other flies, allatectomy would interfere with the synthesis of yolk, but not with other periods of protein synthesis that were independent of allatum hormone, and in that way would affect the yolk peak selectively. In the cockroach *Leucophaea maderae,* egg maturation stimulates food intake if not enough reserves are available for the completion of yolk development (Englemann and Rau, 1965). As with the fly, the implication is that some raw material is being depleted. To test this hypothesis a number of experiments in which protein synthesis was interfered with were undertaken (Belzer, 1970). Some females were sterilized by feeding with hydroxyurea, a specific inhibitor of DNA synthesis. The behavioral effects were qualitatively similar to those following sterilization by allatectomy and by feeding with apholate (Figs. 182 and 183). A more critical test of the role of protein synthesis involved injecting flies with cycloheximide, a specific inhibitor of protein synthesis. Price (1969) had demonstrated that this compound inhibited synthesis in the fat body of *Calliphora* in vitro.

When the inhibitor was injected at age 9 hours and preference tests run at a time when protein hunger normally increases rapidly, that hunger was greatly depressed (Fig. 184). The same dose injected into 6-day-old females (whose protein hunger would already have reached maximum proportions) did not significantly reduce the protein peak (Fig. 185). Thus inhibition of protein synthesis when the protein hunger is growing delays its further development but, once the hunger is established, the inhibitor has no effect upon it. These results support the idea that protein synthesis produces a deficit and that the deficit has to be abolished before the hunger is relieved. Indeed, injection of cycloheximide at the beginning of vitellogenesis delayed the development of eggs and the yolk peak in the feeding cycle (Fig. 186).

An attempt by Belzer to induce protein synthesis artificially by injecting farnesyl methyl ether (a mimic of the allatum hormone) into flies that had been sterilized with apholate (to prevent egg development and protein ingestion) had inconclusive results.

Pinpointing the site of the postulated deficit is a difficulty. Lack of knowledge regarding the timetable of endocrinological and biochemical events. coupled with an incomplete understanding of mechanism of yolk formation (cf. Telfer, 1965), is a very great handicap. Prime candidates for the site of action are the blood and the fat body. The mechanism is not merely a matter of blood pressure or general osmotic relations because injection of water, carbohydrate, or hypotonic salt into the haemocoele fails to alter the pattern of selective ingestion. As Telfer (1954, 1960) demonstrated in saturniid moths, the protein used by the oöcytes is a specific female protein synthesized by the fat body. It is the deficit of this particular protein that would be critical; on the other hand, the explanation of the initial peak of ingestion invokes other materials.

The fat body is an attractive candidate because it synthesizes the protein required by the ovary, it undergoes sequential changes associated with the development of the oöcytes (Thomsen and Thomsen, 1974), it has neural connections with the head region where its input

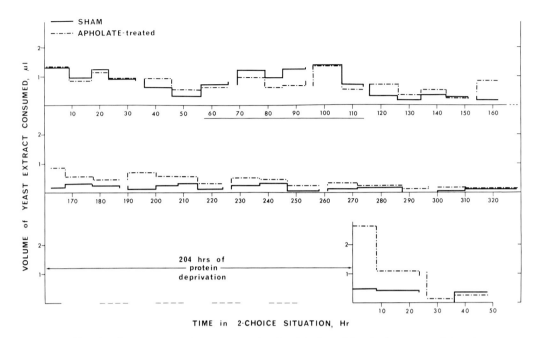

Fig. 182. Average protein consumption by 0.5-day-old apholate-treated female flies and sham-treated flies in a two-choice test. Protein intake was identical in the two groups until hour 78. Between hours 79 and 94 (yolk peak) sham females ate more than experimentals. After hour 115, experimentals ate more than shams (from Belzer, 1970).

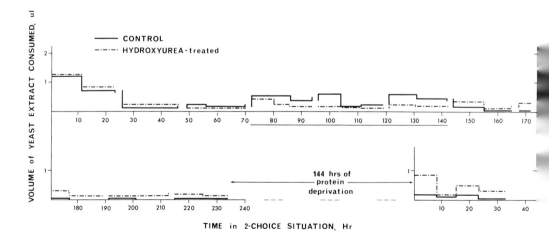

Fig. 183. Average protein consumption by 0.5-day-old hydroxyurea-treated females and shams (from Belzer, 1970).

Food For the Next Generation: Specific Hungers   327

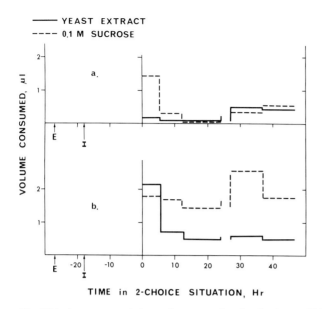

Fig. 184. Average protein and sucrose intake during a 47-hour choice test by (a) 27-hour-old sugar-fed females injected with cycloheximide at 9 hours of age, and (b) by shams (from Belzer, 1970).

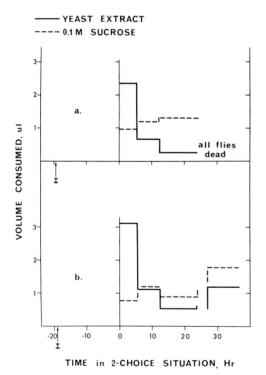

Fig. 185. Average protein and sucrose intake during a standard choice test by (a) 6-day-old-sugar-fed females injected with cycloheximide 19 hours before testing, and (b) by shams. I indicates time of injection (from Belzer, 1970).

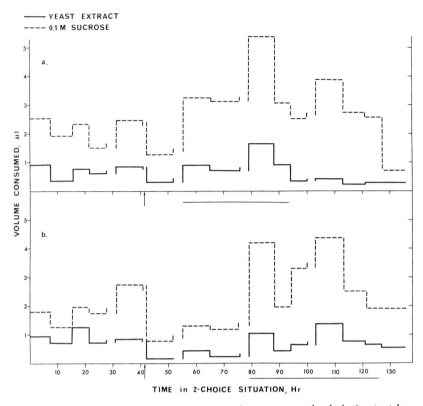

Fig. 186. Average protein and sucrose intake during a standard choice test by (a) 11 newly emerged females which were injected with cycloheximide after the first peak of protein feeding, and (b) by 5 shams. Arrow indicates time of injection. The bar beneath the trace delimits the apparent yolk peak (from Belzer, 1970).

could be integrated with sensory input from the labellum, and it exercises its demand on digestive proteins earlier (at 3 days) than the ovary does. The peak of ingestion correlates better with fat body activity than with ovarian activity. One difficulty, however, is that ovariectomy and allatectomy have different effects on the fat body but similar effects on blood protein levels and feeding behavior. Neither completely abolishes the initial peak. According to Orr (1964b), protein accumulates in the blood after each operation, but whereas the fat body functions normally after ovariectomy (female protein accumulates in the blood because there are no ovaries to remove it), it hypertrophies after allatectomy. How or what protein accumulates in the blood in this instance is a mystery.

If proof should eventually be found that some deficit initiates protein hunger, it does not necessarily follow that satisfying the deficit terminates the hunger by the same pathways. It seems quite clear that supplying blood-borne factors directly does not alleviate the situation. In the case of parabiotic protein-deprived flies, feeding one partner terminated its protein hunger but had no effect on the other fly (Belzer, 1970). Other mechanisms for terminating the specific feeding behavior were sought by Belzer in a long series of diverse experiments. These included: cutting nerves from the abdominal stretch receptors, bloating the abdomen artificially, and cutting the recurrent nerve. The experiments were performed on males and females, the latter in various stages of the reproductive cycle.

When abdominal nerves were cut, both sexes became hyperphagic on sugar when that was the only solution available and on protein when that was the only solution. In a choice situation females increased their intake of both solutions, but the preference for protein remained; males increased only their intake of sugar.

Protein-deprived females that were artificially bloated (to simulate a gravid condition) by being injected with 16.5 μl of a 10% solution of methyl cellulose consumed little protein. If they then had the abdominal nerves cut, they ate. This sequence of events simulated the one that occurs naturally when a female becomes gravid and then oviposits. The normal gravid female takes little or no protein until the eggs are laid. A comparable situation occurs in the cockroach *Leucophaea maderae*, where a restriction of feeding by pregnant females is caused by egg cases in the brood sac (Englemann and Rau, 1965). After oviposition the normal fly takes a long drink of protein. This appears to be a response to relaxed abdominal pressure and indicates that at the time of becoming gravid the female still had a protein appetite that was shut off by abdominal distension before becoming fully satisfied; hence, the rebound after the eggs are laid. There is no comparable sugar-rebound.

Abdominal pressure is not the only pathway for terminating protein feeding, however. Females sterilized by allatectomy, by apholate, or by hydroxyurea continued a slightly elevated consumption of protein 2 to 5 days longer than control females, which by this time were carrying mature eggs and had reduced their intake of protein. Extended protein feeding was also produced during the second reproductive cycle. In this case gravid females with fully matured eggs were allatectomized so that there would be no second crop of eggs. The first clutch was laid normally; *then* the usual protein rebound occurred, but it persisted longer.

If allatectomized females were deprived of their protein after their hunger had returned to normal levels, a rebound in protein hunger occurred and after 5 days of deprivation was vigorous. Thus, after protein hunger has normally been satisfied it can be instituted again by deprivation. This observation agrees with the fact that gravid females, as stated, have an unfulfilled hunger because the pressure of the egg mass shuts it off; they are deprived until the mass is removed; they show a rebound when it is removed.

As Belzer has pointed out, there are obvious adaptive advantages to this arrangement. Oviposition occurs on or near proteinaceous materials. It makes sense for the fly to develop a protein hunger immediately after oviposition, while it still has a bounteous supply. The long drink of yeast extract that the experimental female does in fact take (4 to 7 μl or more) exceeds that contained in the entire fat body of a gravid female.

The recurrent nerve system also plays some part in terminating the ingestion of protein. Section of the nerve caused protein-sated males and females to become hyperphagic on protein as well as on carbohydrate if tested in a no-choice situation. The hyperphagia on protein can be spectacular. This finding raises the question as to whether cutting the recurrent nerve indiscriminately alters feeding (in other words, makes for abnormal feeding), as Evans and Barton Browne (1960) have argued. However, when Belzer placed his flies in choice situations, clear distinctions were observed. Protein-deprived males and females with the recurrent nerve cut both ignored protein in a two-choice situation and became hyperphagic on carbohydrate. Furthermore, when the responses of these operated flies to other kinds of solutions were tested, gustatory discrimination was still operative and different degrees of hyperphagia ensued. He concluded that the inhibitory feedback from the recurrent nerve must be stronger for input from sugar receptors than from the input signaling protein, so that when the feed-back loop is surgically eliminated, a preference for sugar emerges. Belzer also found, in contradiction to the observations of Dethier and Gelperin (1967), that the recurrent nerve exerted a more powerful effect on feeding than did the abdominal nerves. Also, inhibition from the latter does not appear to be selective with respect to the ingestion of sugar and of protein.

From all of these studies a model for the control of protein feeding, patterned after the model of the control of carbohydrate feeding (Dethier and Gelperin, 1967), had been proposed (Belzer, 1970). It is diagrammed in Figure 187. The basic premises are the same for both, namely, an interaction between incoming chemosensory excitation and inhibition mediated by mechanoreceptors in the foregut (I_{FG}) and the abdominal wall (I_{AbN}). An additional inhibition is postulated, that is, a unique inhibitory feedback to protein hunger (I_S), which is the *decrease* in the deficit of some reserve material related to protein synthesis. In the absence of any concrete knowledge protein satiety is assumed to be simply the consequence of relieving the hypothetical deficit. Fluctuations of I_S are postulated as being the source of patterning the ingestion of protein.

Feeding on protein will cease when $E - I \leq 0$. In a protein-deprived fly (Fig. 187), I_S is assumed to be small. A large $I_{FG} + I_{AbN}$ is required to

Food For the Next Generation: Specific Hungers

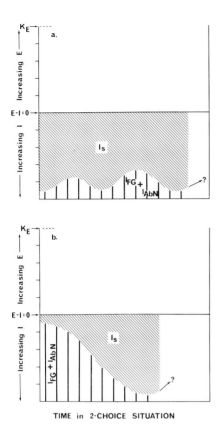

Fig. 187. Model for the generation of the typical patterns of protein feeding seen (a) in newly emerged flies, and (b) in sugar flies (from Belzer, 1970). Explanation in text, Chapter 9.

terminate protein ingestion. Since these I_S are related to meal size, the fly would be expected to take large initial meals (as indeed it does). As protein is consumed, I_S increases and less $I_{FG} + I_{AbN}$ is required to inhibit feeding; accordingly, meals should become progressively shorter. Developing egg masses would further increase I_{AbN}, hence, decrease feeding. As pointed out earlier the gravid female tends to develop a deficit that is held partly in check by I_{AbN}, and this is indicated in the diagram by a question mark. The actual *preference* for protein emerges only because ingestion of sucrose is selectively moderated by the recurrent nerve.

Flies allowed access to protein from the time of emergence take longer and longer drinks as protein synthesis associated with tanning and other developmental events depletes reserves and decreases I_S. With a decline in synthesis, I_S increases, and drinks again become shorter. With the onset of vitellogenesis, I_S again decreases, and drinks become longer. The net result of these interactions could be the biphasic pattern of protein feeding observed by Belzer.

Literature Cited

Adams, T. S., Hintz, A. M., and Pomonis, J. G. 1968. Oöstatic hormone production in house flies, *Musca domestica,* with developing ovaries. *J. Insect Physiol.* 14:983–993.

Balboni, E. R. 1968. The respiratory metabolism of insect flight muscle during adult maturation. *J. Insect Physiol.* 13:1849–1856.

Belzer, W. R. 1970. The control of protein ingestion in the black blowfly, *Phormia regina* (Meigen). Doctoral dissertation, University of Pennsylvania, Philadelphia.

Bennettová-Řežábová, B. 1972. The regulation of vitellogenesis by the central nervous system in the blow fly, *Phormia regina* (Meigen). *Acta Ent. Bohemoslovaca* 69:78–88.

Blake, G. M. 1961. Length of life, fecundity and the oviposition cycle in *Anthrenus verbasci* (L.) (Col., Dermestidae) as affected by adult diet. *Bull. Ent. Res.* 52:459–472.

Bodenstein, D. 1953. Studies on the humoral mechanisims in growth and metamorphosis in the cockroach, *Periplaneta americana.* III. Humoral effects on metabolism. *J. Exp. Zool.* 124:105–115.

Braken, G. K., and Nair, K. K. 1967. Stimulation of yolk deposition in an ichneumonid by feeding synthetic juvenile hormone. *Nature* 216:483–484.

Bradley, R. M. 1972. Duplexity of taste receptor loci. In *Olfaction and Taste,* IV D. Schneider, ed., pp. 273–279. Wissenschaftliche Verlagsgesellschaft MBH, Stuttgart.

Brosemer, R. W., Vogell, W., and Bucher, T. 1963. Morphologische und enzymatische Muster bei der Entwicklung indirekter Flugmuskelin von Locusta migratoria. *Biochem. Zeit.* 338:854–910.

Calabrese, E. J., and Stoffolano, J. G. 1974. The influence of age and diet on respiration in adult male and female black blowflies, *Phormia regina. J. Insect Physiol.* 20:383–393.

Davies, D. M., and Peterson, B. V. 1956. Observations on the mating, feeding, ovarian development, and oviposition of adult black flies (Simuliidae, Diptera). *Canad. J. Zool.* 34:615–655.

Denton, D. A. 1967. Salt appetite. In *Handbook of Physiology,* C. F. Code, ed. Vol. 1, Sec. 6:433–459. Amer Physiol. Soc., Washington, D.C.

Dethier, V. G. 1961. Behavioral aspects of protein ingestion by the blowfly *Phormia regina* Meigen. *Biol. Bull.* 121:456–470.

Dethier, V. G. 1969. Feeding behavior of the blowfly. *Adv. Study of Behavior* 2:111–266.

Dethier, V. G., and Gelperin, A. 1967. Hyperphagia in the blowfly, *J. Exp. Biol.* 47:191–200.

Dethier, V. G., Solomon, R. L., and Turner, L. H. 1965. Sensory input and central excitation and inhibition in the blowfly. *J. Comp. Physiol. Psychol.* 60:303–313.

Englemann, F., and Rau, I. 1965. A correlation between feeding and the sexual cycle in *Leucophaea maderae* (Blattaria). *J. Insect Physiol.* 11:53–64.

Englemann, F., and Wilkens, J. L. 1969. Synthesis of digestive enzyme in the fleshfly *Sarcophaga bullata* stimulated by food. *Nature* 222:798.

Evans, D. R., and Barton Browne, L. 1960. Physiology of hunger in the blowfly. *Amer. Midland Nat.* 64:282–300.

Evans, D. R., and Dethier, V. G. 1957. The regulation of taste thresholds for sugars in the blowfly. *J. Insect Physiol.* 1:3–17.

Finlayson, L. H., and Osborn, M. P. 1968. Peripheral neurosecretory cells in the stick insect (*Carausius morosus*) and the blowfly larva (*Phormia terraenovae*) *J. Insect Physiol.* 14:1793–1801.

Gelperin, A. 1971. Abdominal sensory neurons providing negative feedback to the feeding behavior of the blowfly. *Zeit. vergl. Physiol.* 72:17–31.

Green, G. W. 1964. The control of spontaneous locomotor activity patterns in *Phormia regina* Meigen. II. Experiments to determine the mechanism involved. *J. Insect Physiol.* 10:727–752.

Greenberg, B. 1959. Housefly nutrition. I. Quantitative study of the sugar and protein requirements of males and females. *J. Cell. Comp. Physiol.* 53:169–177.

Gupta, B. L., and Berridge, M. J. 1968. Fine structural organization of the rectum in the blowfly, Calliphora erythrocephala (Meig.) with special reference to connective tissue, tracheae and neurosecretory innervation in the rectal papillae. *J. Morph.* 120:23–82.

Harlow, P. M. 1956. A study of ovarial development and its relation to adult nutrition in the blowfly *Protophormia terraenovae* (R. D.). *J. Exp. Biol.* 33:337–396.

Herold, R. C. 1965. Development and ultrastructural changes of sarcosomes during honey bee flight muscle development. *Devel. Biol.* 21:269–286.

Hewitt, C. G. 1914. *The Housefly Musca domestica* L. Cambridge University Press.

Johansson, A. S. 1955. The relationship between corpora allata and reproductive organs in starved female *Leucophaea maderae* (Blattaria). *Biol. Bull.* 108:40–44.

Johansson, A. S. 1958. Relation of nutrition to endocrine-reproductive functions in the milkweed bugs *Oncopeltus fasciatus* (Dal.). *Nytt. Mag. Zool.* 7:1–132.

Keeley, L. L. 1972. Biogenesis of mitochondria: neuroendocrine effects on the development of respiratory functions in fat body mitochondria of the cockroach, *Blaberus discoidalis*. *Arch. Biochem. Biophysics* 153:8–15.

Larsen, J. R., and Bodenstein, D. 1959. The humoral control of egg maturation in the mosquito. *J. Exp. Zool.* 140:343–381.

Larsen, W. J. 1970. Genesis of mitochondria in insect fat body. *J. Cell. Biol.* 47:373–383.

Lea, A. O., and Thomsen, E. 1962. Cycles in the synthetic activity of the median neurosecretory cells of *Calliphora erythrocephala* and their regulation. In *Neurosecretion*, H. Heller and R. B. Clark, eds., pp. 345–347. Academic Press, New York.

Lennie, R. W., and Birt, L. M. 1967. Aspects of the development of flight muscle sarcosomes in the sheep blowfly, *Lucilia cuprina*, in relation to changes in the distribution of protein and some respiratory enzymes during metamorphosis. *Biochem. J.* 102:338–350.

Levenbook, L., and Williams, C. M. 1956. Mitochondria in the flight muscles of insects. III. Mitochondrial cytochrome c in relation to the ageing and the wing beat frequency of flies. *J. Gen. Physiol.* 39:497–512.

Lowne, B. T. 1890–1895. *The Anatomy, Physiology and Development of the Blowfly*. London.

Maddrell, S. H. P. 1970. Neurosecretory control systems in insects. In *Insect Ultrastructure*, pp. 101–116. *Symp. Roy. Ent. Soc.*, London.

Orr, C. W. M. 1964a. The influence of nutritional and hormonal factors on egg development in the blowfly *Phormia regina* (Meig.). *J. Insect Physiol.* 10:53–64.

Orr, C. W. M. 1964b. The influence of nutritional and hormonal factors in the chemistry of the fat body, blood, and ovaries of the blowfly *Phormia regina* Meig. *J. Insect Physiol.* 10:103–119.

Painter, R. R., Kilgore, W. W., and Gadallah, A. I. 1972. Influence of apholate on the haemolymph proteins of adult house flies. *J. Econ. Ent.* 65:23–27.

Pfaffmann, C., and Bare, J. K. 1950. Gustatory thresholds in normal and adrenalectomized rats. *J. Comp. Physiol. Psychol.* 43:320–324.

Pospíšil, J. 1958. Some problems of the smell of saprophilic flies. *Acta Soc. Entomol. Bohem.* (Csl). 55:316–334.

Power, M. E. 1948. The thoracico-abdominal nervous system of an adult insect, Drosophila melanogaster. *J. Comp. Neurol.* 88:725–752.

Price, G. M. 1969. Protein synthesis and nucleic acid metabolism in the fat body of the blowfly, *Calliphora erythrocephala. J. Insect Physiol.* 15:931–944.

Raabe, M. 1965. Récherches sur la neurosécretion dans la chaîne nerveuse ventrale du Phasme, Clitumnus extradentatus. Les epaississements des nerfs transverses, organes de signification neurohémale. *C. R. Hebd. Séanc. Acad. Sci.* Paris, 261:4240–4243.

Raabe, M., and Ramade, F. 1967. Observations sur l'ultrastructure des organes périsympathiques des phasmides. *C. R. Hebd. Séanc. Acad. Sci.* Paris, 264:77–80.

Richter, C. P., and Barelare, B. 1938. Nutritional requirements of pregnant and lactating rats studied by the self-selection method. *Endocrinology* 23:15–24.

Robbins, W. E., and Shortino, T. J. 1962. Effect of cholesterol in the larval diet on ovarian development in the adult housefly. *Nature* 194:502–503.

Rockstein, M. 1956. Metamorphosis. A physiological interpretation. *Science* 123:534–536.

Rozin, P. 1967. Thiamine specific hunger. In *Handbook of Physiology* C. F. Code, ed., Vol. 1, Sec. 6:411–431. Amer. Physiol. Soc., Washington, D.C.

Strangways-Dixon, J. 1959. Hormonal control of selective feeding in female *Calliphora erythrocephala* Meig. *Nature* 184:2040.

Strangways-Dixon, J. 1961a. The relationship between nutrition, hormones and reproduction in the blowfly *Calliphora erythrocephala* (Meig.). I. Selective feeding in relation to the reproductive cycle, the corpus allatum volume and fertilization. *J. Exp. Biol.* 38:225–235.

Strangways-Dixon, J. 1961b. The relationships between nutrition, hormones and reproduction in the blowfly *Calliphora erythrocephala* (Meigen). II. The effect of removing the ovaries, the corpus allatum and the neurosecretory cells upon selective feeding, and the demonstration of the corpus allatum cycle. *J. Exp. Biol.* 38:637–646.

Telfer, W. H. 1954. Immunological studies of insect metamorphosis. II. The role of sex-limited female protein in egg formation by the cecropia silkworm. *J. Gen. Physiol.* 37:539–558.

Telfer, W. H. 1960. The selective accumulation of blood proteins by the oöcytes of saturniid moths. *Biol. Bull.* 118:338–351.

Telfer, W. H. 1965. The mechanism and control of yolk formation. *Annu. Rev. Ent.* 10:161–184.

Thomsen, E. 1942. An experimental and anatomical study of the corpus allatum of the blow-fly *Calliphora erythrocephala* (Meig.). *Vidensk. Medd. Dansk Naturh. Foren.* 106:319–405.

Thomsen, E. 1948. Effect of removal of neurosecretory cells in the brain of adult *Calliphora erythrocephala* Meig. *Nature* 161:439.

Thomsen, E. 1952. Functional significance of the neurosecretory brain cells and the corpus cardiacum in the female blow-fly, *Calliphora erythrocephala* Meig. *J. Exp. Biol.* 29:137–172.

Thomsen, E., and Møller, I. 1959. Neurosecretion and intestinal proteinase activity in an insect, *Calliphora erythrocephala* Meig. *J. Exp. Biol.* 40:301–321.

Thomsen, E., and Thomsen, M. 1974. Fine structure of the fat body of the female of *Calliphora erythrocephala* during the first egg-maturation cycle. *Cell Tiss. Res.* 152:193–217.

Wigglesworth, V. B. 1936. The function of the corpus allatum in the growth and reproduction of *Rhodnius prolixus*. *Quart. J. Microscop. Sci.* 79:91–121.

Wigglesworth, V. B. 1954. *The Physiology of Insect Metamorphosis*. Cambridge University Press.

Wilkens, J. L. 1968. The endocrine and nutritional control of egg maturation in the fleshfly *Sarcophaga bullata*. *J. Insect Physiol.* 14:927–943.

10
Thirst

No wonder that living creatures find things that are fluid and immersed in moisture friendly to the watery core of their own being.

Santayana, *Praises of Water*

Small terrestrial animals that do not live in a moist habitat where the humidity of the air is high constantly face a pressing problem in water conservation. Because their surface area is large in comparison with their volume the potential for desiccation by evaporation is very high. Several strategies have been evolved in response to this danger: behavioral, structural, and physiological. Among behavioral adaptations is the adoption of nocturnal and crepuscular habits. The times of darkness and twilight are those when atmospheric humidity is normally highest. Leading circumscribed lives under rocks, in rotten logs, and in association with plants whose rates of transpiration ensure a humid microclimate is another behavioral gambit.

Morphological adaptations include the acquisition of a cuticle rendered nonpermeable to water by tanning, by the incorporation of lipids, and by deposition of a hard waxy or soft greasy coating of compounds that frustrate the egress of water. The tremendous importance of epicuticular waxes and greases is strikingly demonstrated by covering the surface with abrasive dusts that disrupt the waxy layer or by exposing the greases to solvents. These treatments cause such an increase in transpiration through the general surface of the body that the insects rapidly succumb to desiccation. Additionally, communication between the interior of the body and the external environment is greatly restricted by the possession of a treacheolar respiratory system in place of gills or lungs, provision for regulating the opening and closing the spiracular openings in relation to the state of hydration of the body (Miller, 1961; Bursell, 1957) or in direct relation to atmospheric humidity, and extension of the protective cuticular covering throughout the respiratory network, and into the foregut and hindgut. It is estimated that 60 to 70% of the water lost by insects finds its way out through the spiracles, a fact testifying, as Edney (1957) pointed out, to the effectiveness of the waterproofing of the rest of the cuticle.

Physiological mechanisms include conversion of excretory products to uric acid, which, being almost insoluble in water and nontoxic in form, can be eliminated without the necessity of wasting water as a solvent. Additionally the Malpighian tubules, the principal excretory organs, salvage water from the wastes in their lumina and restore it to the blood. Similarly the hindgut and rectum reabsorb water, and in some species of insects there are elaborate mechanisms associated with different parts of the gut to extract the last traces of water (see Edney, 1957).

An important part of the water relations of insects is the maintenance of an optimal ionic balance. Insects that live in desiccating environments, or consume such waterless food as dry wood (e.g., some termites and powder post beetles) or stored grain, or live in salt water, are constantly threatened with hypertonicity, whereas those living in fresh water or eating watery food (e.g., aphids) are constantly fighting hydration. Some regulation is obviously required. Even so, insects as a whole are able to tolerate much greater osmotic stress than mammals. Despite the high tolerance of which insects are capable there are still very effective regulatory mechanisms which operate to maintain osmotic homeostasis. One mechanism of regulation depends upon the presence of a balance between amino acid and serum protein, capable of counteracting large osmotic variation occurring when salt content changes as a result of desiccation or hydration. Further regulation is accomplished by the Malpighian-tubule-rectal system, which is analogous to the vertebrate glomerular kidney.

Water is also conserved because it is not employed for heat control as in the sweating and salivation of mammals. Still another means of conserving water, especially in times of acute stress, is to enter into a state of dormancy, either aestivation or diapause, a hormonally and environmentally controlled state in which behavioral activity ceases and metabolic activity is reduced to a minimum. In this state some insects can survive extreme desiccation for long periods of time. The midges described by Hinton (1951) have survived for as long as one and one-half years in a state of nearly absolute desiccation.

Conservation of water is not in itself a sufficient means for maintaining an optimum water balance. Water must be taken into the body. Some insects, as, for example the firebrat, *Thermobia domestica,* and larvae of *Anisotarsus cupripennis* can take up water from subsaturated atmospheres through the surface of the body (Núñez, 1956; Noble-Nesbitt, 1969; Okasha, 1972). With the firebrat the uptake of water is concerned primarily with the regulation of the volume of the body and not with water balance. The net uptake ceases when a particular body volume is reached; it is not dependent upon the content of water. As a matter of fact, the percentage water content increases when the insect is starved because the volume formerly occupied by solids is refilled with water. Regulation of volume may be achieved at the expense of altering the proportion of water, and the animals then become overhydrated. Whether such dropsical insects are or are not in physiological distress is an open question (cf. Mellanby, 1958; Okasha, 1972).

Normally an adequate supply of water is acquired with the usual diet. Not many insects drink for the express purpose of increasing their water content (Leclerq, 1946; Edney, 1957; Mellanby and French, 1958). Among the exceptions are honeybees, some butterflies, and muscoid flies which excrete a liquid urine and must drink often to counterbalance the fluid loss. Honeybees ingest large quantities of water for the purpose, among others, of bringing it to the hive where it is dispensed for evaporation as an air-conditioning process. Few experiments have been conducted on the behavior and physiological control of drinking. The most extensive studies have been made of *Phormia,* the sheep blowfly (*Lucilia cuprina*), the tsetse fly, and *Rodnius.*

The pattern of behavior followed by a thirsty fly consists of the same principal components as those constituting feeding behavior, namely, orientation from a distance, arresting of locomotion upon contact with water, turning of the body to align the proboscis with the water, extension of the proboscis, spreading of the labellar lobes, and sucking. In the blowfly the difference between drinking and eating (since all food is either liquid or liquified by salivation before ingestion) lies in the nature of the stimuli that initiate each behavior and ultimately terminate drinking. The stimulus for initiation is, of course, water, and it is effective both as a vapor and as a liquid.

A unique feature of water as a stimulus is its presence both in the organism and the environment. Its existence on both sides of the integument poses problems that are not encountered in the case of other stimuli (Dethier, 1963). This is particularly true of responses to humidity. Conceivably water may act on an organism independently of what is within (as do, for example, odor molecules), or it may cause an outward movement of internal water so that the insect does not respond directly to a change in external environment but rather to a change induced in its own internal environment. In the first instance humidity receptors would be conceived as responding to a concentration of water molecules. The relationship of water vapor to total atmospheric gases that is relevant in this case is the absolute concentration. If, on the other hand, the insect possesses receptors that are responding to internal osmotic changes or volumetric or other physical changes brought about by evaporation or transpiration, the relevant condition in the atmosphere is the drying power of the air.

Conventionally there are a number of definitions of humidity (Humphreys, 1940). The *absolute humidity* is either the mass of water vapor per unit volume or the gas pressure exerted by the water vapor per unit area. *Relative humidity* is the ratio of the actual mass of water vapor in a small volume to the maximum mass that can exist in the same volume at the same temperature. It is also defined as the ratio of the actual to the maximum pressure of water vapor per unit area that can exist in the presence of a flat surface of pure water at the same temperature. *Saturation deficit* may be defined as: (1) the amount of water vapor in addition to that already present, per unit volume, necessary to produce saturation at the existing temperature and pressure; (2) the difference between actual and saturation pressures; (3) the ratio of the vapor pressure deficit to the saturation pressure at the existing temperature.

Which one of these parameters is critical in enabling an insect to avoid excessively humid areas, excessively dry areas, and to orient to zones of preferred humidity or to sources of water when thirsty is difficult to ascertain. Nor is there much available information on the mechanism of action of hygroreceptors. They could respond directly to water molecules striking the surface; they could permit evaporation of internal water, with a consequent change in chemical composition or osmotic pressure in the milieu of the receptor or by mechanical deformation; they could respond to temperature changes arising from evaporation; or they could utilize special hygroscopic properties to act as hygrometers.

Little is known about the humidity receptors of flies. Oriented responses to water vapor are mediated principally by hygroreceptors in the antennae (Begg and Hogben, 1946; Cragg and Cole, 1956; Perttunen and Syrjämäki, 1958; Syrjämäki, 1962), but additional receptors are clearly involved. The normal orthokinetic response of the tsetse fly *Glossina morsitans* to water vapor is abolished when the branched hairs guarding the thoracic spiracles are removed (Bursell, 1957).

Generally speaking, a water deficit in the body elicits a positive response to high humidity, whereas a maximum amount elicits a dry reaction. Other modifying factors are starvation, age, circadial rhythms, stage or reproductive cycle, stage of development, and sex; for a more complete discussion of humidity responses in general, consult Dethier (1963).

There is no doubt that flies can orient to a source of water from a distance although the critical limits are not known. From a behavioral point of view, orientation to water operates in the same manner as orientation to odors, and in this particular respect water vapor may be

Table 20. The Effect of Desiccation on Water and Sugar Consumption by the Blowfly.[a]

Experiment number	Treatment of flies	Av. wt. of fly minus wings (mg.)	Av. duration (sec.) of sucking of each solution presented successively		
			H_2O	0.1 M sucrose	1.0 M sucrose
1	3-day-old flies fed once on 0.1 M sucrose, starved 24 hr., then desiccated 24 hr.	12.1 (10.2–16.9)	24 (6–52)	46 (23–73)	35 (17–62)
2	3-day-old flies fed once on 0.1 M sucrose, starved 24 hr., then desiccated 24 hr.	11.0 (9.7–17.0)	—	—	54 (40–90)
3	3-day-old flies fed once on 0.1 M sucrose, starved 24 hr., then humidified 24 hr.	23.4 (19.6–27.7)	0	0	38 (20–60)

From Dethier and Evans, 1961.

[a] Each value is based upon tests with 30 individual flies. The figures in parentheses represent ranges.

considered as a special case of olfaction. Also, as with odors, water vapor is an adequate stimulus for extension of the proboscis. All other segments of the drinking pattern are initiated by liquid water stimulating the water receptors of the tarsi, labellar hairs, or oral papillae.

Conditions that might conceivably modify drinking behavior are contaminants, that is, unacceptable substances, in the water; desiccation and state of water balance in general; and state of hunger. Of these only the state of food-deprivation has been found to be unimportant. Experiments involving food-deprived flies are necessarily limited because the average life-span in the absence of food is only 2.5 to 3 days. During this period starving flies take approximately the same amount of water each day (Dethier and Evans, 1961). In another experiment designed to control the effects of desiccation, 30 flies were starved for 24 hours and then placed in a humidifier for 24 hours. Another group of flies was placed in a desiccator. At the conclusion of the treatment each fly was weighed and then allowed to drink, successively, water, 0.1 M sucrose, and 1.0 M sucrose (Table 20). The desiccated flies had lost about half of their body weight by evaporation. When presented with water they sucked for an average of 24 seconds.

They then drank 0.1 M sucrose for 46 seconds, and 1.0 M for 35 seconds. The total period of ingestion of fluid was thus 105 seconds. Similarly, desiccated flies that were presented only with 1.0 M sucrose sucked for 54 seconds. For these flies the concentrated sucrose was the only available source of water. It is interesting that the total amount of fluid ingested was less than that of the flies that also had water available. This observation raises a number of intriguing questions for investigation. The duration of sucking is supposed to be a function of the time required for sensory adaptation, which in turn is a function of the concentration of the stimulus. The difference in duration of sucking in these two samples of desiccated flies could be explained by assuming that the flies that had first had 0.1 M sucrose were already partially adapted when they were presented with the 1.0 M solution — hence the abbreviated sucking of the latter. On the other hand, from an adaptive point of view one might have expected the flies presented solely with 1.0 M sucrose to have consumed more because they had only this solution to compensate for a water deficit as well as a food deficit; however, high concentrations of sugar inhibit stimulation of the water receptor so that from a sensory point of view 1.0 M sucrose is a waterless solution. A host of additional experiments in this area would be fruitful.

The humidified flies drank no water when the opportunity presented itself. Surprisingly, they also refused 0.1 M sucrose. The amount of 1.0 M sucrose ingested was less than that taken by the desiccated flies that were presented with 1.0 M sucrose only (Table 20, experiment 2).

In a more drastic experiment flies were confined in a humidifier until the last had died of starvation five days later. At no time did any drink water. Taken together these experiments show that starvation per se does not induce drinking of water.

There is another aspect of feeding that might be expected to influence drinking behavior, namely, the question of whether the ingestion of dry food enhances drinking. Rats exhibit a type of behavior known as prandial drinking, that is, drinking that serves to assist mechanically the ingestion of dry food and is unrelated to water balance (cf. Teitlebaum and Epstein, 1962). Prandial drinking apparently is not indulged in by flies. A comparison of two groups of flies, one of which had free access to water but no food, and the other, dry food but no water, showed no difference in drinking behavior (Table 21). The ingestion of dry food did not cause increased drinking even though the process of eating solid sugar requires copious regurgitation.

A powerful determinant of the volume of water drunk is the composition of the water. By analogy with human experience one might anticipate less finickiness as the degree of thirst increases. This expec-

Table 21. The Effect of NaCl on Water Intake by the Blowfly.[a]

Materials available to the fly	Mean volume of water consumed in 3 days (μl)	Mean volume of NaCl consumed in 3 days (μl)	Mean total fluid intake in 3 days (μl)
Water (no food)	29 (12–39)	—	29
Water and dry sucrose	16 (12–61)	—	16
0.1 M NaCl and dry sucrose	—	23 (9–36)	23
0.5 M NaCl and dry sucrose	—	19 (12–38)	19
0.5 M NaCl, dry sucrose, and water	12 (5–18)	6 (4–7)	18
1 M NaCl and dry sucrose	—	12 (3–24)	12
1 M NaCl, dry sucrose, and water	14 (6–18)	4 (0–5)	18
2 M NaCl and dry sucrose	—	3 (0–12)	3
2 M NaCl, dry sucrose, and water	15 (3–24)	4 (0–7)	19

From Dethier and Evans, 1961.

[a] Each value is based on tests with 10 individual flies. The figures in parentheses represent ranges.

tation was borne out by a series of experiments in which sodium chloride was used as a contaminant (Dethier and Evans, 1961). The first experiment utilized the technique of preference-aversion tests, the choice in this case being between water and salt. The salt solutions tested ranged from 10^{-5} M to 5 M (Fig. 188). Below 10^{-4} M the data were very variable, no clear-cut choice was discernible, and it is probably safe to assume that the salt solutions in this range are below detection threshold. Whether or not there is an actual preference for very low concentrations of salt is not certain. Bimodal curves relating choice of salt solutions to concentration have been noted for mammals, including man, for whom low concentrations of salt are said to be sweet. Also, it has been reported that the cockroach *Periplaneta americana* accepts low concentrations of salt when it is water-satiated and rejects high (Frings, 1946). Additional experiments with *Phormia* would be enlightening, especially in view of the possibility that response curves to certain sugars are also bimodal.

There is some electrophysiological evidence bearing on this point. As mentioned earlier, it had always been assumed that a fly accepts (not prefers) dilute solutions of salt because of the activity of the water receptor, which is not inhibited by low salt concentrations. The experiments described earlier however, show that in the absence of activity

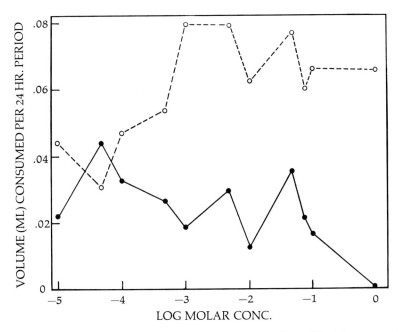

Fig. 188. Volume of different concentrations of sodium chloride ingested per fly per 24 hours in a two-choice situation. Solid line, sodium chloride; broken line, water (from Dethier and Evans, 1961).

from the water receptor low-frequency action potentials from the salt receptor alone initiate acceptance rather than rejection.

Regardless of the uncertainty of interpreting responses to low concentrations, there is no question that there is a decisive rejection of sodium chloride at concentrations above 10^{-4} M. Rejection at 10^{-4} M, together with the possibility of acceptance at lower concentrations, is particularly interesting in view of what is currently known about the chemoreceptors. It will be recalled that no electrophysiological responses have been recorded from the labellar hairs at concentrations of sodium chloride below 0.1 M. Here is another instance of failure to detect responses to concentrations to which the intact fly is known to respond. The other case is that of sugar, to which very low behavioral responses have been recorded.

An obvious answer is that receptors other than those tested electrophysiologically are the ones that mediate the behavioral response. There is also the possibility, albeit remote since no interaction among receptors has been demonstrated, that massive stimulation, that is, stimulation of a whole receptor field, somehow engenders an interaction whereby subthreshold depolarization of hundreds of receptors unstabilizes the system.

In any case, there is enough evidence to suggest that at concentrations where both salt and water receptors are active, the nature of the behavioral response may depend not on an all-or-none response of either receptor but upon subtle interactions in the central nervous system that quantitatively assess the contributions of each. Rees (1970, 1972) remarked that the water receptor is in fact a receptor that responds to low concentrations of salt (including zero concentration) and that, since distilled deionized water does not occur in nature, the fly can in fact sense an extraordinarily great range of electrolyte concentrations by allocating one receptor to the low range (the water receptor) and another to the high range (the salt receptor). Two modalities would thus be possible. By utilizing an across-fiber patterning in that range of concentrations where both fibers are active, still another modality could be differentiated. To push the concept to its limit one could introduce the activity of the fifth cell. Whatever its adequate stimuli may be, it does respond to salts. Utilization of an across-fiber pattern involving all three receptors—water, salt, and fifth cell—could yield a great deal of information about different concentrations of salt and provide the means for several different behavioral patterns correlated with various concentrations.

The rejection of salt when it is placed beside water becomes more pronounced as the concentration increases to 5 M. Concomitantly the intake of water rises, reaching a maximum when the salt is 10^{-3} M. The total fluid intake remains approximately constant and may be considered to be the maintained response to the particular ambient conditions of the experiment.

The deterrent effects of contaminants on drinking are further exemplified in a series of experiments in which flies received only salt solutions as their supply of fluid (Table 21). When the material available for ingestion was dry sucrose and 0.1 M sodium chloride, the mean volume of fluid drunk in 3 days was 23 μl; when 0.5 M salt was provided instead, the intake dropped to 19 μl. (Table 21). In a similar series of experiments where water was available in addition to a salt solution (a typical two-choice situation), the intake of water increased as the intake of salt decreased, and the total fluid intake was nearly constant and approximately equal to that taken when only water and dry sucrose were available.

In another way of investigating the relation between thirst and contaminants, flies were placed in a desiccator where they had a supply of dry sugar as food but no fluid. Periodically their behavioral responses to 1.0 M sodium chloride were tested. At the beginning of the experiment, before being placed in the desiccator, the flies were given water ad libitum; after this none of them responded when the tarsi were touched to the salt solution. As water loss increased on successive days, their behavior changed. First they extended the proboscis when the tarsi were stimulated but did not open the lobes of the labellum when the aboral hairs touched the solution; later they opened the labellar lobes but did not drink; still later they drank briefly; gradually with prolonged desiccation the duration of drinking increased. In short, as water loss increased the rejection threshold of the tarsi increased; this increase was followed by a rise in the rejection threshold of the labellum.

All of these behavioral experiments agree in showing that the response to water and the toleration of contaminants are powerfully influenced by the state of water balance. None of these observations should come as any surprise because they follow the same pattern as that of mammalian behavior under similar circumstances. Implicit in both cases is the idea that sensory discrimination is important in initiating drinking, that there is some mechanism for terminating drinking at the proper stage of acquisition of water, and that some mechanism associated with water balance causes the central nervous system to assess differently the sensory information that normally deters drinking. Very little is known about the last-mentioned phenomenon (cf. Dethier, 1968).

The control of responsiveness to potable water has been the subject of considerable experimentation. The most obvious variant that could theoretically control drinking is the osmotic concentration of the blood. This was assessed in *Phormia* by injecting various substances into the haemocoele. The results are summarized in Table 22 (Dethier and Evans, 1961). Injection of water into flies that responded positively to water (that is, were thirsty) abolished all responses to water (experiments 1–4). The percent that failed to respond increased as the volume injected was increased. The effects of repeated injections were additive (experiment 3), and the effect was immediate (experiment 4). More surprising was the discovery that injections of even enormous volumes of highly concentrated solutions also rendered thirsty flies unresponsive to water (experiments 5–7). The fact that these hypertonic solutions abolished responsiveness to water rather than increasing it, as might have been expected, indicated that volumetric, not osmotic or ionic factors, were the controlling feature. Even mineral oil, before its toxic effects were apparent, rendered thirsty flies unresponsive to water (experiment 12). Responsiveness to water was also abolished by a meal of nearly saturated glucose (experiment 13).

Although these results implicated volume or pressure in the haemocoele, the effects could have been nonspecific, that is, they could have inhibited ingestion of anything. Experiments 3, 5, and 6 did not support this idea because responsiveness to sugar was unaffected. A decisive experiment would be one in which alteration of volume or pressure in a negative direction induced responsiveness in water-sa-

Table 22. The Effect of Injections on Responses of the Blowfly to Water.

Experiment number	Number of flies	Response before treatment	Injected[a]	Percent negative after treatment
1	35	+	2.5 µl water	6
2	52	+	8 µl water	85
3	27	+	2 µl water	7
			2 µl water	22
			2 µl water	70[b]
4	26	+	3 µl water immediately	58
			at 10 min.	58
			at 60 min.	54
5	120	−	3 µl 2 M glucose	82[b]
6	116	−	7 µl 2 M glucose	96[b]
7	29	−	2.4 µl 4 × saline	100
8	53	+	3 µl 4 × saline	55
9	69	+	3 µl 2 M glucose	55
10	48	+	3 µl 2 M glucose in saline	56
11	40	+	6 µl 2 M glucose in saline	85
12	92	+	4 µl mineral oil, moribund at 15 min.	66
13	14	+	Fed 2 M glucose 0–60 min.	100

From Dethier and Evans, 1961.

[a] Water indicates distilled water; the saline was Bodenstein's; 4 × indicates saline four times more concentrated; experiments 1–9 from Evans (1961).

[b] Responded subsequently to 0.1 M sucrose.

tiated flies. Bleeding actually accomplished this. It was reasoned that an effect could be most conclusively demonstrated by placing flies just at the threshold of thirst rather than having them maximally satiated. Accordingly a group was mildly desiccated until a few responded to water. Bleeding now caused the unresponsive flies to react vigorously to water within a very few seconds. It had little effect on flies that were already responsive, and in only a few (experiment 4 or 13) that were given water to repletion did it re-establish responsiveness. Apparently there is a threshold such that the volume of blood must be reduced below some critical level before bleeding becomes effective.

If the increase in volume in the haemocoele inhibits drinking by activating abdominal stretch receptors, it might be expected that filling the crop would achieve the same end. It might also be argued that cutting the recurrent nerve would induce polydipsia because the foregut receptors are mechanoreceptors and hence insensitive to chemical and osmotic characteristics. When in fact the recurrent nerve was cut in 80 flies that were unresponsive to water before the operation, 60%

became bloated. The same operation on 50 water-satiated flies produced a bloated state in 50%. Sensory control of drinking (that is, sensory input to drive and adaptation to stop) still operates in these flies. An operated fly becomes bloated by repeated rather than continuous drinking as is the case with ingestion of sugar. After a while the fly no longer imbibes even though kept in contact with water; nevertheless, feeble responses continue for more than 24 hours. If flies in this state are presented with sugar, they resume vigorous sucking until the crop and abdomen burst. As mentioned in the discussion on control of feeding, there is a strong presumption that back pressure in the crop and/or the inhibition that it generates prevents the sensory input from driving the pump. The stimulus of water is not intense enough to effect sucking under these circumstances; but sugar, a stronger stimulus, can do so. Evans and Barton Browne (1960) also reported that flies with cut recurrent nerves repeatedly responded to water. Whether or not they ingested abnormal amounts was not stated. In a more recent report Dethier and Gelperin (1967) stated that flies in which the recurrent nerve had been cut did not respond abnormally to water. Again no quantitative comparison between operated and normal flies was made and no criterion for polydipsia was established.

One possible explanation of the confusion attending reports about polydipsia is afforded by some experiments that Belzer (1970) conducted in his investigation of hunger for proteins. He compared successively the responses to various solutions by flies in which the recurrent nerve had been cut at the frontal ganglion. Such flies, offered 1.0 M NaCl, refused it; offered 0.25 M NaCl, they drank 13 μl per fly and stopped; then offered 0.1 M sucrose, they each drank an additional 23 μl. Flies in another experiment offered water became polydipsic (no value for intake was given) but eventually refused water; offered 0.1 M sucrose, they drank an additional 8.3 μl. It would appear that any liquid in the foregut elicits activity from the mechanoreceptors, which then transmit impulses to the brain where they counteract the effect of incoming messages from water receptors and sugar receptors, and complex combined messages that are codes representing protein. The different solutions have different excitatory values in terms of sensory input. Thus, 1.0 M NaCl already supplies enough inhibition via the sensory route so that removal of recurrent nerve inhibition still does not leave the way open for any excitation. Water or 0.25 M NaCl has a low excitatory level; hence removal of recurrent nerve inhibition permits more than normal to be taken. Sucrose has a high excitatory level, so that more is taken than normal when the recurrent nerve is cut. In short, the recurrent nerve regulates the intake of water as well as of other fluids, but there are other inhibitory systems also operating on ingestion as, for example, the abdominal nerves and the specific protein inhibitors postulated earlier. It is equally reasonable to suppose that there is an additional system for water. It now seems clear that a polydipsia is in fact established. By this is meant an abnormally large ingestion of water but not necessarily a volume equivalent to the volume of sugar that would be taken under the same circumstances.

Day (1943) and Thomsen (1952) had earlier observed polydipsia in some flies after allatectomy. Allatectomy frequently involves a variable degree of injury to the recurrent nerve. This could account for the low incidence (ca. 10%) of bloating observed by these workers. We were not able to produce polydipsia by allatectomy. Removal of the median neurosecretory cells of the brain also occasionally resulted in bloating. A hormonal involvement is not necessarily indicated, as distinct from a neural one, because these cells are connected to the recurrent nerve. If alterations in drinking behavior can indeed be brought about by removal of the corpus allatum or median neurosecretory cells, an indirect effect cannot be ruled out because in some species the corpus allatum affects the production of urine (Altmann, 1956), and the median neurosecretory cells produce a diuretic hormone (Núñez, 1963) as does also the mesothoracic ganglionic mass (Maddrell, 1958). The effect could be indirect, in the sense that the hormone, rather than acting on the brain to augment or depress the sensory effect of water, could alter the state of water balance, which in turn acts on the brain in some as yet unknown manner. Altmann (1956) had demonstrated that injection of extract of corpus allatum into the haemocoele of honeybees causes an increase in the intake of water and that an extract of corpus cardiacum causes a decrease. That the effect is in fact indirect is suggested by the in vitro observation that extract of corpus allatum increases the rate of urine production by the Malpighian tubules, and extract of corpus cardiacum decreases it.

The experiments carried out with *Phormia* favor a volume hypothesis. Experiments conducted by Barton Browne (1964) and Barton Browne and Dudzinski (1968) with the sheep blowfly, *Lucilia cuprina,* favor a concentration hypothesis. The mechanism must be quite specific because, as already pointed out, insects are able to tolerate very wide variations in water content. In well-hydrated individuals tissues are bathed in a liberal volume of fluid, whereas in desiccated individuals the tissues may be nearly dry (Mellanby, 1937, 1958). The tolerated range in osmotic pressure is impressive; in the housefly it varies from 4 to 20 atmospheres, depending upon the period of water deprivation (Bolwig, 1953).

Barton Browne and Dudzinski found that the responsiveness of sheep blowflies to water was poorly correlated with the volume of water in the crop and with the volume and osmotic pressure of the blood. On the other hand, a good correlation existed between responsiveness to water and the concentration of chloride and sodium ions in the blood. Injections of sodium chloride solutions of concentrations greater than 0.3 M and of Bodenstein's solution of equivalent molar concentration caused negative flies immediately to become positive in their response and markedly increased the intake of already positive flies (Barton Browne, 1964). Injections of water, sugar solutions, and solutions of nontoxic sodium salts other than sodium chloride, irrespective of concentration, made flies less ready to respond to water. No dramatic results followed bleeding. Removal of 2 mg of haemolymph caused no immediate increase in intake, but there was an indication that flies that remained indifferent to water for abnormally long periods after drinking would occasionally drink immediately after bleeding. Interpretation of these results was confounded by the fact that "stabbed" controls as well as bled flies became abnormally responsive to water for a prolonged period following treatment.

In an attempt to marshall evidence in support of his hypothesis that the water and the sugar receptors in the D hairs of the tarsi of *Calliphora vicina* (= *erythrocephala*) are one and the same, and that the distinction between water and sugar is accomplished by across-fiber patterning, van der Starre (1972) has applied new interpretations to all of the available data relevant to water and sugar stimulation. He questioned the validity of threshold measurements obtained with fixed flies and the attempts to relate behavioral and electrophysiological results, overlooking perhaps the frequently expressed admonition that the interpretations of thresholds rest on an impeccably sound base only when derived from comparisons within any given set of circumstances. More particularly, in assessing the experiments relating to the ingestive behavior of humidified and desiccated flies he decided that the conclusions to be drawn are that starvation and desiccation affect flies' abilities to distinguish between water and sugar (desiccated starved flies discriminate better than humidified starved flies), and therefore that water and sugar thresholds are interdependent. He further questioned the conclusion that the ingestion of water and of sugar are regulated independently. In a series of experiments in which he attempted to show that water can bring about adaptation to sugar he rejected a role of central excitatory states on the grounds that it is more complicated than a peripheral role. The evidence given is not compelling.

Whatever may be the situation with respect to discrimination between water and sugar via tarsal and labellar hairs, the nature of the mechanisms controlling thirst is unclear. Van der Starre focuses on some weaknesses in the current hypothesis. Beyond a doubt more experimental work is required before our understanding of the factors involved in the ingestion of water become clear.

Some change in the blood, reflecting the state of water balance or water content, seems to be a highly probably factor. Whether the change is in volume or concentration, or differs from one species of fly to the next, remains to be seen. At which point these changes interact with the central nervous system is also a mystery. If change in concentration is the critical factor, this change could directly affect receptors in the brain analogous to the osmoreceptors in the hypothalamus of mammals. Or it could act indirectly by controlling the rate at which the crop empties, as is the case with sugar. If this is the mechanism, the final action would take place via the mechanoreceptors in the foregut and would be responsive to cutting of the recurrent nerve. If the characteristic of the blood that is important is its volume, one might expect the stretch receptors in the abdomen to be involved, whereupon transection of the ventral nerve cord should result in polydipsia.

The implication of hormones in the regulation of water balance is clear. Whether or not they are directly involved in the behavioral responsiveness to water is another matter. In mammals, angiotensin powerfully affects drinking behavior in a direct manner (Epstein, Fitzsimons, and Simons, 1969; Epstein, Fitzsimons, and Rolls, 1970; Fitzsimons, 1969). An investigation of the direct effects of hormones on the brain of the fly would open new fields for speculation. In the final analysis it is the brain that integrates all incoming information and makes the decisions.

Literature Cited

Altmann, G. 1956. Die Regulation des Wasserhaltes der Honigbiene. *Insectes Sociaux* 3:33–40.

Barton Browne, L. 1964. Water regulation in insects. *Annu. Rev. Ent.* 9:63–82.

Barton Browne, L., and Dudzinski, A. 1968. Some changes resulting from water deprivation in the blowfly, *Lucilia cuprina*. *J. Insect Physiol.* 14:1423–1434.

Begg, M., and Hogben, L. 1946. Chemoreception of *Drosophila melanogaster*. *Proc. Roy. Soc. London* B133:1–19.

Belzer, W. R. 1970. The control of protein ingestion in the black blowfly, *Phormia regina* (Meigen). Doctoral dissertation, University of Pennsylvania, Philadelphia.

Bolwig, N. 1953. On the variation of the osmotic pressure of the haemolymph in flies. *S. African Ind. Chemist.* 7:113–115.

Bursell, E. 1957. The effect of humidity on the activity of tsetse flies. *J. Exp. Biol.* 34:42–51.

Cragg, J. B., and Cole, P. 1956. Laboratory studies on the chemosensory reactions of blowflies. *Ann. Appl. Biol.* 44:478–491.

Day, M. F. 1943. The function of the corpus allatum in muscoid Diptera. *Biol. Bull.* 84:127–140.
Dethier, V. G. 1963. *The Physiology of Insect Senses.* Methuen, London.
Dethier, V. G. 1968. Chemosensory input and taste discrimination in the blowfly. *Science* 161:389–391.
Dethier, V. G., and Evans, D. R. 1961. Physiological control of water ingestion in the blowfly. *Biol. Bull.* 121:108–116.
Dethier, V. G., and Gelperin, A. 1967. Hyperphagia in the blowfly. *J. Exp. Biol.* 47:191–200.
Edney, E. B. 1957. *The Water Relations of Terrestrial Arthropods.* Cambridge University Press.
Epstein, A. N., Fitzsimons, J. T., and Simons, B. J. 1969. Drinking caused by the intracranial injection of angiotensin into the rat. *J. Physiol.* (London) 200:98–100 P.
Epstein, A. N., Fitzsimons, J. T., and Rolls, B. J. 1970. Drinking induced by injection of angiotensin into the brain of the rat. *J. Physiol.* (London) 210:457–474.
Evans, D. R. 1961. Control of the responsiveness of the blowfly to water. *Nature* 190:1132–1133.
Evans, D. R., and Barton Browne, L. 1960. The physiology of hunger in the blowfly. *Amer. Midland Nat.* 64:282–300.
Fitzsimons, J. T. 1969. The role of a renal thirst factor in drinking induced by extracellular stimuli. *J. Physiol.* (London), 201:349–368.
Frings, H. 1946. Gustatory thresholds for sucrose and electrolytes for the cockroach, *Periplaneta americana* (Linn.). *J. Exp. Zool.* 102:23–50.
Hinton, H. E. 1951. A new chironomid from Africa, the larva of which can be dehydrated with injury. *Proc. Zool. Soc. London* 121:371–380.
Humphreys, W. J. 1940. *Physics of the Air,* 3rd ed. McGraw-Hill, New York.
Leclerq, J. 1964. Des insectes qui boivent de l'eau. *Bull. Ann. Soc. Ent. Belge* 82:71–75.
Maddrell, S. H. P. 1958. A diuretic hormone in *Rhodnius prolixus* Stal. *Nature* 194:605–606.
Mellanby, K. 1937. Water and fat content of tsetse flies. *Nature* 139:883.
Mellanby, K. 1958. Water content and insect metabolism. *Nature* 181:1403.
Mellanby, K., and French, R. A. 1958. The importance of drinking water to larval insects. *Ent. Exp. Appl.* 1:116–124.
Miller, P. L. 1961. Spiracle control in dragonflies. *Nature* 191:621–622.
Noble-Nesbitt, J. 1969. Water balance in the firebrat, *Thermobia domestica* (Packard). Exchanges of water with the atmosphere. *J. Exp. Biol.* 50:745–769.
Nũnez, J. A. 1956. Untersuchungen über die Regelung des Wasserhaushaltes bei *Anisotarsus cupripennis* Germ. *Zeit. vergl. Physiol.* 38:341–354.
Nũnez, J. A. 1963. Probable mechanisms regulating water economy of *Rodnius prolixus. Nature* 197:312.
Okasha, A. Y. K. 1972. Water relations in an insect, *Thermobia domestica.* II. Relationships between water content, water vapor from subsaturated atmospheres and water loss. *J. Exp. Biol.* 57:285–296.
Perttunen, V., and Syrjämäki, J. 1958. The effect of antennectomy on the humidity reactions of Drosophila melanogaster (Dipt., Drosophilidae). *Ann. Ent. Fennica* 24:78–83.
Rees, C. J. C. 1970. The primary process of reception in the type 3 ("water") receptor cell of the fly, *Phormia terranovae. Proc. Roy. Soc. London* B174:469–490.

Rees, C. J. C. 1972. Responses of some sensory cells probably associated with the detection of water. In *Olfaction and Taste,* IV, D. Schneider, ed., pp. 88–94. Wissenschaftliche Verlagsgesellschaft MBH, Stuttgart.

van der Starre, H. 1972. Tarsal taste discrimination in the blowfly, *Calliphora vicina* Robineau-Desvoidy. *Netherlands J. Zool.* 22:277–282.

Syrjamaki, J. 1962. Humidity reception in *Drosophila melanogaster. Ann. Zool. Soc. Zool.-Botan. Fennicae Vanamo* 23:1–74.

Teitlebaum, P., and Epstein, A. N. 1962. The lateral hypothalamic syndrome. *Psychol. Rev.* 69:74–90.

Thomsen, E. 1952. Functional significance of the neurosecretory brain cells and the corpus cardiacum in the female blowfly, *Calliphora erythrocephala* Meig. *J. Exp. Biol.* 29:137–172.

11
Winter: Diapause

A sunbeam on a winter's day,
Is all the proud and mighty have
Between the cradle and the grave.
John Dyer, *Grongar Hill*

The homeostatic mechanisms that ensure an optimum quality and quantity of food to meet the exigencies of daily living and the special cyclic demands of reproduction are tuned to operate efficiently against a wide but not unlimited environmental background. The limits are exceeded seasonally in many parts of the world. The most disruptive influence on the life style of animals inhabiting temperate latitudes is wrought by winter. Temperatures then drop below the allowable limits for development and reproduction. In the tropics impediments to normal living may be drought, excessive moisture, excessive heat, or seasonal absence of food.

At these times of environmental stress some animals migrate, some aestivate or hibernate in states of dormancy, others enter into a state of reduced or arrested development called diapause (Beck, 1968). Each of these states is preceded and accompanied by alterations in feeding behavior. Migrating birds and hibernating mammals embark upon a period of excessive eating that results in obesity. Fat constitutes a reserve that migrating birds draw upon to satisfy the energy demands of long journeys; hibernating mammals also draw on this reserve to maintain basal metabolism during the dark months of winter. Diapausing insects also exhibit aberrations in feeding habits, tuned to the special needs of that state.

Diapause differs from dormancy initiated directly by heat or cold in that it occurs in *anticipation* of the environmental stress against which it will protect the insect. This capacity, or its absence, is genetically determined in each species or race. In diapausing individuals there must be intrinsic physiological mechanisms to switch diapause on and off, and they must be coupled to and responsive to environmental signals that inevitable and reliably precede seasonal stress.

In some insects diapause is obligatory. Intrinsic events dominate these individuals; when some invariably scheduled signal is given, the machinery for establishing diapause starts regardless of whether intolerable environmental conditions ensue. Diapause continues, even if the period of stress has come to an end or never occurred, until some other prearranged environmental signal is given. Thus an insect with obligatory winter diapause will enter into this state even in a warm laboratory admirably appointed to its needs and will remain dormant until the normal time appointed for awakening arrives.

In facultative diapause, on the other hand, extrinsic factors dominate. Whether or not it occurs depends on environmental conditions existing at some critical stage in the insect's life. Extremes of temperature and humidity and suboptimal diet are among the conditions believed to be effective inductors.

The most reliable harbinger of winter and summer is the changing length of day. It is not surprising, therefore, that day-length is the anticipatory signal that initiates diapause. In insects that experience obligatory diapause the advent of a critical length of day starts the physiological machinery. Long days start it for some insects, and short days for others. Still others respond neither to long nor short days but to days of intermediate duration (ca. 8 hours). A rare type of diapause also exists that is the reverse of that just mentioned; some moths are induced to diapause by long *and* by short days, and are free of diapause only at some intermediate length of day.

Facultative diapause also is initiated by critical photoperiods; however, other environmental conditions existing at the time determine whether or not the length of day at that particular moment is critical. Accordingly, the time of diapause in nature depends upon prevailing weather conditions.

Some states of diapause terminate spontaneously. The termination of many is hastened by exposure to low temperatures for several weeks or months. Cold reactivation is generally the rule in the temperate latitudes where the range 0° to 12°C is most effective. Then as the temperature rises higher, normal development and behavior resume. As Danilevskii, Goryshin, and Tyshchenko (1970) have pointed out, this is a sensible arrangement because in temperate zones the yearly course of temperature is greatly delayed in relation to increasing length of day. Were insects to terminate their diapause in response to critical photoperiod, the days would still be too cold.

Phormia regina is a temperate-zone spring-fall fly. It thrives under cool conditions, but those parts of the world that offer these conditions are also those areas where winters are too severe for survival in the usual mode. It has been known for many years that the species passes the winter in the adult stage (Roberts, 1930; Hall, 1948; Mail and

Schoof, 1954; Wallis, 1962; Dondero and Shaw, 1971; Siverly, 1972). As winter approaches flies creep into nooks and crannies in walls, houses, trees, and rocks. On particularly warm and sunny days some emerge and lounge around in a semiactive state. One set of observations showed that *Phormia regina* would come out of hiding on such days between the hours of 11:00 A.M. and 2:00 P.M. and rest on the vertical brick walls of buildings (Wallis, 1962). On cloudy days at the same or higher temperatures no flies appeared. A more northern relative, *Protophormia terraenovae* exhibits the same behavior (Roubaud, 1927; Cousin, 1932; Danilevskii, 1965; Nuorteva, 1966). Only recently, however, has it been proven that a true facultative diapause occurs in these two species (Stoffolano, Greenberg, and Calabrese, 1974). The diapause is induced by a combination of low temperature and short photoperiod. It is characterized by negative phototaxis, a tendency to crowd into crevices (positive thigmotaxis?), failure to feed, failure to mate, hypertrophy of the fat body, lack of ovarian development, absence of sperm in the spermathecae, and markedly lower respiratory rates in both sexes (Calabrese and Stoffolano, 1974a, 1974b).

The fact that the reduction in respiratory rate is directly correlated with a reduction in protein synthesis in some insects (Slama, 1964a, 1964b) suggests that this relation may exist also in *Phormia*. The absence of vitellogenesis and the lack of development of the accessory reproductive glands in *P. regina* (Stoffolano, in press) and in *Pyrrhocoris apterus* (Slama, 1964a, 1964b), the degeneration of the flight muscles in the adult Colorado potato beetle (Stegwee, 1964), and the great reduction in the incorporation of amino acids into the pupal fat body of cecropia (Wyatt, 1963) are all measures of the severity of the reduction of protein synthesis in diapausing insects.

Considering all of the behavioral and physiological changes that occur during diapause, it is hardly surprising that there is a change in feeding behavior too. Many insects entering diapause, and in diapause, cease drinking and eating or eat only intermittently. Female mosquitoes, for example, entering diapause do not bite when given the opportunity (Tate and Vincent, 1932, 1936; de Buck and Swellengrebel, 1934; Washino, 1970; Washino and Bailey, 1970; Schaefer, Miura, and Washino, 1971; Washino, Gieke, and Schaefer, 1971). Face flies (*Musca autumnalis* De Geer) feed but little, and their tarsal acceptance thresholds to sugars and proteins are elevated (Stoffolano, 1968). The same is true of *Phormia*. A comparison of tarsal acceptance thresholds for sucrose of diapausing and nondiapausing flies of identical ages (39 days) showed that whereas 54% of the nondiapausing flies responded to 1.0 M sucrose, only 38% of those in diapause did (Fig. 189) (Stoffolano, 1974). Another interesting comparison was made. Flies that were 41 days old and in diapause were placed in conditions

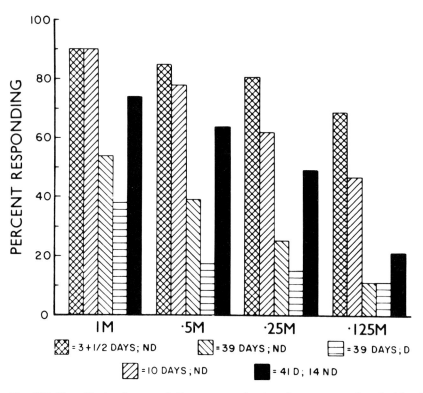

Fig. 189. The effects of age and diapause on the tarsal acceptance thresholds of *Phormia regina* to various concentrations of sucrose (from Stoffolano, 1974).

that do not induce diapause (27°C and a 16-hour day) for 14 days and tested when 55 days old. Their thresholds were lower than those of 39-day-old diapausing individuals. One inference is that some condition of diapause causes threshold to be elevated and nondiapausing conditions remove this constraint.

As might be expected, the volumes of fluid imbibed by the two categories of flies are different. Nondiapausing individuals take much longer drinks, and a correspondingly large increase occurs in the volume imbibed (Table 23). When diapausing flies are returned to nondiapausing conditions they increase their intake. Thus the threshold data and intake data complement one another.

One interpretation of these results is that some physiological condition specifically associated with diapause acts upon the central nervous system to modify its threshold to sensory input. Stoffolano proposed, for face flies, that hypertrophy of the fat body stretched the abdomen and activated the stretch receptors that send negative feedback to the central nervous system. Tests of this hypothesis as it applied to mosquitoes indicated, however, that their biting rate was not influenced by abdominal stretch (Chen, 1969).

Table 23. Mean Intake and Duration of One Uninterrupted Drink of 0.125 M Sucrose Solution by Diapausing and Nondiapausing *Phormia regina*.

Physiological condition of flies when tested	Age (days)	Intake (μl)	Duration (sec.)
Nondiapause	40	24.7	140.6
Diapause	40	6.0	37.5
Nondiapause[a]	49	22.0	168.1

From Stoffolano, 1974.
[a] These flies had been in diapause for 41 days, and were then returned to nondiapause for 8 days before being tested.

Stoffolano (1974) then began a long series of experiments designed to reveal the mechanism underlying the rise in tarsal thresholds in *Phormia* and hence the disinclination to feed (Fig. 189). One specific question was whether the elevation in behavioral threshold stemmed from peripheral or from central neural changes. The first step was to measure electrophysiologically the sensitivity of the receptors of diapausing and nondiapausing flies; however, since flies in diapause are normally older than nondiapausing flies at the peak of their activity, the study had to be preceded by an investigation of the effect of age itself. Because it is easier to record from labellar than from tarsal hairs the former were chosen for study. The results indicate that it is probably safe in this instance to extrapolate behaviorally and electrophysiologically from one system to the other. Rees (1970) had reported that there is an increase in the mean frequency of action potentials generated by the salt receptor to a given concentration up to and including the age of 3 days, after which there is a gradual decline. The maximum frequency of response to 1.0 M NaCl at day 3 was 41 impulses per second for *Phormia regina* and 63 impulses per second for *Protophormia terraenovae*. The number of hairs that were responsive also decreased with age. When flies attained the age of 25 days only 60% of the hairs responded to salt solutions. An increase in the frequency of impulses up to the third day of adult life was also recorded for the water, sugar, and salt receptors by Omand (1971). Neither of these studies mentioned the sex of the experimental animals. In the first, only largest hairs numbers 10 and 11 were examined. In the second study the identity of the hairs was not specified.

A more detailed analysis of the effect of aging on labellar hairs gave slightly different results (Stoffolano, 1973). Tests of largest hairs numbers 1 to 4 and 8 to 11 revealed that from day 3.5 to day 25 age-related changes occurred in the frequency of impulses from the salt receptors of males and females but that the change with age did not

Table 24. Effect of Age and Sex on the Mean Impulse Frequency and Percentage of Inoperative Labellar Chemoreceptor Sensilla in Nondiapausing *Phormia regina*; Test Solutions Were 1 M NaCl and 0.5 M Sucrose.

No. flies tested		Age (days)	Frequency (impulses/sec.)		% Sensilla inoperative		No. Sensilla recorded from	
			salt	sugar	salt	sugar	salt	sugar
Males:	10	3.5[a]	41.2	40.7	12.5	16.3	140	134
	9	18	48.7	37.9	34.7	36.8	94	91
	12	25	51.7	31.4	51.7	54.6	92	89
Females:	6	3.5[a]	41.2	36.3	5.2	6.3	91	90
	10	15	59.2	48.4	21.9	24.4	125	121
	10	26	47.8	39.4	26.3	28.1	118	115

From Stoffolano, 1973.

[a] Flies 3.5 days old were offered only water. All other flies were given free access to sugar and liver until 1 day before testing.

progress in an orderly fashion. Frequency responses of the sugar receptor was significantly influenced by the age of the fly, its sex, and the combination of age and sex. In males the frequency decreased progressively with age; in females the frequency varied in an irregular pattern (Table 24). The number of hairs that were responsive also changed with age. As the flies aged, fewer salt and sugar receptors were operative. Details of the progressive failure differed in males and females (Table 24).

Although the three studies are not in complete agreement, they all lead to the conclusion that age exerts a pronounced effect on the number of sensilla that remain responsive and on the vigor of response of those that do. Clearly any comparison of diapausing and nondiapausing flies must be made with flies of identical ages. When this was done, no difference was found in the frequency of action potentials generated by the sensilla of the two groups of flies at age 60 days (Table 25). Incidentally, a difference in both salt and sugar receptors was found between flies fed on sugar and those fed on sugar and liver (Table 25). A significant difference was also evident between the diapausing and nondiapausing flies with respect to the number of hairs that were responsive, the number being less in the diapausing flies. Thus these flies emerging from their nooks and crannies in the spring are handicapped with a large percentage of nonfunctional receptors. Despite this impediment they regain normal thresholds and regulation of intake.

The difference in the behavioral responsiveness of diapausing and nondiapausing flies could be explained on the basis of the degen-

Table 25. Effect of Diapause and Diet on the Mean Impulse Frequency and Percentage of Inoperative Chemoreceptor Labellar Sensilla in Female *Phormia regina*; Test Solutions Were 1 M NaCl and 0.5 M Sucrose.

No. flies tested	Condition	Age (days)	Frequency (impulses/sec.)		Inoperative sensilla		No. sensilla recorded from	
			salt	sugar	salt	sugar	salt	sugar
6	Nondiapause	60[a]	53.4	34.0	24.0	27.1	73	70
9	Diapause	63[b]	59.2	39.7	20.1	21.5	115	113
11	Diapause	60[a]	49.7	27.3	51.7	60.2	85	70

From Stoffolano, 1973.
[a] Flies fed sugar and liver.
[b] Flies fed sugar only.

eration of receptors in the former (Table 25); however, the return to behavioral sensitivity following the termination of diapause argues against a peripheral determinant. Were the periphery involved, it would be necessary to postulate that the receptors regained their sensitivity or that receptors that had not deteriorated became more sensitive (i.e., fired at a higher frequency) and so compensated for the deficit in numbers. No information is available on the first point, and evidence on the second point is contradictory (cf. Gillary, 1966; Rees, 1970).

An alternative hypothesis is the one proposed by Stoffolano (1968) for the face fly, namely, that some central inhibitory influence is present during diapause. This is presumed to arise from stretch derived from hypertrophy of the fat body. It has already been shown that the presence of mature eggs in the ovaries can inhibit ingestion, and Stoffolano (1968) has demonstrated that in gravid females of *Musca autumnalis* the tarsal threshold was high. It is not an unreasonable supposition, therefore, that an enlarged fat body could perform the same service, but experimental proof is lacking. Should the fat body eventually be incriminated, the diagram in Figure 187 could be modified to include under the heading "stretch receptors" not only full crop but also mature ovaries and hypertrophied fat body.

Whatever the mechanism, it operates in harmony with other aspects of the fly's physiology to enable the insect to survive the winter. As fall with its shortened days and reduced temperatures approaches, olfactory behavior changes so that flies ignore sources of protein (Schoof and Savage, 1955). Ingestion of carbohydrates continues. As Stoffolano described the situation, environmental factors have produced a shutdown of the endocrines involved in ovarian development (see also Mordue et al., 1970; de Wilde, 1970). If ingestion of protein were to occur, it might force the female into a nondiapause state. Experiments with mosquitoes have shown that cold tolerance and survival are reduced in mosquitoes that take a blood meal just before overwintering (Wallis, 1959; Hall, 1967). In general, cold hardiness is greater when no feeding has occurred and the gut is empty (Salt, 1961).

With the arrival of spring and the rise in temperature and longer days that accompany it, flies become ever more active. Their accelerated activity gradually depletes their energy stores, the fat body shrinks, negative feedback from stretch receptors is removed, behavioral thresholds for taste are lowered. The fly is once again attracted to protein, and the usual mechanisms regulating feeding resume control.

Literature Cited

Beck, S. D. 1968. *Insect Photoperiodism.* Academic Press, New York.

de Buck, A., and Swellengrebel, N. H. 1934. Behaviour of Dutch *Anopheles atroparvus* and *messeae* in winter under artificial conditions. *Riv. Malariol.* 13:404–416.

Calabrese, E. J., and Stoffolano, J. G. 1974a. The influence of diapause on respiration in adult male and female black blowflies, *Phormia regina. Ann. Ent Soc. Amer.* 67:715–717.

Calabrese, E. J., and Stoffolano, J. G. 1974b. The influence of age and diet on respiration in adult male and female blowflies, *Phormia regina. J. Insect Physiol.* 20:383–393.

Chen, S. S. 1969. The neuroendocrine regulation of gut protease activity and ovarian development in mosquitoes of the *Culex pipiens* complex and the induction of imaginal diapause. *Diss. Abstr.* B30(9):4187–B.

Cousin, G. 1932. Étude expérimentale de la diapause des insectes. *Bull. Biol. Franc. Belg. Suppl.* 15:1–341.

Danilevskii, A. S. 1965. *Photoperiodism and Seasonal Development of Insects.* Oliver and Boyd, London.

Danilevskii, A. S., Goryshin, N. I., and Tyshchenko, V. P. 1970. Biological rhythms in terrestrial arthropods. *Annu. Rev. Ent.* 15:201–244.

Dondero, L., and Shaw, F. R. 1971. The overwintering of some muscoidean Diptera in the Amherst area of Massachusetts. *Proc. Ent. Soc. Washington* 73:52–53.

Gillary, H. L. 1966. Stimulation of the salt receptor of the blowfly. I. NaCl. *J. Gen. Physiol.* 50:337–350.

Hall, D. G. 1948. *The Blow Flies of North America.* Thomas Say Foundation, Ent. Soc. Amer., Washington, D.C., Vol. 4.

Hall, D. W. 1967. Factors associated with hibernation of *Culex pipiens* Linnaeus in Central Indiana. M. S. thesis, Purdue University.

Mail, G. A., and Schoof, H. F. 1954. Overwintering habits of domestic flies at Charleston, West Virginia. *Ann. Ent. Soc. Amer.* 47:668–676.

Mordue, W., Highnam, K. C., Hill, L., and Luntz, A. J. 1970. Environmental effects upon endocrine-mediated processes in locusts. In *Hormones and the Environment,* G. K. Benson and J. G. Phillips, eds., pp. 111–136. *Mem. Soc. Endocrin.* No. 18. Cambridge University Press, London.

Nuorteva, P. 1966. The flying activity of *Phormia terrae-novae* R. D. (Dipt., Calliphoridae) in subarctic conditions. *Ann. Zool. Fenn.* 3:73–81.
Omand, E. 1971. A peripheral sensory basis for behavioral regulation. *Comp. Biochem. Physiol.* 38A:265–278.
Rees, C. J. C. 1970. Age dependency response in an insect chemoreceptor sensillum. *Nature*, 227:740–742.
Roberts, R. A. 1930. The wintering habits of muscoid flies in Iowa. *Ann. Ent. Soc. Amer.* 23:784–792.
Roubaud, E. 1927. Sur l'hibernation de quelques mouches communes. *Bull. Soc. Ent. France* 24–25.
Salt, R. W. 1961. Principles of insect cold-hardiness. *Annu. Rev. Ent.* 6:55–74.
Schaefer, C. H., Miura, T., and Washino, R. K. 1971. Studies on the overwintering biology of natural populations of *Anopheles freeborni* and *Culex tarsalis* in California. *Mosqu. News*, 31:153–157.
Schoof, H. F., and Savage, E. P. 1955. Comparative studies of urban fly populations in Arizona, Kansas, Michigan, New York, and West Virginia. *Ann. Ent. Soc. Amer.* 48:1–12.
Siverly, R. E. 1972. Overwintering of the black blowfly, *Phormia regina*. *Environ. Ent.* 1:526.
Slama, K. 1964a. Hormonal control of respiratory metabolism during growth, reproduction, and diapause in female adults of *Pyrrhocoris apterus* L. (Hemiptera). *J. Insect Physiol.* 10:283–303.
Slama, K. 1964b. Hormonal control of respiratory metabolism during growth, reproduction and diapause in male adults of *Pyrrhocoris apterus* L. (Hemiptera). *Biol. Bull.* 127:499–510.
Stegwee, D. 1964. Respiratory chain metabolism in the Colorado potato beetle. II. Respiration and oxidative phosphorylation in "sarcosomes" from diapausing beetles. *J. Insect Physiol.* 10:97–102.
Stoffolano, J. G. 1968. The effect of diapause and age on the tarsal acceptance threshold of the fly, *Musca autumnalis*. *J. Insect Physiol.* 14:1205–1214.
Stoffolano, J. G. 1973. Effect of age and diapause on the mean impulse frequency and failure to generate impulses in labellar chemoreceptor sensilla of *Phormia regina*. *J. Gerontol.* 28:35–39.
Stoffolano, J. G. 1974. Control of feeding and drinking in diapausing insects. In *Experimental Analysis of Insect Behaviour*, L. Barton Browne, ed., pp. 32–47. Springer-Verlag, New York.
Stoffolano, J. G., Greenberg, S., and Calabrese, E. 1974. A facultative, imaginal diapause in the black blowfly, *Phormia regina*. *Ann. Ent. Soc. Amer.* 67: 518–519.
Tate, P., and Vincent, M. 1932. Influence of light on the gorging of *Culex pipiens* L. *Nature* 130:366–367.
Tate, P., and Vincent, M. 1936. The biology of autogeneous and anautogenous races of *Culex pipiens* L. (Diptera: Culicidae). *Parasitol.* 28:115–145.
Wallis, R. C. 1959. Diapause and fat body formation by *Culex restuans* Theobald. *Proc. Ent. Soc. Washington* 61:219–222.
Wallis, R. C. 1962. Overwintering activity of the blowfly, *Phormia regina*. *Ent. News* 73:1–5.
Washino, R. K. 1970. Physiological condition of overwintering female *Anopheles freeborni* in California (Diptera: Culicidae). *Ann. Ent. Soc. Amer.* 63:210–216.
Washino, R. K., and Bailey, S. F. 1970. Overwintering of *Anopheles punctipennis* (Diptera: Culicidae) in California. *J. Med. Ent.* 7:95–98.

Washino, R. K., Gieke, P. A., and Shaefer, C. H. 1971. Physiological changes in the overwintering females of *Anopheles freeborni* (Diptera: Culicidae) in California. *J. Med. Ent.* 8:279–282.

de Wilde, J. 1970. Hormones and insect diapause. In *Hormones and Environment,* G. K. Benson and J. G. Phillips, eds., pp. 487–514. *Mem. Soc. Endocrin.* No. 18. Cambridge University Press, London.

Wyatt, G. R. 1963. Metabolic regulation in the development of insects. In *Control Mechanisms in Respiration and Fermentation,* B. Wright, ed., pp. 179–188. Ronald Press, New York.

12
Microscopic Brains

And still they gaz'd, and still the wonder grew,
That one small head could carry all he knew.
Goldsmith, *The Deserted Village*

Among primates the size of the head tends to reflect the size of the brain, and large cranial capacity is equated with superior intelligence. Insects tend to have very large heads relative to the rest of the body, but the proportionately large size is partly an illusion because the head is usually mounted on a thin neck which accentuates its appearance of massiveness. Furthermore, the large convex compound eyes, which occupy most of the head, also magnify its apparent size. In any case, the dimensions of the head give no indication of the size of the brain within because most of the head's interior is filled with muscle or air sacs. Proportionately and absolutely the insect brain is very small. Among the smallest brains are those of the culicoid midge or no-see-um and the African ant *Oligomyrmex*. The no-see-um, with a brain that is only 200 microns in its greatest dimension, is capable of behavior equal in every respect to that of *Phormia*, and the ant, with a brain only 150 microns at its greatest, being a social insect, probably surpasses *Phormia* in its abilities.

The brain of a blowfly weighs on the average 0.84 milligram (wet weight) (Dethier, 1965). Its maximum linear dimension is 1583 microns (Fig. 193). It probably contains not more than 100,000 cells. In gross appearance it is a laterally elongated body devoid of conspicuous prominences other than the optic areas. The paired optic lobes, which constitute 80% of the brain's mass, are fan-shaped structures to which the whole organ owes its elongate appearance. Since the brain is situated dorsad of the alimentary canal and the rest of the nervous system ventrad, the two must be connected in the vicinity of the oesophagus. In primitive insects the circumoesophageal connectives are clearly visible as paired cords extending from the brain (the supraoesophageal ganglion) to the suboesophageal ganglion, the first of a linear series of ventral segmental ganglia fused to different extents in different insects. The fusion of the two head ganglia is so extensive in *Phormia* that all of the neural material in the head appears as a single mass, through the center of which the oesophagus passes via a small tunnel.

The brain proper, that is, the supraoesophageal ganglion, consists of three neuromeres; the protocerebrum (forebrain), the deutocerebrum (midbrain), and the tritocerebrum (hindbrain). The right and left halves of the fore- and midbrain are connected by commissures that lie dorsal to the oesophagus. They, together with the structures that they innervate (labrum, eyes, ocelli, and antennae), are thus preoral and homologous with the prostomium of annelids. The halves of the hindbrain are connected by a commissure that is ventral to the oesophagus; therefore the hindbrain is in fact the first ganglion of the primitive ventral nerve cord. In *Phormia* much of its mass is situated laterad and even slightly ventrad of the oesophagus. It innervates the oral region (labium, maxillae, palpi, hypopharynx). A unique feature, as Snodgrass (1935) pointed out, is its connection both with the protodeutocerebral mass and the stomatogastric system (the analogue of the vertebrate autonomic system) via the frontal ganglion.

Usually the external form of the insect brain exhibits lateral lobes that delineate the three neuromeres. In *Phormia* their external contours are invisible, so complete is the fusion. The different regions are most easily identified externally by their nerves (Figs. 37, 38, 190). The brain proper has five nerves: nervus opticus, nervus ocellarius, nervus ganglii occipitalis, nervus antennalis, and nervus labrofrontalis. The remaining large cephalic nerve, the nervus labialis, is associated with the suboesophageal ganglion. The first three nerves belong to the forebrain. This neuromere is the dorsal and largest part of the brain and contains most of the so-called higher integrative centers, including the massive optic lobes.

The nervous ocellarius is attached in a median, dorsal, posterior location; the nervus opticus, within the optic lobes; and the nervus ganglii occipitalis on the posterior wall. The nervus antennalis is connected with the anterior part of the midbrain. This region contains little more than antennal centers. The nervus labrofrontalis is attached anteriorly to the hindbrain. This section is the smallest of the three neuromeres.

The head of *Phormia* is hypognathous; consequently, the primitive anterior-posterior relationships are in fact ventral-dorsal relationships. The hindbrain, for example, which is ontogenetically posterior to the forebrain, is anatomically ventral to it. For convenience, as Power (1943a) did in his study of the brain of *Drosophila,* the planes of section and axes of orientation are described relative to the body proper.

Microscopic Brains

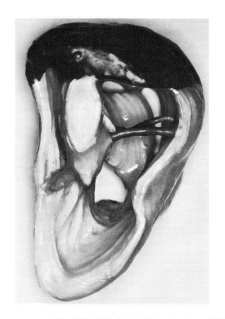
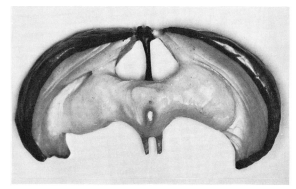

Fig. 190. Model of the brain of *Phormia* reconstructed from serial sections cut in the sagittal plane. Anterior surface on right (courtesy of M. Wilczek).

Fig. 191. Model of the brain reconstructed from serial sections cut in the frontal plane viewed from posterior (courtesy of M. Wilczek).

Thus, the horizontal plane is the one parallel to the top of the head; the sagittal plane is parallel to the side; the frontal plane is parallel to the face (frons and clypeus).

The topographical relation of the brain to the rest of the head can best be appreciated by viewing the set of scale models prepared by Wilczek. These are serial reconstructions in three dimensions (Figs. 190 and 191). In the frontal aspect (Fig. 191) the brain has an outline with a striking resemblance to a fly that is flying head on toward the observer. The central part of the brain would correspond to the body and the optic lobes to the blur of the wings in full sweep. The transverse axis lies slightly below the midpoint. The median ocellar stalk extends vertically from the top and the two longitudinal connectives passing back to the ventral abdominal ganglia extend from the posterior area of the bottom. The oesophageal canal lies slightly below the transverse horizontal axis. In this plane the brain, exclusive of the optic lobes, is much thinner. The transverse axis passes through the posterior area of the head. The most conspicuous feature is the pair of antennal nerves. The sagittal view also shows that the brain lies posteriorly in the head, is displaced slightly ventrally (if only the head capsule and not the labellar complex is considered), and is thicker from top to bottom than from front to back (Figs. 192, 193, and 194).

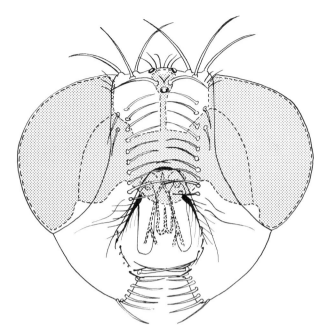

Fig. 192. Frontal view of the head of a female *Phormia regina* with a silhouette of the brain showing its position in the head and the relation of its parts to external features of the head.

The gross histological structure is immediately apparent in any section. The greater mass is a central feltwork of nerve fibers, the neuropile, interspersed with greater and lesser fiber tracts, the special groups of cells, compact bodies consisting of dense aggregations of interneurons and dense masses of terminal arborizations (glomeruli) (e.g., corpora pedunculata, the central body, etc.) It is overlain cortically with neural cell bodies. The whole is encased in a sheath of nonneural connective tissue cells or glial elements. The most striking organizational feature of the insect brain at this level is the sharp separation of cell body regions and synaptic fields. In contrast to the vertebrate brain, where cell bodies may lie in intimate association with dendritic fields, most of the cell bodies of the insect brain lie far removed at the periphery. Synaptic contact is restricted to the neuropile (Smith, 1967); however, it is difficult, often well nigh impossible, to identify synapses by light microscopy. There are no axosomatic synapses because, as already noted, the somata of motor neurons and interneurons are situated in the cortex, and those cell bodies that are present in the neuropile are ensheathed by glial elements (Wigglesworth, 1960). There are also many narrow glial elements extending well into the neuropile from cell bodies in the cortex (Smith, 1967). It is this organizational feature and the haystack nature of the neuropile

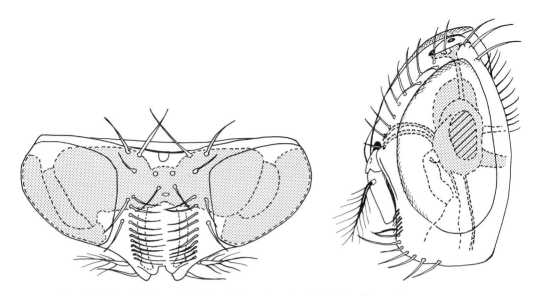

Fig. 193. Dorsal view of the head showing the brain in silhouette.

Fig. 194. Sagittal view of the head showing the brain in silhouette.

that has been the despair of neurophysiologists endeavoring to trace and study central connections and integrating mechanisms. Even though there seems to be some semblance of order in the neuropile (cf. Maynard, 1962), the complexity of the system and the small size of the cells as compared, for example, with that of molluscs is awesome.

Maynard, on the basis of gross fiber configurations, has divided neuropile into two classes: structured and unstructured. Unstructured neuropile can be subdivided into plexiform and diffuse; structured, into glomerula and stratified. Plexiform is homogeneous and netlike. Diffuse, referred to as "tangled confusion" in classical neurological studies, is the kind characteristic of the brain of insects. Most of its neurons are either monopolar interneurons and motor neurons or peripheral bipolar neurons. Their processes are tortuous, extensively branched, and ramifying; however, each characteristically has its own domain of arborization and each is identifiable in every individual. Glomerular neuropile is characterized by knotlike clusters of tightly ramifying pre- and postsynaptic structures. The antennal glomeruli and the corpora pedunculata are examples of this type. Stratified neuropile, as exemplified by the optic lobes, is an orderly three-dimensional lattice constructed of precisely oriented neuronal processes. As Maynard has described it, it's basic pattern consists of vertically or radially oriented fibers passing at right angles through horizontal or tangential layers of ramifying processes. In short, although neuropile is complex, it is not haphazard.

Even though architectural details at the cellular level are at present beyond analysis, the topographical relations of the principal association areas and the courses of the main fiber tracts have been established (Larsen, Dethier, and Broadbent, 1975). Comparing these with descriptions of the brains of related Diptera can convey some idea of the organization of the brain of *Phormia*. The earlier works included an investigation of the internal structure of the brain of *Somomya erythrocephala* (Cuccati, 1888), a brief description of the brain of *Drosophila melanogaster* (Hertwick, 1931), the classical detailed studies of *D. melanogaster* by Power (1943a, 1943b), studies of species of *Calliphora, Sarcophaga, Tubifera,* and *Eristalis* (Jarnicka, 1959), and investigations of the postembryological development of the brain of *Calliphora vomitoria* L. (Gieryng, 1964, 1965).

As an introduction to the gross internal anatomy of the brain of *Phormia*, the detailed scale reconstruction of Wilczek is very helpful (Fig. 195). In this model only the large aggregates of fibers and glomeruli (the so-called bodies) and a few major fiber tracts are shown.

The protocerebrum consists of the following structural parts: the central complex, the corpora pedunculata, the protocerebral bridge, the optic lobes, the protocerebral lobes, the accessory protocerebral lobes, and several large commissures. The center of the protocerebrum, which is also the center of the whole brain, is occupied by the central complex, which consists of the central body, the ellipsoid body, and a pair of ventral tubercles. The complex is bounded dorsally by the accessory protocerebral lobes and the dorsally ascending posterior branches of the median root of the corpora pedunculata, ventrally by the central commissure and the oesophageal canal. Posterolaterally it is bounded by the paired olfactoria-globularis tracts and the accessory protocerebral lobes. Anteriorly it is bounded by the accessory protocerebral lobes and the median roots of the corpora pedunculata.

The largest unit of the central complex is the central body (Central körper, fan-shaped body). It has a convexoconcave shape, the convex surface being directed posteriorly. The ellipsoid body (ringförmiger Basalteil des Fächerkörpers, Koper elliptischer Sektion, nodulus, kleine ovale schalenartige Gebilde) is considerably smaller and kidney-shaped — it is also kidney-shaped in *Calliphora vomitoria* L. (Jarnicka, 1959; Gieryng, 1964) — whereas in *Drosophila* it is ellipsoidal (Power, 1943a). It dovetails into the concavity of the central body and is flattened anteroposteriorly. In most insects these two parts of the central complex are situated one above the other, but in *Phormia* (and *Calliphora*) the ellipsoid body is situated anterior to the central body. The central complex has many connections with various parts of the neuropile of the protocerebral and accessory protocerebral lobes. Two small bundles extend laterally between the lateral anterior area of the

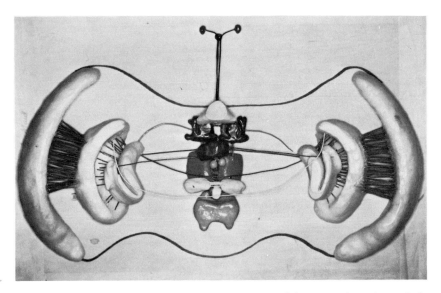

Fig. 195. A detailed model of the brain reconstructed from serial sections. Only a few of the major commissures are indicated.

protocerebral lobes and the central complex. The fibers lie posterior to the median root of the corpora pedunculata and ventral to the fibers connecting the central complex and the accessory protocerebral lobes. Fibers also run between the medial areas of the protocerebral lobes and the central complex (Fig. 197h). They branch to the opposite side of the central complex as well as fanning out to enter the posterior side of the central body. There are fibers passing laterally from the anterior area of the ellipsoid body to the posterior side of the accessory lobe and finally to the lateral side. No connections were observed between the central body and the median root of the corpora pedunculata even though the two lie in close proximity (Fig. 197i). Ventral to these two lies a pair of spherical globuli, the ventral tubercles. They are bilobed ventrally, hence in some sections appear as four bodies (Fig. 197f). In *Calliphora,* according to Satija (1958), there are no connections between the tubercles any any other part of the brain except the central body. In *Phormia* there are many connections. The most prominent are large branching nerves to the central body. Some of these fibers extend to the ipsilateral side, and others spread to the contralateral side (Fig. 196d). Other connections occur here in the form of a large plexus of fine fibers.

At a point where the ventral tubercles are bilobed there are two large accessory commissures passing between the accessory protocerebral lobes. They lie anterior to the tubercles and give off numerous fine fibers to the tubercles (Figs. 197f, 197g, 196e). No connections were observed between the tubercles and the median root of the

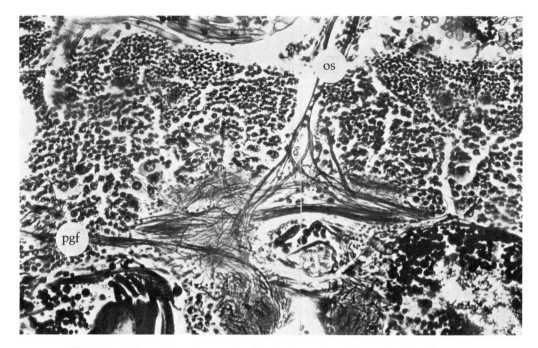

Fig. 196a–i. Frontal sections of the brain, in serial order, stained with Bodian's protargol technique.

Symbols for Figure 196

AP, accessory protocerebral lobe
ACF, antennal coarse fibers
AFF, antennal fine fibers
AG, antennal glomeruli
AN, antennal nerve
ABO, anterior branch of anterior optic tract
AOT, anterior optic tract
ACP, anterior root of corpora pedunculata
BS, brain stem
BAN, branch of antennal nerve
BCC, branch of central commissure
BMB, branch of median bundle
CCP, calyx of corpora pedunculata
CB, central body

CC, central commissure
CCM, central complex
CHC, chiasmic commissure
CAL, commissure between accessory lobes
CFF, commissure of fine fibers for antennal glomerulus
CMG, commissure of middle glomerulus
CP, corpora pedunculata
DC, deutocerebral commissure
DG, deutocerebral glomerulus
DPCP, dorsal posterior commissure of protocerebrum
DB, dorsal posterior part of brain
EB, ellipsoid body

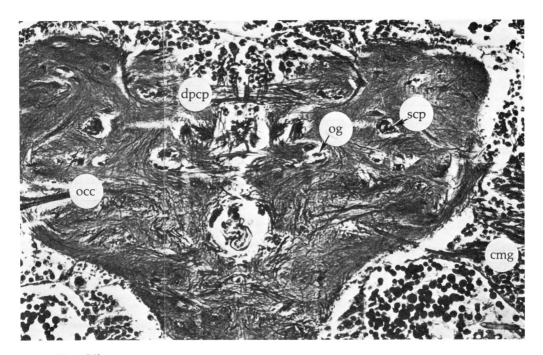

Fig. 196b

LN, labial nerve
LBN, labrofrontal nerve
MB, median bundle
MBB, median bundle branching
MCP, median root of corpora pedunculata
NS, neurosecretory cells
ON, ocellar nerve
OS, ocellar stalk
OE, oesophageal canal
OG, olfactorio globularis
OCC, optical component of central commissure
PAN, posterior antennal center
PBO, posterior branch of optic track
PFF, posterior fine fibers
PGF, posterior giant fibers
PB, protocerebral bridge
PL, protocerebral lobe
RN, recurrent nerve
RAG, root of antennal glomerulus
RAN, root of antennal nerve
RCP, root of corpora pedunculata
RPB, root of protocerebral bridge
SCP, stalk of corpora pedunculata
SOE, suboesophageal area
US, unnamed commissure
VSCP, ventral branch of stalk of corpora pendunculata
VPCC, ventral protocerebral-central complex tract
VT, ventral tubercle

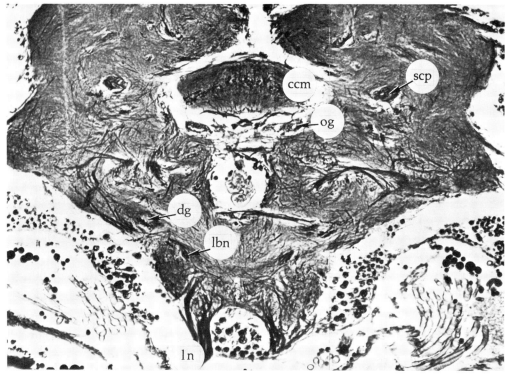

Fig. 196c

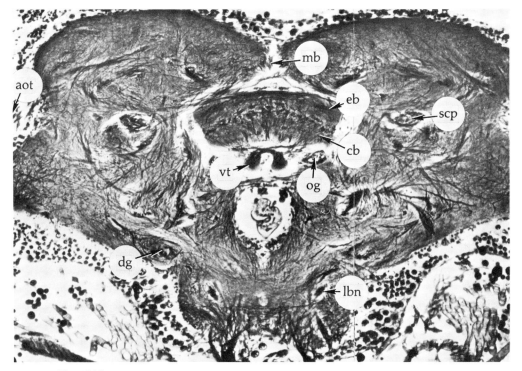

Fig. 196d

Microscopic Brains

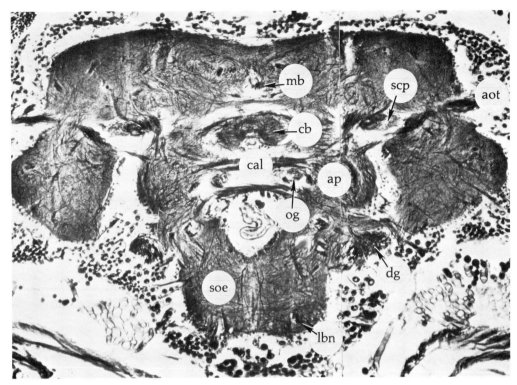

Fig. 196e

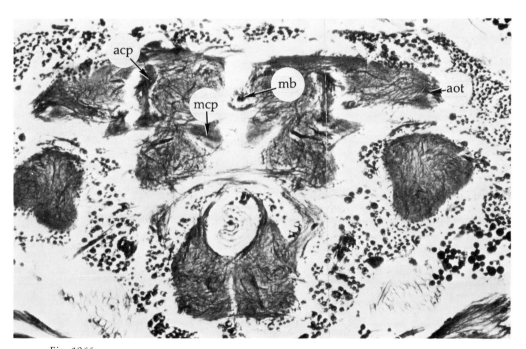

Fig. 196f

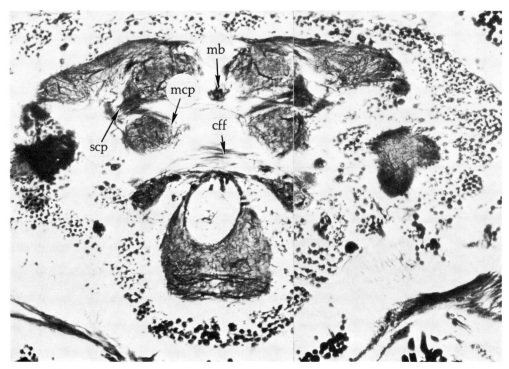

Fig. 196g

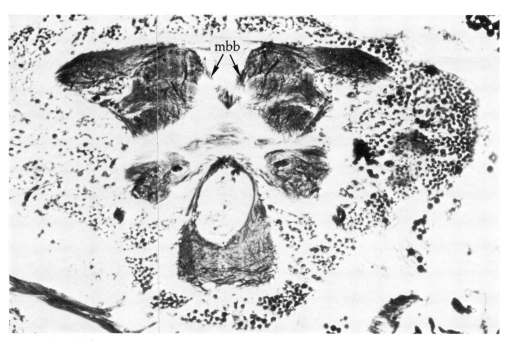

Fig. 196h

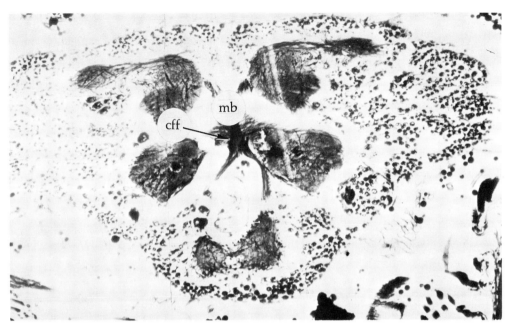

Fig. 196i

corpora pedunculata, but there are connections with the stalk of the pedunculata. Small fibers also connect the ventral tubercles with the anterior branch of the olfactorioglobularis as it passes toward the antennal glomeruli (Figs. 196c, 196d, 202c, 202d).

The paired corpora pedunculata (mushroom bodies), usually associated with learning (Vowles, 1954, 1961), are poorly developed in *Phormia* as in other Diptera. They are neither as large nor as complex as those in Hymenoptera. Yet, they are extensive, conspicuous structures extending from the dorsal posterior region of the brain to the midanterior region (Fib. 198b). They lie dorsal to the central complex and by virtue of their complicated branching extend to either side and in front of the central complex. They are above the central commissure and the oesophagus.

The complexity of their shape makes them difficult to visualize in three dimensions. Each member of the pair consists of three stalks. Power's (1943a) description of the spatial relationships cannot be excelled. "The three stalks unite at a common antero-lateral point and extend from this junction in the three principal planes: dorso-ventrad, antero-posteriorly, and latero-mesad. They thus bear the same spatial relationship to one another as do the planes of the semicircular canals in the inner ears of vertebrate animals."

The most conspicuous of the three parts is the posterior stalk (pedunculus, Stiele, Haupstiel). From the point of junction of the three parts it extends posteriorly and dorsally, terminating in the pedunculate bodies (calyx cups), which are approximately twice as large as the diameter of the stalk (Figs. 198a, 198b). It was the manner of attachment of the terminal globuli cells to the cups of the calyx in Hymenoptera that was suggestive of the stipe and pileus of a mushroom. This stalk appears longitudinally in sagittal sections and is cut transversely in horizontal and frontal sections. The calices are connected by the dorsal posterior commissure lying immediately anterior to them (Fig. 197p). Near the anterior end of the posterior stalk a small branch separates and passes ventroposteriorly. It sends fibers to the central commissure and the ventral posterior commissure. The stalk itself has numerous fibers connecting with various parts of the protocerebrum. In the region of the central complex a major trunk enters the anterior accessory commissure. There is also a branch passing to the central body and the ventral tubercles.

From the point of junction of the three stalks two, comprising the median bundle (tubercule interne, Mediananschwellung des Stieles, corpus callosum, inner root, Balken), pass in a general anterior direction. Of the two stalks of this bundle, one is directed dorsally (the α-lobe of Vowles); the other goes medially (the median or β-lobe of Vowles). The posterior or dorsal stalk (tubercle exterieur, Cylinder des Stieles, rückläufiger Stiel) extends dorsally where it comes to lie immediately laterad of the accessory protocerebral lobes. It appears in cross section in the horizontal plane of the head, and longitudinally in sagittal and frontal planes. It is slightly bifurcate at the end.

The median stalk lies anterior to the central complex (Figs. 196a, 197i, 198c). It appears in cross section in the sagittal plane and longitudinally in the horizontal and frontal planes. It is rod-shaped with a bulbous swelling at the medial termination. Each stalk of the median bundle of one side meets the corresponding stalk of the opposite side at the midline of the brain but does not fuse.

At the same level as the central complex, in the posterior region of the brain is located the protocerebral bridge (pons cerebralis, le pont du lobes cerebraux, gabelförmiger Körper, fibrillar arch, Ocellarnervenbrüche, dorsal posterior commissure, Hirnbrüche). As in *Drosophila*, it is at the same level as the terminations of the posterior stalks of the corpora pedunculata but more medial (Fig. 198h). It is cylindrical in shape, and both ends bend posteriorly so that it is slightly horseshoe-shaped (Figs. 197o, 197p). The concavity faces posteriorly. In *Drosophila* the pons is divided into lateral parts by a median cleft (Power, 1943a). In *Calliphora* it is undivided (Cuccati, 1888; Jarnicka, 1959). The situation in *Phormia* is unclear; some horizontal sections have shown the bridge as divided. The ocellar nerves pass just posterior to the pons but are not connected in *Phormia*. The fibers in the pons are of two types (thick and fine). Two main bundles of fibers connect the pons with the neuropile: the olfactorio globularis (Figs. 197o, 198o) and the root of the pons (Fig. 197o). The roots proceed ventrally, curving gradually in an anterior direction. At the point of maximum curvature they enter the neuropile and continue in the direction of the central complex. The roots of the pons lie lateral of the olfactory glomerulus. In this region the axons of some large neurosecretory cells enter the roots of the pons (Figs. 197l).

The optic lobes form the lateral winglike extensions of the brain mass. Each consists of the following principal areas in the nomenclature of Snodgrass (1935): lamina ganglionaris, medulla externa, medulla interna. The lamina ganglionaris, as the term is employed here, corresponds to the "external glomerulus and auxiliary layers" of Power and the "intermediate retina" of Cajal and Sanchez (1915). Various subdivisions are known as Moleculärschichte der Retina, Palissaden-Schicht des Retina-Ganglion, periopticon, lame ganglionnare, zona plexiform externa, outer ganglion. It is the most distal synaptic region of the optic lobes and shaped like an oblong convex box with the longest axis dorsoventral. Several structural layers are discernible within it. The outermost is that in which axons from the visual cells leave the basement membrane either directly en route to the medulla externa or indirectly after having first crossed laterally with other axons to form local chiasmata. Proximad of this layer is a layer of monopolar neurons (centripetal cells). This layer is followed by a layer of centrifugal cells. The fibers leaving the lamina ganglionaris pass onward as discrete bundles, the optic cartridges. Each optic cartridge consists of a long retinal fiber, several short retinal fibers, a giant monopolar fiber, and a centrifugal fiber. For more details one should consult Cajal and Sanchez (1915), Bretschneider (1921), and Power (1943a). Fibers leaving the lamina approach the focus of its arched shape and cross to enter the opposite side of the medulla externa. The crossing is from anterior to posterior (and vice versa) and not from dorsal to ventral; consequently, the external chiasma may be seen only in sections that are cut in a horizontal plane.

The medulla externa (middle glomerulus of Power) is essentially the same as that in *Drosophila,* where it has the shape of an oblong saucer so situated that its lateral convex curvature is approximately parallel with the curvature of the cornea.

The medulla interna—the lobula and lobula plate of Bishop, Keehn and McCann (1968)—is composed of two unequal parts. The larger part (anterior inner glomerulus) is oblong, bent so that the convex bow is approximately parallel to the curvature of the cornea. The smaller part (posterior inner glomerulus) is bent posteriorly. In horizontal sections the area between these two parts is V-shaped. Within this triangular area is located the internal chiasma; however,

Symbols for Figure 197

AP, accessory protocerebral lobe
ACF, antennal coarse fibers
AFF, antennal fine fibers
AG, antennal glomeruli
AN, antennal nerve
ABO, anterior branch of anterior optic tract
AOT, anterior optic tract
ACP, anterior root of corpora pedunculata
BS, brain stem
BAN, branch of antennal nerve
BCC, branch of central commissure
BMB, branch of median bundle
CCP, calyx of corpora pedunculata
CB, central body
CC, central commissure
CCM, central complex
CHC, chiasmic commissure
CAL, commissure between accessory lobes
CFF, commissure of fine fibers for antennal glomerulus
CMG, commissure of middle glomerulus
CP, corpora pedunculata
DC, deutocerebral commissure
DG, deutocerebral glomerulus
DPCP, dorsal posterior commissure of protocerebrum
DB, dorsal posterior part of brain
EB, ellipsoid body
LN, labial nerve
LBN, labrofrontal nerve
MB, median bundle
MBB, median bundle branching
MCP, median root of corpora pedunculata
NS, neurosecretory cells
ON, ocellar nerve
OS, ocellar stalk
OE, oesophageal canal
OG, olfactorio globularis
OCC, optical component of central commissure
PAN, posterior antennal center
PBO, posterior branch of optic track
PFF, posterior fine fibers
PGF, posterior giant fibers
PB, protocerebral bridge
PL, protocerebral lobe
RN, recurrent nerve
RAG, root of antennal glomerulus
RAN, root of antennal nerve
RCP, root of corpora pedunculata
RPB, root of protocerebral bridge
SCP, stalk of corpora pedunculata
SOE, suboesophageal area
US, unname commissure
VSCP, ventral branch of stalk of corpora pedunculata
VPCC, ventral protocerebral-central complex tract
VT, ventral tubercle

Microscopic Brains 379

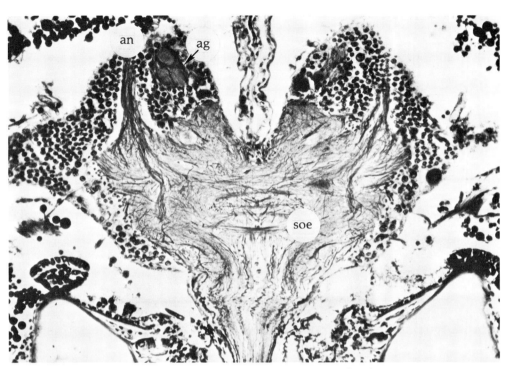

Fig. 197a–p. Dorso-ventral sections of the brain, in serial order.

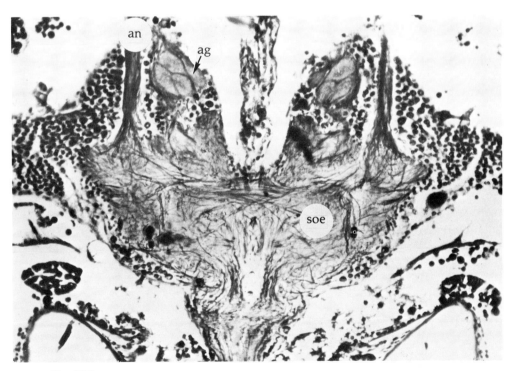

Fig. 197b

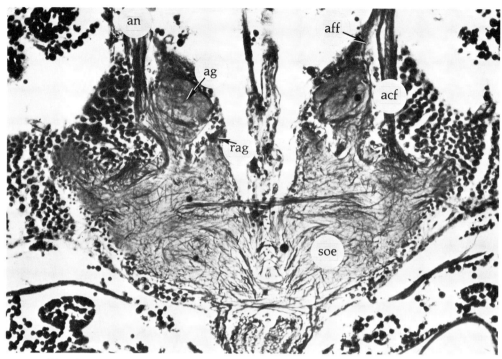

Fig. 197c

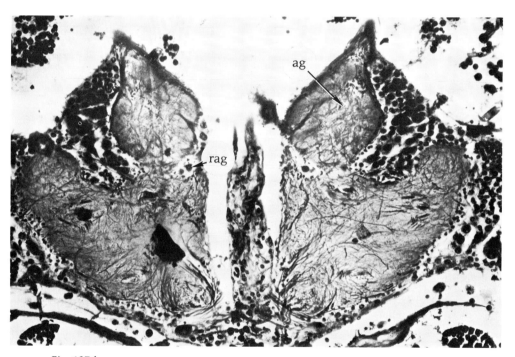

Fig. 197d

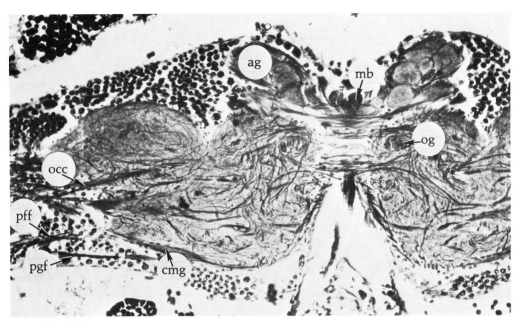

Fig. 197e

chiasma is a misnomer because, as is also true in *Drosophila*, the fibers that connect the external and internal medullae do not branch to the opposite sides.

The deutocerebrum is the olfactory part of the brain. It consists chiefly of the paired antennal glomeruli and the fiber tracts associated with them. The pear-shaped glomeruli lie in the anterior region of the brain. They consist of several spherical regions fused together in a grapelike cluster. The interior is a dense neuropile interdigitating with the cortex (Fig. 197d). A thick bundle of fibers connects the glomeruli to the suboesophageal areas (Figs. 198d, 198e). Two types of fibers connect the glomeruli and the antennae (Fig. 197c). The fine fibers (presumed to be sensory) from the antennae enter the glomeruli directly. Some traverse the glomeruli and enter the antennal commissure. The thick fibers (presumed to be motor) actually lie ventral to the glomeruli and appear to originate in the suboesophageal region. In horizontal sections (Fig. 197b) the coarse fibers can be traced all the way from the brain stem to the antennae. Some form a distinct plexus in the suboesophageal region (Fig. 203a).

Except for the "stalk" of fibers connecting the antennal glomeruli to the suboesophageal regions of the brain, the largest connecting tract is the commissure (commissure of fine fibers) joining the right and left members of the pair (Figs. 197g, 198h-l). It leaves the posterior portion of each glomerulus and loops posterior to the median bundle (Figs. 196h, 196i, 197g). No connections between the two bundles were observed. They lie at right angles to each other.

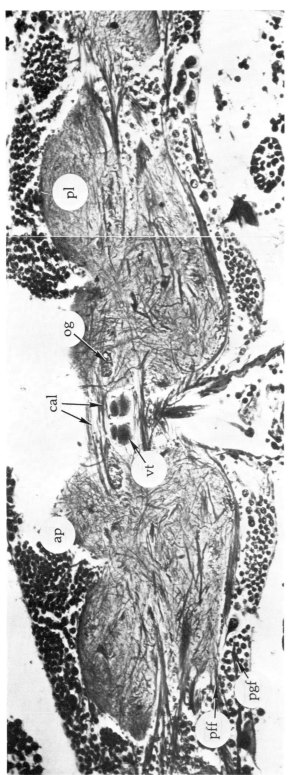

Fig. 197f

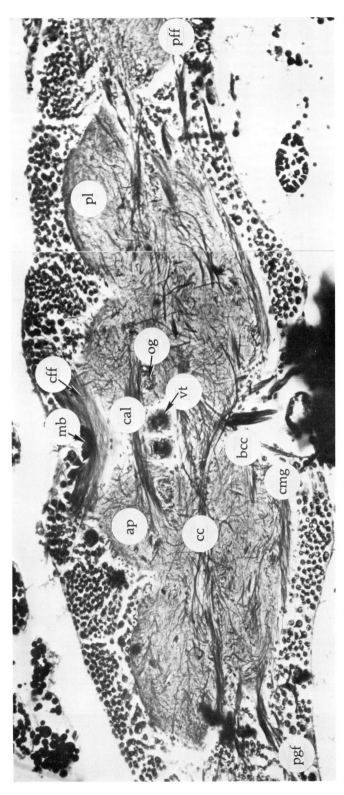

Fig. 197g

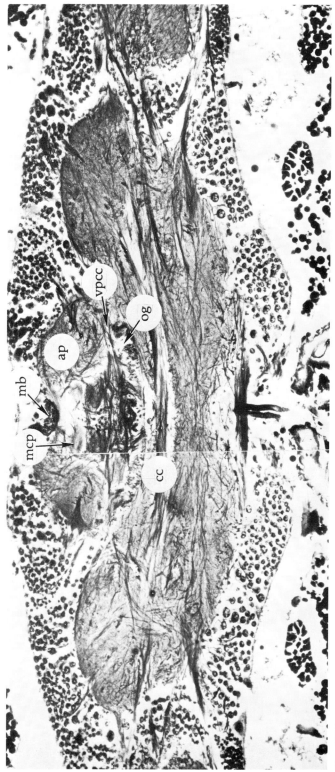

Fig. 197h

Microscopic Brains

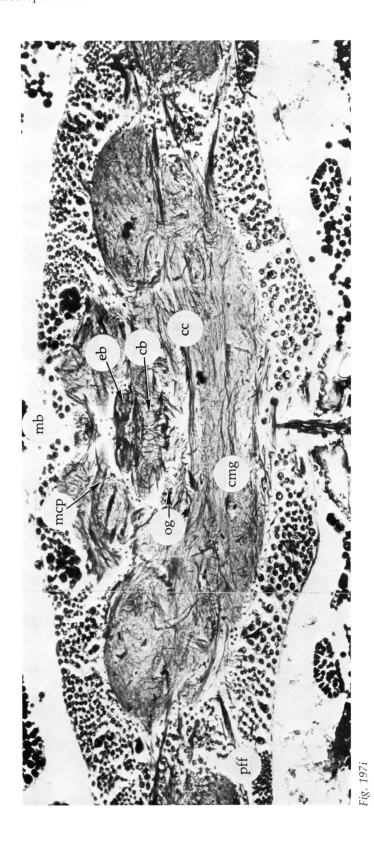

Fig. 197i

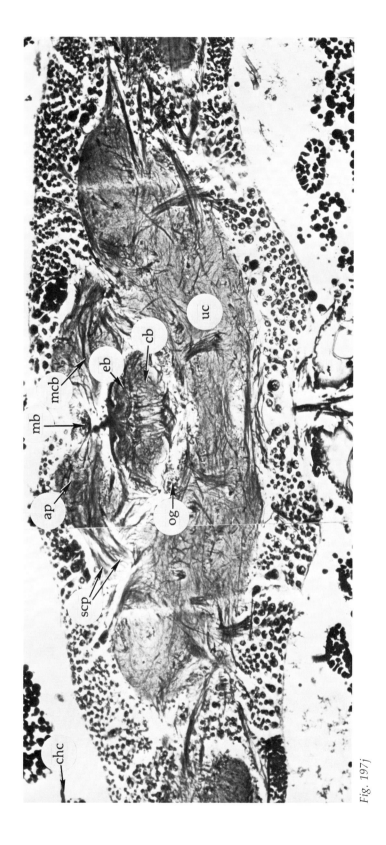

Fig. 197j

Microscopic Brains

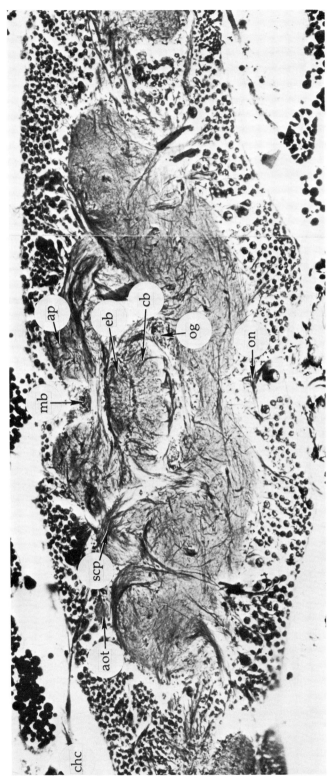

Fig. 197k

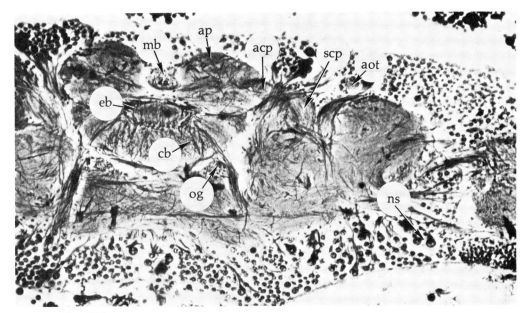

Fig. 197l

Although the glomeruli are situated at a considerable distance from the main centers of the brain and the larger commissures, there are many connections via small tracts. There is, for example, a small bundle of fibers passing between the glomeruli and the central commissure (Fig. 198c). A few small fibers also connect with the corpora pedunculata. The olfactoria globularis, a large commissure, terminates in the antennal glomeruli (Figs. 198e, 198f, 198o).

In insect brains generally there are three principal commissures (in addition to other smaller ones). These are: the optic tract connecting the medullary bodies of opposite eyes, the commissure that connects the ventral bodies beneath the roots of the pedunculata, and the deutocerebral commissure traversing the ventral part of the brain between the antennal bodies. A fourth is a tritocerebral connective passing beneath the stomodaeum (Snodgrass, 1935). Thus the right and left halves of each neuromere, the protocerebrum, deutocerebrum, and tritocerebrum, are connected. There are also a number of prominent commissures that are not transverse. The principal commissures in *Phormia* are: the optic tract, the central commissure, the anterior accessory commissure, the dorsal posterior commissure, the anterior optical tract, the posterior tract of giant fibers, the antennal fine fiber commissure, the olfactorio globularis, and the median bundle. Of these the major commissures of the protocerebrum exclusive of the optic tracts are: the central commissure, the anterior accessory commissure, and the dorsal posterior commissure. The major commissures of the deutocerebrum are the antennal commissures and the olfactorio globularis.

Microscopic Brains

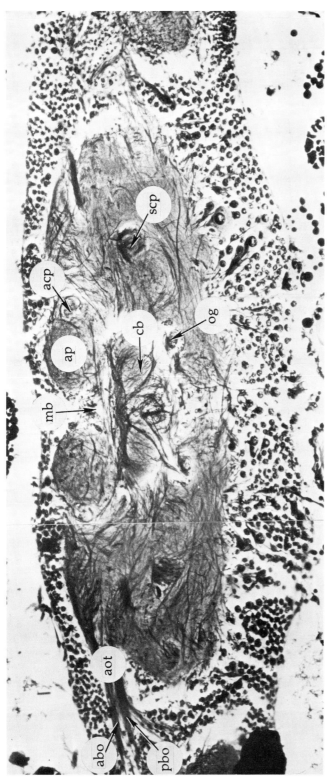

Fig. 197m

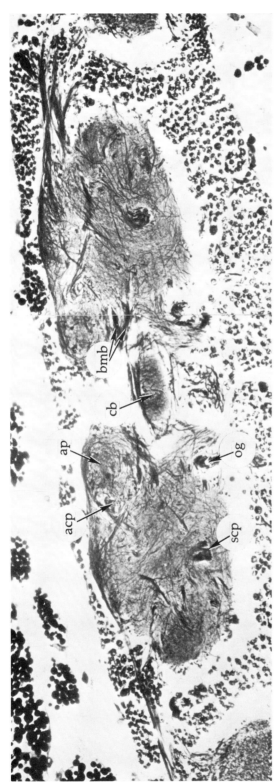

Fig. 197n

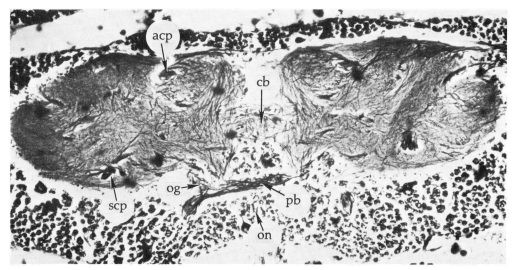

Fig. 197o

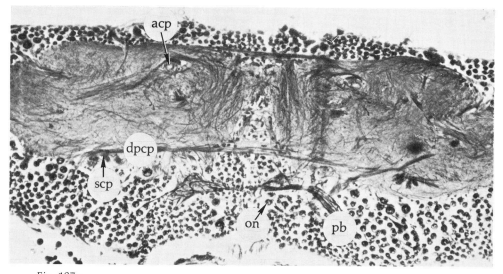

Fig. 197p

The optic tract (central commissure S of Cuccati, commissure of middle glomerulus of Power, optic tract of Snodgrass) connects the right and left medullae externae. It leaves the anterior edge of the medulla externa, passes posteriorly close to the medulla interna, then traverses the forebrain posterior to and in the plane of the ventral tubercles above the oesophagus, and again curves forward to enter the anterior edge of the opposite medulla interna.

Other optical tracts include the anterior optical tract and the posterior tract of giant fibers. The anterior optical tract leaves the medial edge of the smaller part of the medulla interna and sends one branch medially and anteriorly to the central complex, while the other continues to the opposite medulla interna. In *Drosophila* it becomes associated with ocellar fibers in the median line and passes ventrally and posteriorly into the suboesophageal ganglion. There is also in *Drosophila* and *Calliphora* a tract of fine fibers (referred to by Cuccati as sr, op), which leaves the posterior medulla interna at the same place and extends anteriomedially into the central brain areas.

The central commissure lies in the same plane as the ventral tubercles, posterior to them and the central body and slightly dorsad. It is ventrad of the protocerebral bridge. It extends transversely through the midsection of the brain. Fibers from the protocerebrum as well as from the optic lobes contribute to this tract (Figs. 197f-i). Two major tracts, which are optical components of this commissure, connect the right and left medullae externae. The anterior one (Power's "optical component of the central commissure") passes from the dorsal medial side of the larger part of the medulla externa, then bends ventrally and extends medially under the pedunculata and above the antennal glomeruli to reach the medulla interna of the opposite side. The posterior commissure of the medullae internae (Power's "commissure of anterior inner glomerulus") lies in the brain at the level of the ventral tubercles and posterior to them. It leaves the medial edge of the larger part of the medulla interna and runs under the central body. Near the midregion of the brain the central commissure splits (Fig. 198a). From each side a branch extends into the opposite posterior region of the protocerebral lobe. The ends of this branch lie near the optic tract; however, no direct connections between the two have been observed. There are connections between the central commissure and the posterior roots of the corpora pedunculata and the central complex.

The anterior accessory commissure connects (Fig. 197f) the accessory protocerebral lobes and lies immediately anterior to the olfactorio globularis and ventral tubercles, posterior to the commissure of the antennal glomeruli, and ventral to the median bundle.

The dorsal posterior commissure connects the dorsal posterior halves of the brain (Figs. 197b, 197p). It lies in the same plane as the protocerebral bridge but in a more anterior location. The two do not connect. The commissure connects the two lobes of the corpora pedunculata in the region of the anterior sides of the calyx cups.

In the deutocerebrum, aside from the antennal commissure already described, the most conspicuous fiber tract is the oblique olfactorio globularis. This structure extends ventrally from the protocerebral bridge as a single tract. It bifurcates posterior to the central body near the central commissure (Figs. 198e, 198g, 198m-o). One branch continues ventrally into the suboesophageal area as a plexus of fine fibers.

Large fibers (the ventral branch) enter the brain stem (Fig. 198g). An anterior branch continues forward until it enters the antennal glomerulus (Fig. 198f). An anterior subdivision of the ventral branch extends in the suboesophageal area to the vicinity of the labrofrontal nerves. A branch can be observed in a saggital plane arising on the dorsal surface of the protocerebral lobe and joining the main trunk of the olfactorio globularis at a point dorsal and posterior to the central body (Fig. 198f). In the region of the central body numerous small fibers connect that structure with the globularis. Fibers also connect the globularis with the central commissure at this point. Where the globularis meets the antennal glomerulus there are fibers connecting with the ventral tubercles (Fig. 198g).

Finally, there is the median bundle, a very broad tract lying vertically in the anterior median region of the brain. It is the major connection between the suboesophageal area and the protocerebrum. It is bounded on each side by the accessory protocerebral lobes. Its ventral portion divides and encircles the oesophagus to enter the suboesophageal area as a plexus. At the point of bifurcation around the oesophagus a single branch of fibers extends from the posterior suface to the dorsal surface of the oesophagus (Fig. 198j). In the same plane, fibers can be seen connecting the bundles to the stalk of the corpora pedunculata. At its dorsal end the median bundle also bifurcates. Each branch enters the corresponding lateral protocerebral lobe (Fig. 196h). At the point of dorsal bifurcation two small bundles of fibers leave the median bundle at a 90° angle and penetrate the protocerebrum (Fig. 196g.). Sectioning horizontally from the dorsal surface one can discern the two anterior branches of the median bundle that are associated with the accessory lobe of the protocerebrum. A few fibers are exchanged between the median bundle and the accessory lobe. Proceeding ventrally in the same plane one can see the point of bifurcation (Fig. 197m) just dorsal to the main stalk of the bundle.

Fibers connecting the bundle with the suboesophageal area completely encircle the oesophagus en route to their destinations. They enter the suboesophageal area in the same plane as the labial (Fig. 198h) and labrofrontal nerves. There are close associations with fibers of the latter. In the suboesophageal ganglion there is also a close association of the fibers of the median bundle with fibers of the ventral branch of the olfactorio globularis. Fibers of the median bundle and the commissure of the antennal glomeruli are juxtaposed at a 90° angle but apparently do not share fibers at this point. The bundle passes anterior to the commissure (Fig. 197e).

Such details as are known of the interconnections of the various sense organs and association centers can now be discussed. Our principal concern is with the chemoreceptive system; however, as complete a picture as possible will be given. Compare also Power's (1948) description of the thoracico-abdominal system.

As mentioned earlier, there are five pairs of nerves associated with the brain proper. The labial nerve and the brain stem, the connection with the thorax, connect with the suboesophageal ganglion. The nerves are: the ocellar stalk, the optic lobes, the antennal nerves, the nervus ganglii occipitalis, the labrofrontal nerves, and the labial nerves.

Symbols for Figure 198

AP, accessory protocerebral lobe
ACF, antennal coarse fibers
AFF, antennal fine fibers
AG, antennal glomeruli
AN, antennal nerve
ABO, anterior branch of anterior optic tract
AOT, anterior optic tract
ACP, anterior root of corpora pedunculata
BS, brain stem
BAN, branch of antennal nerve
BCC, branch of central commissure
BMB, branch of median bundle
CCP, calyx of corpora pedunculata
CB, central body
CC, central commissure
CCM, central complex
CHC, chiasmic commissure
CAL, commissure between accessory lobes
CFF, commissure of fine fibers for antennal glomerulus
CMG, commissure of middle glomerulus
CP, corpora pedunculata
DC, deutocerebral commissure
DG, deutocerebral glomerulus
DPCP, dorsal posterior commissure of protocerebrum
DB, dorsal posterior part of brain
EB, ellipsoid body

LN, labial nerve
LBN, labrofrontal nerve
MB, median bundle
MBB, median bundle branching
MCP, median root of corpora pedunculata
NS, neurosecretory cells
ON, ocellar nerve
OS, ocellar stalk
OE, oesophageal canal
OG, olfactorio globularis
OCC, optical component of central commissure
PAN, posterior antennal center
PBO, posterior branch of optic track
PFF, posterior fine fibers
PGF, posterior giant fibers
PB, protocerebral bridge
PL, protocerebral lobe
RN, recurrent nerve
RAG, root of antennal glomerulus
RAN, root of antennal nerve
RCP, root of corpora pedunculata
RPB, root of protocerebral bridge
SCP, stalk of corpora pedunculata
SOE, suboesophageal area
US, unname commissure
VSCP, ventral branch of stalk of corpora pendunculata
VPCC, ventral protocerebral-central complex tract
VT, ventral tubercle

Microscopic Brains

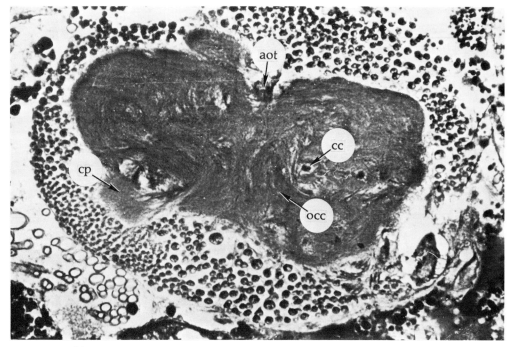

Fig. 198a

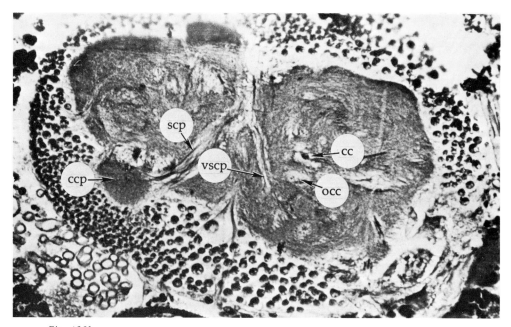

Fig. 198b

Fig. 198a–p. Sagittal sections of the brain, in serial order.

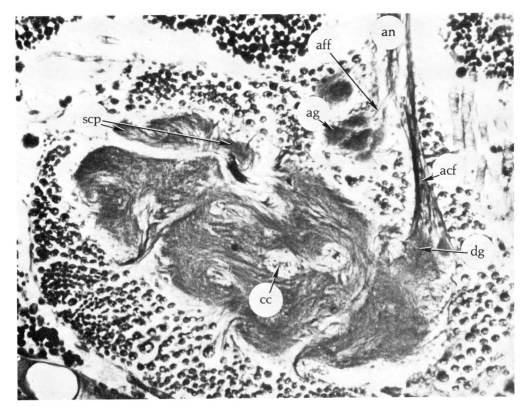

Fig. 198c

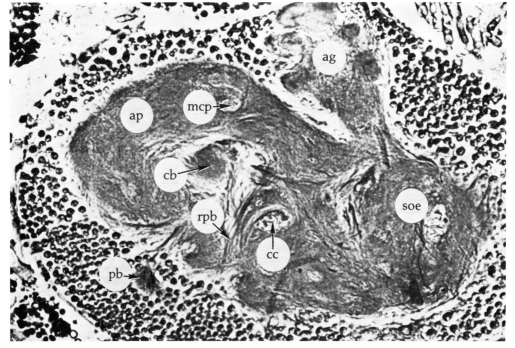

Fig. 198d

Microscopic Brains

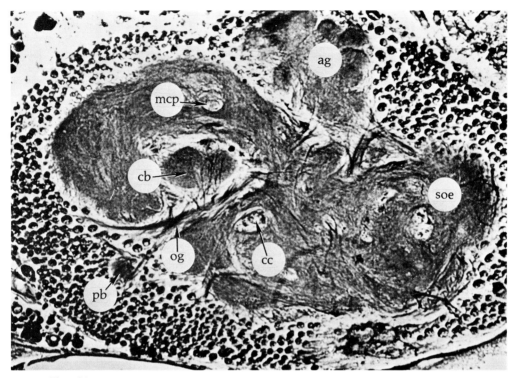

Fig. 198e

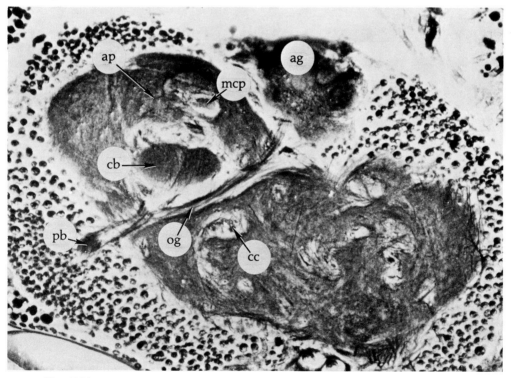

Fig. 198f

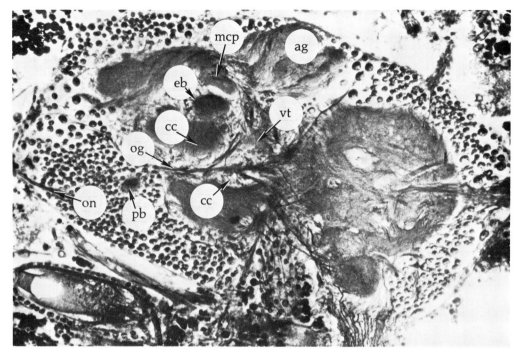

Fig. 198g

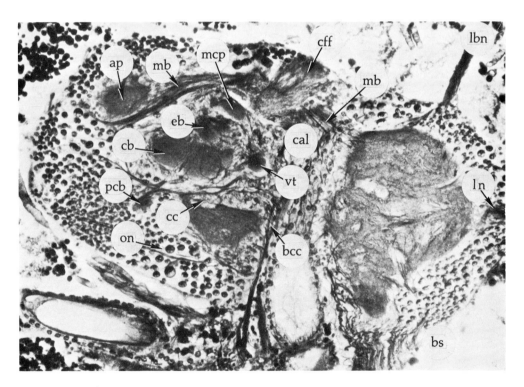

Fig. 198h

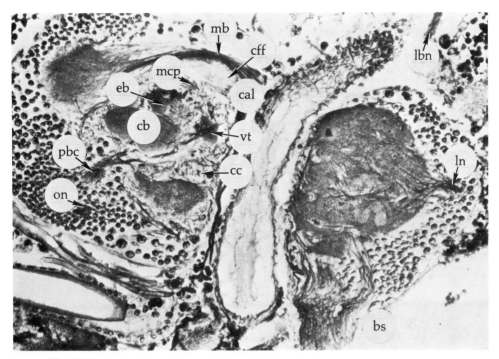

Fig. 198i

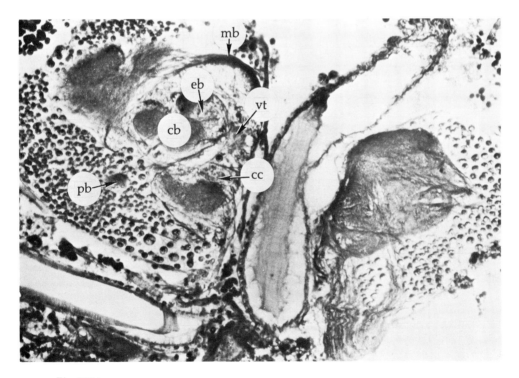

Fig. 198j

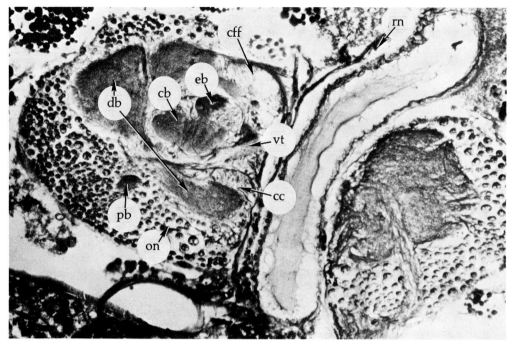

Fig. 198k

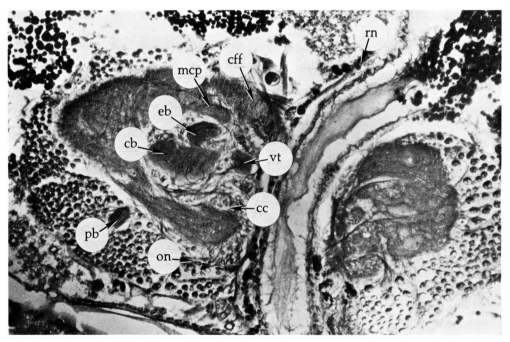

Fig. 198l

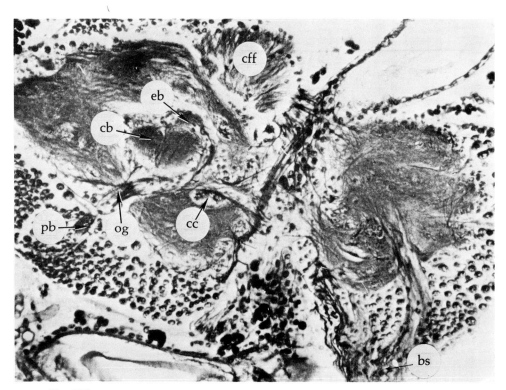

Fig. 198m

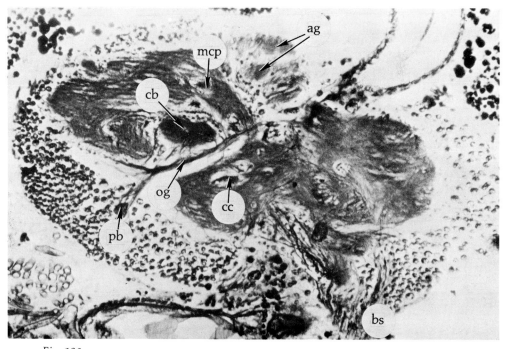

Fig. 198n

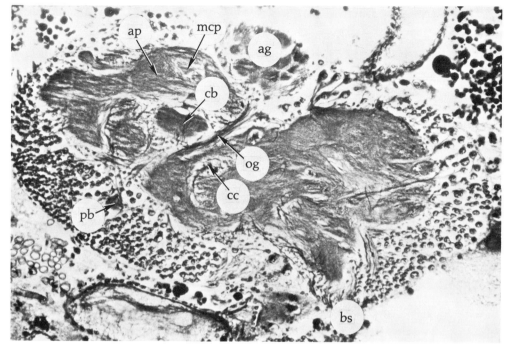

Fig. 198o

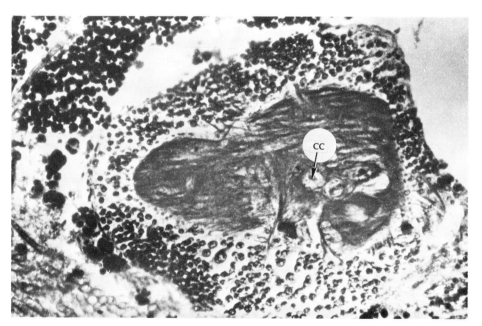

Fig. 198p

The ocellar stalk attaches to the dorsal side of the forebrain and the fibers cross ventrally to the posterior side until they reach the level of the oesophageal canal, at which point they turn anteriorly and enter the neuropile. This is the smallest cerebral nerve and is difficult to observe except in frontal sections. The fibers are paired as they approach the brain and branch before entering (Fig. 196a). The stalk is not, strictly speaking, a nerve; it is an extension of the brain comparable to the optic lobes. The true ocellar nerves are the short axons of the ocellar retinal cells. They synapse in the ocellar pedicel at the distal ends of the stalk.

The optic system is undoubtedly the most complex component, and the internal organization of the optic lobes themselves has received much attention since the time of Cajal. For details of the optic ganglia one is referred to Cajal and Sanchez (1915), Zawarzin (1914), Bretschneider (1921), Autrum (1958), and later work. The routes by which visual information reaches the integrative centers of the brain are the four tracts leaving the medulla interna. One bundle of fibers goes to the ventral body, one to the central body and pons, one to the pons and pedunculata. The commissure passing beneath the central body sends fibers to the central body. Another bundle is connected directly with the suboesophageal ganglion.

The antennal nerves attach to the midbrain at approximately the level of the oesophageal canal and constitute the largest bundles of nerves in the brain. Their initial connection is lateroventrally with the antennal glomeruli. They are mixed nerves containing sensory fibers and motor fibers serving the antennal muscles in the head and scape. The sensory fibers enter the glomeruli directly. Some pass through and enter the antennal commissure.

The nervus ganglii occipitalis is a pair of short connectives passing from the posterior area of the protocerebrum to the recurrent nerve as it emerges posteriorly from the passage in the head containing the oesophagus. It has been demonstrated in other insects that it arises from neurosecretory cells in the dorsal part of the protocerebrum. After coming to lie with the recurrent nerve the fibers enter the hypocerebral ganglion. These fibers constitute an important pathway between the brain and the endocrine complex (corpus allatum-corpora cardiaca).

The labrofrontal nerves extend from the brain from positions slightly ventrad of the antennal nerves (Figs. 198g, 198h) (Dethier, 1959). They are smaller in diameter than the antennal nerves. They pass down the proboscis on either side of the oesophagus. Each one branches almost immediately. The branch nearer the midline divides again, and part of it curves anteriorly and dorsally, meeting the corresponding branch of the opposite side and uniting with it in a single nerve. This nerve loops to continue ventrally for a short distance along the oesophagus before looping once again to reverse its direction and continue dorsally and posteriorly up the oesophagus as the recurrent nerve (Figs. 38, 198k, 198l). The recurving branches are the frontal connectives. In some specimens the frontal connectives leave the main trunk before it branches. The exact disposition of the loops varies, depending on whether the proboscis is retracted or extended; however, the first recurved dorsal loop is constant even though its tightness varies. It is suspended from the anterior walls of the proboscis and the anterodorsal wall of the head capsule in the region of the remnants of the ptilinum by strands of fine tracheae and fat body, both of which by their elasticity permit great flexibility and shock-absorption of the nerves as the mouth parts move. The most ventral extension of the nerve, before it retraces its dorsal course, gives rise to a small nerve that continues for a short distance down the oesophagus (Fig. 38). It also gives off to the surface of the oesophagus at least two lateral pairs of very fine nerves.

The exact position of the frontal ganglion is difficult to ascertain because it is not conspicuously ganglionlike in appearance. After the two frontal connectives fuse, there is a swelling in the region of the first loop and another in the region of the second loop. It is probable that the first is the frontal ganglion. A histological examination of this region would be desirable because the suggestion has been made that there is no frontal ganglion in the groups of flies to which *Phormia* belongs (Langley, 1965).

The recurrent nerve continues along the anterior surface of the oesophagus as it traverses the brain. Emerging posteriorly from the brain, it proceeds along the dorsal surface of the oesophagus and joins the hypocerebral ganglion just anterior to the junction of the crop and proventriculus (Figs. 126, 131). Here small nerves branch to the corpus allatum and corpora cardiaca. The largest nerves from the hypocerebral ganglion extend to the crop and to the proventriculus.

After the frontal connectives branch from the labrofrontal nerves, the two medial nerves unite on the anterior surface of the oesophagus in the region of the insertions of the retractors of the fulcrum. The fused nerve then extends down the proboscis, passing between the paired dilators of the cibarial pump to which small branches are sent.

The lateral branches of the labrofrontal nerves pass ventrally on either side of the oesophagus, each giving a branch to the retractors of the fulcrum, passing laterad of these muscles, entering the fulcrum, and extending to the labrum, These extensions constitute the innervation of the labrum-epipharynx. The central projections of the labrofrontal nerve have not been traced.

The labial nerves (Figs. 37, 38) are suboesophageal (Fig. 198h). Each member of the pair divides into two large branches soon after leaving the suboesophageal ganglion. The medial branch extends ventrally along the posterior portion of the proboscis. Almost immediately it gives off a small branch which, extending laterally and dorsally, innervates the accessory retractors of the rostrum. The principal branch then continues ventrally to the labellum. It innervates the retractors of the furca, the retractors of the paraphyses, and the transverse muscles of the haustellum. It is the principal sensory trunk from the labellum.

The second principal branch of the labial nerve, the lateral branch, immediately subdivides. The more dorsal branch sends a small twig to the accessory retractors of the rostrum, then continues up into the cranial cavity to the tracheae and fat body.

The remaining branch passes anterior to the retractors of the rostrum, then curves posterior to the flexors of the haustellum and the accessory retractors of the rostrum, thence anterior to all three muscles and ventrally down the proboscis. Near the proximal cornua of the fulcrum it gives off a small branch to the retractors of the rostrum. Shortly thereafter it subdivides to send sensory fibers to the maxillary palpi and motor fibers to the extensors of the haustellum and the adductors of the apodemes. To this point at least the main nerve is demonstrated to be a compound maxillary-labial nerve.

The central projections of the labial nerve are but imperfectly known. The axons from the labellar chemoreceptors, after entering the suboesophageal region of the brain mass, fan out and extend a short distance dorsally in the neuropile. Some fibers cross to the contralateral side. The motor fibers originate in a few large cortical cells in the ventral and ventrolateral region of the ganglion.

Distressingly little is known about the functional organization of the insect brain—nothing, as far as Diptera are concerned. One can only piece together bits of information derived from experiments with other insects (principally honeybees, ants, locusts, and crickets) and extrapolate to *Phormia,* with full knowledge of the perils this entails. Of all areas the optic is undoubtedly the most well understood (Bishop, Keehn, and McCann, 1968). The first synaptic level in the optic lobes occurs in the medulla externa. Here are to be found interneurons sensitive to nonselective ipsilateral motion. In the anterior region of the medulla interna are found units that detect motion in specific directions. Some have ipsilateral visual fields and others contralateral fields. The latter are obviously connected via optic fibers of the central commissure. In the corpora pedunculata there are units with monocular contralateral and ipsilateral fields and binocular units, all sensitive to selective motion. Some of these units perform summations of paired combinations of activity from certain of the units in the medulla interna. Presumably these units are connected to each other via the fibers going from the central commissure to the pedunculate body. As Wiersma (1967) has pointed out, there are many similarities between the arthropod retina and that of vertebrates, in which moving objects are specific stimuli for certain optic neurons (Lettvin et al., 1959; Maturana and Frank, 1963; Barlow, Hill, and Levick, 1964; Hubel, 1960; Hubel and Wiesel, 1959, 1960, 1962).

The corpora pedunculata are the only other areas of the brain that have been studied extensively by electrophysiological stimulation and lesions. The many small cells, the somata of which make up the calyx cups, send fibers down the stalk and into the α and β lobes. Afferent interneurons synapse with these and stimulate the calyx cells. The interneurons are probably first-order interneurons, and Maynard believed that there probably exists only one interneuron between receptors and "higher" centers. In any event, there must be a very close connection between the antennae and the corpora pedunculata because ablation of the latter reduces an insect's ability to utilize antennal input and initiate appropriate behavior (Maynard, 1967).

Other roles of the corpora pedunculata have also been demonstrated. Mention has already been made of interneurons in this area that inhibit activity in the suboesophageal ganglion, and interneurons in the corpus centrale that excite the oesophageal ganglion (Huber, 1960). That the corpora pedunculata are also concerned with learning processess (long a speculation) was actually demonstrated by Vowles (1961), who found that lesions in this area interfered profoundly with the ability of ants to run a maze that they had learned.

Gustatory information from the tarsi passes first into the thoracic ganglion of the ventral nervous system. Whether or not it passes directly through to the suboesophageal ganglion is not known. Arriving in the suboesophageal ganglion in some form, it reaches the brain via the median bundle. There are connections between the median bundle and the commissure passing beneath the central body. Access to the central complex is clearly possible. Since the labial nerves, those carrying axons from the labellar hairs and oral papillae, also enter the suboesophageal ganglion, it is possible that they are closely associated with axons from the tarsi in their passage to the central complex. One might expect, therefore, to find interneurons associated with gustation in either the central body or in the pedunculate bodies.

The other nerve that is of particular interest in connection with feeding is the labrofrontal nerve. Unfortunately the pathways within the tritocerebrum, where this nerve enters, have not been identified.

An area of particular interest to endocrinologists is the pars intercerebralis. Here are to be found the median neurosecretory cells whose secretory products play such vital roles in regulating metamorphosis, diapause, protein metabolism, and reproductive development.

Scharrer and Scharrer (1944) have likened this part of the insect brain plus the corpus allatum and corpora cardiaca to the hypophyseal-hypothalamic complex of the vertebrate. The intricate endocrine and neural interrelations amply justify the comparison.

There seems little reason to doubt that the central complex is the most important integrative center in the central nervous system. It is probably here that most of the sensory information is processed and most of the motor commands given. Here also may be located command interneurons that initiate whole patterns of behavior. Most of our knowledge of these matters is derived from the work (with crickets) of Huber and his associates (Huber, 1955, 1959, 1960, 1962). Their analyses reveal in the cricket two categories of behavior: (1) movement patterns in which the interplay of effectors can be modulated by different inputs from the periphery (for example, walking, copulation, egg-laying); (2) movement patterns that are almost irrevocably set by the central nervous system (for example, grooming, singing, flying). Local stimulation of the brain elicits from specific points ordered and coordinated patterns. At any point, however, there are changes in latency, threshold, and activation (from activation to inhibition). In some cases (e.g., acoustic behavior) in which action depends on momentary endogenous states there are changes in threshold related to these. In other cases (e.g., copulation oviposition) only the first local stimulation is effective because a certain constellation of stimuli from the periphery is required. The behavior pattern evoked may be very complex, and its various phases may come into action sequentially, depending on the order of their various thresholds. For example, local stimulation in the brain of the cricket may produce the following actions in this order: increase in respiration, antennal and head movements, walking, jumping.

These analyses have also shown that such a seemingly simple action as walking is in reality part of several complex behavioral situations. As Huber (1960) has shown, one complex category of behavior of which walking is an element is the category comprising flight, hole inspection, and attack. Walking is also an element of food searching, burrow construction, and postmating behavior. Brain stimulation in some animals caused locomotion, coupled with orienting movements of the antennae and palpi, and feeding when food was encountered. A change in threshold occurred with satiation. Huber suggested that the searching movements followed by the taking of food support the hypothesis that activation of an eating drive has occurred.

Literature Cited

Autrum, H. 1958. Electrophysiological analysis of the visual systems in insects. *Exp. Cell Res. (Suppl.)* 5:426–439.

Barlow, H. B., Hill, R. M., and Levick, W. R. 1964. Retinal ganglion cells corresponding selectively to direction and speed of motion in the rabbit. *J. Physiol.* 173:377–407.

Bishop, L. G., Keehn, D. G., and McCann, G. D. 1968. Motion detection by interneurons of optic lobes and brain of flies *Calliphora phaenicia* and *Musca domestica*. *J. Neurophysiol.* 31:509–525.

Bretschneider, F. 1921. Über das Gehirn des Wolfsmilchschwärmers (*Deilephila euphorbiae*). *Jena Zeit, Naturwiss.* 57:423–462.

Cajal, S. R., and Sanchez, D. 1915. Contribucion al conocimiento de los centros nerviosos de los insectos. *Trabajas del Lab. de Investig. Biologicas Univ. Madrid,* 13:1–167.

Cuccati, J. 1888. Über die Organization des Gehirns der Sonomya erythrocephala. *Zeit. wiss. Zool.* 46:240–269.

Dethier, V. G. 1959. The nerves and muscles of the proboscis of the blowfly *Phormia regina* Meigen in relation to feeding responses. In *Studies in Invertebrate Morphology,* Smithson. Misc. Coll. 137:157–174.

Dethier, V. G. 1965. Microscopic brains. *Science* 143:1138–1145.

Gieryng, R. 1964. Post-embryonal development of the central nervous system of Diptera. (Part I. Calliphora vomitoria L.). *Ann. Univ. M. Curie-Sklod; Sec. D.,* 19:377–393.

Gieryng, R. 1965. Veränderungen der histologischen Struktur des Gehirns von Calliphora vomitoria L. (Diptera) während der postembryonalen Entwicklung. *Zeit. wiss. Zool.* 171:81–96.

Hertwick, M. 1931. Anatomie und Variabilität des Nervensystems und der Sinnesorgane von *Drosophila melanogaster* (Meigen). *Zeit. wiss. Zool.* 139:559–663.

Hubel, D. H. 1960. Single unit activity in lateral geniculate body and optic tract of unrestrained cats. *J. Physiol.* 150:91–104.

Hubel, D. H., and Wiesel, T. N. 1959. Receptive fields of single neurones in the cat's striate cortex. *J. Physiol.* 148:574–591.

Hubel, D. H., and Wiesel, T. N. 1960. Receptive fields of optic nerve fibres in the spider monkey. *J. Physiol.* 154:572–580.

Hubel, D. H., and Wiesel, T. N. 1962. Receptive fields, binocular interaction and functional architecture in the cat's visual cortex. *J. Physiol.* 160:106–154.

Huber, F. 1955. Sitz und Bedeutung nervöser Zentren für Instinkthandlungen beim Männchen von *Gryllus campestris* L. *Zeit. Tierpsychol.* 12:12–48.

Huber, F. 1959. Auslösung von Bewegungsmustern durch elektrische Reizung des Oberschlundganglions bei Orthoptera (Saltatoria: Gryllidae Acridiidae). *Verh. dtsch. zool. Ges. Münster, Zool. Anz.* (Suppl. 23) 248–269.

Huber, F. 1960. Untersuchungen über die Funktion des Zentralnervensystems und insbesondere des Gehirnes bei der Fortbewegung und der Lauterzeugung der Grillen. *Zeit. vergl. Physiol.* 44:60–132.

Huber, F. 1962. Vergleichende Physiologie der Nervensysteme von Evertebraten. *Fortschr. Zool.* 15:165–213.

Jarnicka, H. 1959. On the structure of the brain in some Diptera. *Ann. Univ. M. Curie-Sklod.,* Sec. C., 14:161–167.

Langley, P. A. 1965. The neuroendocrine system and stomatogastric nervous system of the adult tsetse fly *Glossina morsitans*. *Proc. Zool. Soc. London* 144:415–424.

Larsen, J. R., Dethier, V. G., and Broadbent, A. H. 1975. The brain of the black blowfly, *Phormia regina* Meigen. In press.

Lettvin, J. Y., Maturana, H. R., McCulloch, W. S., and Pitts, W. H. 1959. What the frog's eye tells the frog's brain. *Proc. Inst. Radio Engineers.* 47:1940–1951.

Maturana, H. R., and Frank, S. 1963. Directional movement and horizontal edge detectors in the pigeon retina. *Science* 142:977–979.

Maynard, D. M. 1962. Organization of neuropil. *Amer. Zoologist.* 2:79–96.

Maynard, D. M. 1967. Organization of central ganglia. In *Invertebrate Nervous Systems*, C.A.G. Wiersma, ed., pp. 231–255. University of Chicago Press, Chicago.

Power, M. E. 1943a. The brain of Drosophila melanogaster. *J. Morph.* 72:517–559.

Power, M. E. 1943b. The effect of reduction in numbers of ommatidia upon the brain of Drosophila melanogaster. *J. Exp. Zool.* 94:33–71.

Power, M. E. 1948. The thoracico-abdominal nervous system of an adult insect, Drosophila melanogaster. *J. Comp. Neur.* 88:347–409.

Satija, R. C. 1958. A histological study of the brain and thoracic nerve cord of *Calliphora erythrocephala* with special reference to the descending nervous pathways. *Res. Bull. Punjab Univ.* 142:81–96.

Smith, D. S. 1967. The organization of the insect neuropile. In *Invertebrate Nervous Systems*, C.A.G. Wiersma, ed., pp. 79–85. University of Chicago Press, Chicago.

Scharrer, B., and Scharrer, E. 1944. Neurosecretion VI. A comparison between the intercerebralis-cardiacum-allatum system of insects and the hypothalamo-hypophyseal system of vertebrates. *Biol. Bull.* 87:242–251.

Snodgrass, R. E. 1935. *Principles of Insect Morphology*. McGraw-Hill, New York.

Vowles, D. M. 1954. The structure and connections of the corpora pedunculata in bees and ants *Quart. J. Micr. Sci.* 96:239–255.

Vowles, D. M. 1961. Neural mechanisms in insect behaviour. In *Current Problems in Animal Behaviour*, W. H. Thorpe and O. L. Zangwill, eds., pp. 5–29. Cambridge University Press, London.

Wiersma, C. A. G. 1967. Visual central processing in crustaceans. In *Invertebrate Nervous Systems*, C.A.G. Wiersma, ed., pp. 269–284. University of Chicago Press, Chicago.

Wigglesworth, V. B. 1960. The nutrition of the central nervous system in the cockroach *Periplaneta americana* L. The role of the perineurium and glial cells in the mobilization of reserves. *J. Exp. Biol.* 37:500–512.

Zawarzin, A. 1914. Histologische Studien über Insekten. IV. Die optischen Ganglion der Aeschna-Larven. *Zeit. wiss. Zool.* 108:175–257.

13
Profiting by Experience

Ils n'ont rien appris, ni rien oublie.
Talleyrand

Paradoxically, the most enduring characteristic of the environment is change. To survive animals must be able to cope with change. In a sense this is what behavior is all about—response to change. But environmental changes can have quite different time courses, and the manner in which organisms respond is adjusted to the temporal as well as to the qualitative aspects of change. If the time course greatly exceeds the life-span of an organism, there is little that the individual can do about it. Changes proceeding at geological rates affect generations and are coped with genetically. Natural selection and mutation are the influential mechanisms. Species of animals become extinct, and others replace them.

Environmental changes occuring within the lifetime of an individual are countered by changes within the animal. Relatively gradual accomodations are accomplished by physiological alterations. Animals become acclimated to changes in temperature, in altitude, in diet, etc. Physiological changes develop comparatively slowly and are reversed slowly. They are tuned to rather long-lasting alterations in the environment. More rapid, unstable, and fluctuating environmental variability demands more rapid countermeasures. The two systems that function effectively in this respect are the endocrine and nervous systems. Again there is a division of responsibility dictated by the rapidity with which a response must occur. Hormonal responses are quicker than physiological responses, and neural responses quicker then hormonal. The division among these three is somewhat arbitrary because all are physiological, and there is considerable overlap and co-operation between endocrine and neural systems. The distinction between neural responses and the others (and again this is blurred at the edges) is that most neural responses involve some overt movement or suppression of movement on the part of the animal. At a gross descriptive level the distinction is obvious, and although it is perfectly true that hormones affect and modify behavior, it is usually the nervous system that actuates it.

In its simplest guise the nervous system is a stimulus-response system. The chemoreceptive system of *Phormia* can serve as an example: a change occurs in the environment, for example, sugar on the tarsus (the stimulus), and the proboscis is extended (the response). This is not to imply that behavior is to be explained as a derivative solely or even primarily of simple reflexes, but merely to indicate at this point that behavior is basically a matter of stimulus and response. Given this basis, a question of profound importance is: do antecedent events affect response? One must distinguish here between antecedent events whose effects are still actively operative at the moment that response is being observed, and those events whose effects are momentarily inoperative. For example, the behavioral taste threshold of a fly is changed by a meal (an antecedent event); however, the effects of this meal continue to influence the nervous system via the stretch receptors of the foregut and abdomen long after eating has ceased. On the other hand, it might be imagined that the consumption of a particular meal predisposed the fly the next time it fed to choose that particular food over others, if given a choice. This predisposition would imply learning because it involves an effect somewhere in the animal that is stored until the propitious moment. Learning of this kind has not been demonstrated in *Phormia*.

Learning has been defined by Thorpe (1963) as "that process which manifests itself by adaptive changes in individual behavior as a result of experience." From a purely operational point of view it is advantageous to classify learning into several categories. It must be emphasized that these categories are based on descriptive behavior and do not imply a commonality of mechanism within a category from one taxonomic group to the next. The following is Thorpe's classification: habituation; associative learning, of which there are two types, conditioned reflex Type I (classical conditioning) and conditioned reflex Type II (trial and error learning, instrumental conditioning, operant conditioning); latent learning; and insight learning. These categories intergrade to a considerable extent and contain examples of very different levels of complexity. For detailed discussions the reader should consult Hinde (1970) and, for studies on nonlaboratory animals and invertebrates, Thorpe (1963), Dethier and Stellar (1970), Hinde (1970), and Manning (1972).

The simplest, most universal form of learning is habituation. It may be as fundamental a characteristic of life as DNA. It differs from other forms of learning in that it is a loss of habitual responses rather than the acquisition of new ones. A habituating animal begins to ignore repeated or continuous stimuli that are neither harmful nor rewarding. The adaptive value is obvious; an animal can expend much time and energy uselessly by responding to meaningless stimuli. Like boy's cry "Wolf!" the call from the sense organs eventually goes unheeded. There is of course, as the fable relates, the off-chance that there really is a wolf.

Habituation is defined as a long-term stimulus-specific waning of a response due to failure to release the consummatory act (Thorpe, 1963). It is further characterized by its tendency to be terminated immediately by a novel stimulus (dishabituation) and eventually by habituation to the novel stimulus (cf. Thompson and Spencer, 1966). Insofar as the definition is descriptive, the difficulty of establishing habituation as a distinctive unitary phenomenon is immense. As soon as one begins to probe underlying mechanisms it becomes apparent that a number of different mechanisms can manifest themselves as waning responses to specific stimuli. The usual criteria for establishing the existence of habituation are waning, stimulus specificity, a long time-course, dishabituation, and central rather than peripheral neural involvement. The criterion of central involvement was presumably established to rule out sensory adaptation; however, experiments with mollusks, especially *Aplysia,* have revealed behavioral phenomena that seem to represent true habituation and yet are mediated peripherally (Lukowiak and Jacklet, 1973).

The fly exhibits any number of waning responses. Most of them do not satisfy the criteria for habituation, and analyses of underlying causes indicate that they do not constitute genuine learning. Consider, for example, fatigue. A fly can be driven away from a landing site repeatedly by a wave of the hand. It persistently lands. Anyone who has ever attempted to catch a fly that lands on him time and time again discovers that there is no waning of response to the moving hand. That particular visual stimulus is neither rewarding nor harmful because the hand always misses. Were habituation occurring in this case, it should eventually become possible to capture the fly. On the other hand, if attempts to catch the fly are pursued to extravagant lengths, the fly ultimately ceases to fly. It can be demonstrated in this case that failure to respond arises from exhaustion of energy reserves. As was described in Chapter 1 a fly can be flown to the point where no stimuli will induce further flight, but an injection of carbohydrate in the blood or a meal of carbohydrate immediately restores its responsiveness to a moving hand.

Fatigue can readily be ruled out experimentally in situations involving responses to chemical stimulation, and various other stimuli can be shown to effect a waning response. A fly repeatedly stimulated with sugar soon stops extending its proboscis if permitted to eat at each presentation. Its acceptance threshold rises. The failure to respond persists for many hours. As shown earlier, however, repeated sensory stimulation is being continuously countered by inhibition from stretch receptors.

A rise in acceptance threshold still occurs upon repeated stimulation even if feeding (the consummatory act) is not allowed to take place. This waning is of short duration. Even here more than one mechanism can be operating. Sensory adaptation can easily account for waning lasting up to 3 minutes. As Getting (1971) demonstrated, stimulus satiation can also occur at the first synapse where several sensory fibers converge. The duration of the refractoriness is hardly long enough to warrant calling it habituation (cf. Getting, 1971).

Another level of adaptation was demonstrated in experiments involving stimulation of contralateral legs (Dethier, 1952). The waning of response in these instances was also of very stort duration. For sucrose it was 1 to 13 seconds. These experiments showed quite clearly that adaptation could occur at some level in the central nervous system. The importance of these experiments lies in demonstrating that from the brevity of an effect, alone, does not warrant our excluding it from the central nervous system.

Thus the hypothesis that habituation involving the contact chemical senses occurs in flies is not irrefutable, and the more penetrating the analysis the less the resemblance of various kinds of waning behavior to habituation. Further experiments, especially some related to stimulus specificity would be fruitful. For example, in some animals response to a given intensity of light gradually wanes upon repeated stimulation, but subsequent stimulation by a *lower* intensity will regenerate the response. The specificity here is intensity, and the fact of its being lower rules out sensory disadaptation. A comparable experiment utilizing chemical stimuli could be designed for the fly.

A better case can be made with the olfactory system (Thorpe, 1938). Thorpe (1939) showed with *Drosophila melanogaster* that a natural aversion to the odor of cedarwood oil could be transformed into a high degree of tolerance by exposing newly emerged adults to an airstream bearing the odor. With the odor of peppermint, aversion could even be converted to attraction in some cases. A possible explanation for the conversion depends on the presence in the oil of two components: menthol, which is repellent, and an ester that is attractive. Habituation to the repellent component, which completely masks the ester, would eventually allow the ester to exert its attractiveness. Thorpe also found that some kinds of stimuli experienced during larval life influenced the responses of adults. For example, adult *D. melanogaster* reared in a medium containing 0.5% peppermint, and thoroughly washed as larvae and pupae, behaved differently from the controls. Control flies never exposed to peppermint were repelled to the extent that only 35% of those placed in a Y-tube olfactometer entered the arm containing the odor. Of the flies exposed as larvae to peppermint, 70% entered the arm containing the odor. Comparable results were obtained with *D. guttifera* (Cushing, 1941) and, by means of a different technique, with *Calliphora* (Crombie, 1944). Thorpe termed this phenomenon "olfactory conditioning" and believed that it constituted evidence of latent learning. Employing techniques similar to Thorpe's, Hershberger and Smith (1967) confirmed the original observations but interpreted the results as demonstrating associative learning (the flies showing reduced aversiveness to the odor were associating it with food).

The whole problem was reinvestigated by Manning (1967), who pointed out that in all of the earlier tests each fly was tested only once so that there were no data relating to the time-course of the phenomenon or its regularity. Using geraniol as a test odor he repeated and confirmed Thorpe's observations. He reared eight generations of flies (controls) in the absence of odor and an identical number of generations exposed to odor. In a Y-tube olfactometer, 87.9% of the controls rejected the odor, whereas only 53.3% of the experimentals did. As he pointed out, each test provided a group of flies which, having been reared in a geraniol-scented medium, "chose" to move toward that odor when moderately food-deprived. The conditioning hypothesis would suggest that flies turning toward geraniol on the first run were doing so because they associated the scent with food. It would lead to the prediction that on a second run a higher proportion would go toward the scent. The habituation hypothesis would suggest that turning toward geraniol on the first run represented random choice. It would predict that on the second run there would also be a random choice. The percentages in both runs should not differ significantly from random choice. This was the result obtained. Manning concluded that his experiments, and earlier ones as well, demonstrated habituation rather than conditioning. Furthermore, he was unable to condition flies to odor by rewarding with food or by association with deprivation. Crombie's results with blowflies were similar. *Calliphora* raised on sterile medium containing menthol or exposed as postfeeding larvae to the odor of menthol became habituated to 2% but not to 100%.

Probably the most straightforward example of habituation for flies that we have is the behavior exhibited by *Musca domestica* toward novel stimuli. Mourier (1965) discovered that flies were strongly attracted to neutral novel objects placed in their cage but that the attractiveness waned with time. Counts were made of the number of visits made to two black squares, 15×15 cm^2, placed symmetrically on the

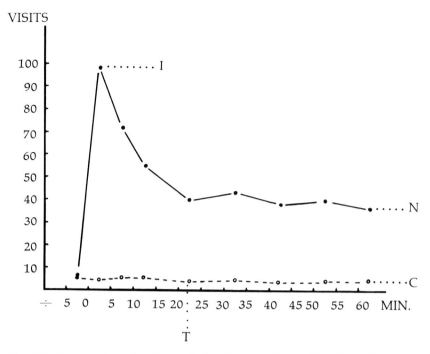

Fig. 199. The number of visits made by flies to a "new object" as a function of time. I = the number during the first 5 minutes. N = the final attractiveness of the object as compared with the floor of the cage (C) (from Mourier, 1965).

floor of a cage 1 m³. One square was always present and constituted the control. The other square was newly introduced at the start of the experiment. Each count was the sum of 20 momentary counts taken at intervals of 15 seconds during counting periods of 5 minutes. Each period of observation lasted 70 minutes and consisted of nine 5-minute counting periods. In the graph showing the number of visits with time (Fig. 199), C represents number of visits to the control square and is a measure of the density of the flies on the bottom of the cage, I is the number of visits to the novel square when it is first introduced, and N is the number of visits after the square is no longer novel. Therefore N is a measure of the normal attractiveness of the object compared with its surroundings. The novel object is 2.5 times more attractive when it is first introduced than its normal attractiveness. The novelty wanes and is lost after approximately 20 minutes. This can be considered genuine habituation because the stimulus is neutral. The waning is not due to sensory adaptation or fatigue, and it is stimulus-specific. These conclusions are based on the observation that introduction of a *different* novel stimulus re-establishes the response immediately (personal observation). This observation satisfies the criterion for dishabituation.

Recovery occurs rather rapidly (compared with that in vertebrates). When a novel object to which the flies had become thoroughly habituated was withdrawn and then reintroduced into the cage after an interval, a 10-minute absence was sufficient to produce some novelty effect, but one hour was required to produce the full novelty effect. Thus it may be concluded that recovery was complete within an hour. The novelty effect is clearly affected by a number of physiological conditions. Flies deprived of food and water, flies deprived of protein, and females, all are more strongly attracted to novel objects than are standard flies (3-day-old flies fed on dry sucrose and water and placed in the observation cage 2 days before the experiment). Age also exerts an influence; novelty increases up to an age of 3 days, after which it remains constant. Unfortunately there are no data describing the course of waning in these flies so that the effect of the various physiological states on habituation cannot be fully assessed.

An experiment involving the behavioral response of freely flying flies to a "target" of odor has been described by Wright (1974) as closely parallel to Mourier's experiment and is interpreted as memory. The investigation has not been carried far enough, however, to determine whether the results merely reflect sensory adaptation or whether habituation is in fact occurring.

Some experiments involving the neural location of "novel movement detection" by locusts are suggestive of where relevant mechanisms might be located in the fly (Horn and Rowell, 1968; Rowell and Horn, 1968). Recordings from the tritocerebrum led to the discovery of interneurons that were only weakly responsive to stationary visual stimuli but were strongly responsive to specific directional movement in the contra-lateral field. When the stimulus object was moved repeatedly along the same axis, the response waned and required from 15 minutes to several hours for complete recovery; however, if the same stimulus object was moved along a parallel axis, the response that had waned not only reappeared but reappeared at a higher level. In this case it would appear that habituation as observed behaviorally reflected decay in responsiveness of a particular interneuron.

Since habituation is most universally manifest in relation to mild and generalized warning stimuli, perhaps the search with *Phormia* should turn in that direction. For example, the fact that *Phormia* mounted by the wings on wax-tipped sticks tend to respond better to chemical stimuli if they have been allowed to rest in harness for several hours strongly suggests habituation.

From time to time reports surfaced in the literature purporting to demonstrate in Diptera learning abilities more complex than habituation. In 1941 Frings reported that a blowfly, *Cynomyia*, could be conditioned to extend its proboscis in response to the vapor of coumarin

(1,2-benzopyrone, an odorous component found in sweet clover, woodruff, etc.) when rewarded with sugar. In these experiments each test fly was first presented with the vapor of coumarin. At the end of one second of olfactory stimulation it was stimulated on the tarsi with sugar. The odor was present during tarsal stimulation. The proboscis was extended in response to the sugar. Responses to coumarin alone, that is, in the absence of sugar, occurred 30% of this time. With simultaneous stimulation with coumarin and sugar the response rose to a mean of 73% for the first set (ten) of trials, 88% for the sixth set (six), and 90% for the tenth (six) set. It was assumed that conditioning had occurred.

Repetitions and extensions of these experiments were undertaken with *Phormia* in our laboratory by Block and by Nelson. Although they obtained somewhat similar results, they also found, that the same result could be obtained by presenting sugar in advance of olfactory stimulation, or by omitting all olfactory stimulation until the flies had undergone several sets of stimulation with sugar alone, and only at the end of the series tested with coumarin alone. Something was obviously happening but what was it? In the meantime two other reports of learning had appeared. Murphy (1967) reported that *Drosophila* could learn a maze and Ilse (1949) reported that drone flies, *Eristalis taenax*, learned to associate the color yellow with food. Murphy's results could not be confirmed and may have been influenced by odor trails in the apparatus (Yeatman and Hirsch, 1971). Ilse's anecdotal report is difficult to interpret because of uncontrolled elements in the experimental design.

The tantalizing report of Frings (1941) led Dethier, Solomon, and Turner (1965) to embark on a long series of investigations of the effect of antecedent events on the response of the chemoreceptors of *Phormia*. Two well-known phenomena provided the starting point, spatial summation and temporal summation. The activity of sugar receptors can be summed in the central nervous system; that is, the concentration of sugar required to elicit extension of the proboscis decreases as the number of receptors stimulated is increased (Dethier, 1955; Arab, 1959). Temporal summation can be demonstrated in several ways: (a) a concentration of sugar normally too weak to elicit extension of the proboscis will do so if two hairs are stimulated successively within an interval of a few seconds; (b) stimulation of a water receptor, normally ineffective in eliciting a response from a water-satiated fly, will be effective if preceded by stimulation of a sugar receptor.

These observations, particularly the second, suggested that at some point in the pathway between the sensory neuron and the motor neurons that activate proboscis extension, a change in excitatory state occurs and that this change has a finite duration in time. The inhibitory counterpart of this state was revealed by the demonstration that stimulation of a salt receptor (which normally prevents or reverses proboscis extension) raises the threshold to subsequent sugar stimulation. Thus the possibility was opened of manipulating independently the intensity of central excitatory and inhibitory processes so as to modify the effectiveness of external stimuli in producing the proboscis extension. This presented the opportunity of making a new type of assessment of excitation and inhibition in the central nervous system (CNS).

The following testing procedure illustrates the logic and strategy of the series of studies. A hungry but thoroughly water-satiated fly was stimulated with a drop of water touched to one labellar hair. There was no behavioral response; i.e., the proboscis remained retracted. Then a drop of sugar was placed on another labellar hair. The fly responded with a vigorous proboscis extension. After removal of the sugar, the proboscis retracted. Retraction was either full and immediate, or partial and slow. Fifteen seconds after the sugar was removed, a drop of water was applied to the same labellar hair that had previously been stimulated with water. In contrast to the outcome of the first stimulation by water, the proboscis extended. This behavioral response to the second water test was not due to reactivation of sugar left on the hair, because the water was always placed on one hair and the sugar on another. Furthermore, electrophysiological evidence has shown that a sugar fiber stops firing as soon as sugar is withdrawn. Therefore the behavioral response to the second water test was not due to summation from simultaneous input from the water and sugar receptors. Finally, electrophysiological evidence has revealed no difference in the nature of the response of the water receptor before and after the stimulation of the sugar receptor.

Therefore the change in responsiveness to water must reflect an excitatory change somewhere in the CNS. Several characteristics of this central excitatory state (CES) can be described: (a) the decay of excitation in time; (b) the influence of stimulus intensity on the rate of decay; (c) the effect of food and water deprivation; and (d) the influence of inhibiting stimuli.

The experimental procedure consisted of test trials with the following standard components. First, there was a water pretest in which a drop of water on a needle was touched to the tip of one of the longest anterior labellar hairs for 5 seconds. Because the fly was highly water-satiated, it usually did not extend its proboscis. In those very rare instances when a response did occur, the fly was allowed to drink as much as it liked. Next, after a negative water pretest was obtained, a drop of sucrose solution was placed on the tip of one of the longest posterior labellar hairs for 5 seconds. This sucrose stimulation almost invariably resulted in proboscis extension.

In order to quantify the magnitude of proboscis extension an arbitrary scale was constructed. This scale is illustrated in Figure 200. For example, a movement from Position 1 to Position 3 would be recorded as 1–3. Several measures can be derived from this scale. First, any increase in extension can be tabulated as a positive event and counted. This yields a measure of percentage (relative frequency) of responding or failure to respond. Second, a measure of the net excursion of the movement can be obtained by subtracting the measure of resting level from the point of greatest extension made. Third, the maximum extension can be used as an index. Although these three response measures are not independent, each has been found to be a useful index of responsiveness. Some qualitative attributes of the response were also recorded. These included trembling of the proboscis without extension, and repeated extension and withdrawal rather than a simple response.

After the sugar was removed, the second water test (the water posttest), was conducted after decay times of 15, 30, 45, 60, 120, 360, 480, or 600 sec.

The water posttest, coming at the end of the decay time, consisted of touching a drop of water for 5 seconds to the same long anterior labellar hair that had been stimulated in the water pretest. This marked the end of a trial. Thus the water pretest and posttest procedures were identical. An intertrial interval of 2 minutes followed.

Some flies were tested with decay times of 15-60 seconds, others with decay times of 120-600 seconds. In both groups the decay times were varied from trial to trial in a predetermined sequence designed to minimize potential order effects. Each fly was tested at least three times on each decay time. The molar concentrations for the sucrose stimulation for short times were 0.008-1.0; for longer times only 1.0 was used.

The results of a typical experiment with decay times of 15-60 seconds are summarized in Figure 201, which shows that the percentage (relative frequency) of water posttests yielding a proboscis extension declined in an orderly fashion as decay time was increased, and that both the mean net excursion and the mean distance of maximum mean net excursion of the proboscis also decreased with decay time. Extension of the proboscis decreased with decay time.

The results obtained for the decay times of 120-600 seconds were highly unreliable and are not presented. It is important to note, however, that some flies gave an occasional proboscis extension response to water even as long as 600 seconds after sucrose stimulation.

Clearly, stimulation of a labellar hair with sucrose increases the re-

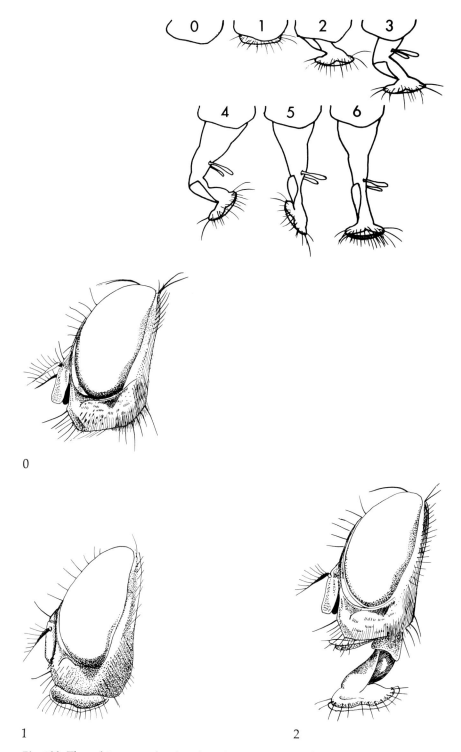

Fig. 200. The arbitrary scale of proboscis extension used to quantify response magnitude (after Dethier, Solomon, and Turner, 1965).

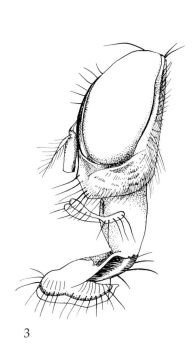

3

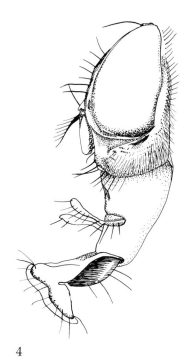

4

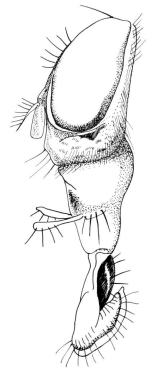

5

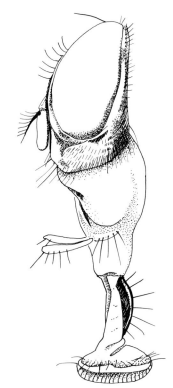

6

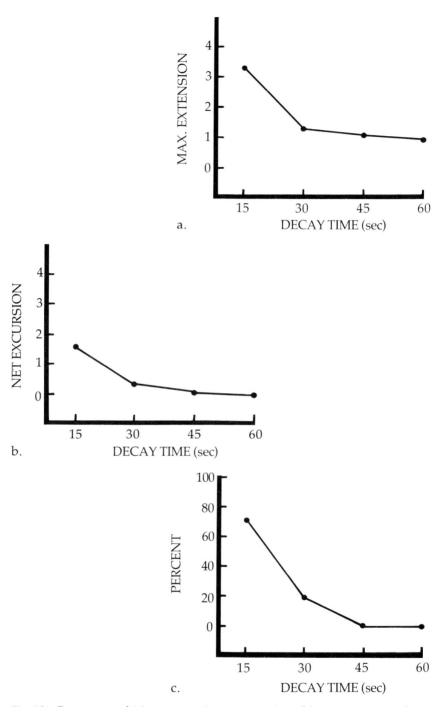

Fig. 201. Percentage of (a) mean maximum extension, (b) mean net excursion, and (c) extension of the proboscis during the water posttest as a function of decay time. The sucrose stimulus was 0.032 M. Each point represents the mean of at least 21 observations from each of seven flies (from Dethier, Solomon, and Turner, 1965).

sponsiveness of the water-satiated fly to water applied later to a different hair. The amount of excitability decays slowly with time.

Because different concentrations of sucrose are known to alter the frequency of nerve impulses passing from the sucrose receptor in the hair to CNS, it might be supposed that these frequency differences would affect the intensity and duration of the CES. In order to test this supposition the concentration of sucrose was varied over a wide range and the responsiveness during the water posttest was measured. Several decay times were sampled for each fly.

The results of this experiment for the 45-second decay time are shown in Figure 202. The effect of sucrose concentration on responsiveness to water in the posttest was large and orderly. The higher the sucrose concentration, the greater the responsiveness to water after a given decay time. This can be interpreted to mean: the greater the frequency of sensory nerve impulses, the greater the intensity of the CES after a given decay time. When all of the data on various sucrose concentrations and different decay times were combined, the family of functions shown in Figure 203 was obtained. Responsiveness to the water posttest was a joint function of sucrose concentration and decay time, such that increased concentration tended to enhance responsiveness and to delay the decline of responsiveness during the decay time.

In the experiments reported above, all flies had been food-deprived for 60 hours when testing was started. Because it has been demonstrated that behavioral thresholds to sucrose stimulation decrease markedly with increasing food deprivation, it seemed possible that the effectiveness of any given concentration of sucrose in arousing the CES would increase with prolonged deprivation.

In experiments that followed the same maintenance and testing procedures used in the previous experiments, flies were tested after varying hours of food deprivation. Each fly was tested five times at one concentration (0.250 M), and one decay time (15 seconds). The deprivation times were 0-80 hours and each fly was run at one deprivation time only.

The results are shown in Figure 204. The responsiveness during the water posttest increased regularly from 0-70-hours food deprivation, after which it fell off rapidly at 80 hours, because of the moribund condition of these flies.

The striking effect of food deprivation could arise from any of several physiological changes occurring during deprivation. The most obvious of these is change in the distension of the crop as food is utilized. The contents of the crop were weighed after varying deprivation periods that followed a 24-hour period of unrestricted 0.1 M sucrose feeding. After 20 hours the crop was nearly empty. Because the

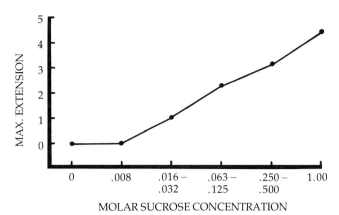

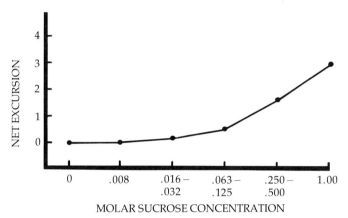

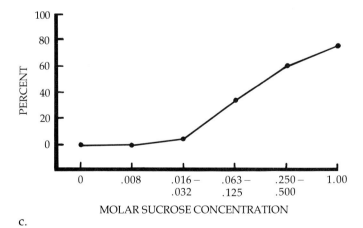

Fig. 202. Percentage of (a) mean maximum extension, (b) net excursion, and (c) extension of the proboscis during the water posttest 45 seconds after sucrose stimulation as a function of sucrose concentration (from Dethier, Solomon, and Turner, 1965).

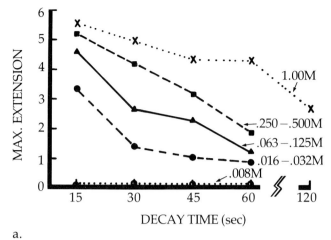
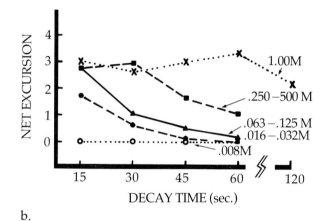

Fig. 203. Percentage of (a) mean maximum extension, (b) mean net excursion, and (c) extension of the proboscis during the water posttest as a joint function of decay time and sucrose concentration (from Dethier, Solomon, and Turner, 1965).

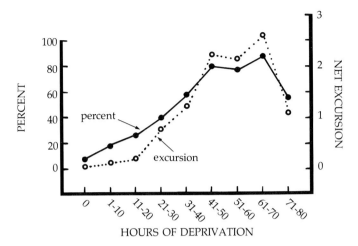

Fig. 204. Percentage of proboscis extension (solid line) and mean net excursion of proboscis (dotted line) during the water posttest as a function of hours of food deprivation. Posttest 15 seconds after stimulation by sucrose (from Dethier, Solomon, and Turner, 1965).

behavioral responsiveness continues to increase for 70 hours, crop factors cannot explain the outcome.

Reduction in blood sugar level during deprivation is another possible mechanism, and to test for its involvement in responsiveness, sets of parabiotic flies were prepared.

Both members of each pair had been food-deprived for 60-70 hours at the time of the operation. One member of each pair was then fed to satiation on 1.0 M sucrose, after which five regular trials were run on both members of each pair. A trial consisted, as usual, of water pretest, sucrose stimulation with 0.250 M solution, and water posttest. The decay time was 15 seconds. Tests were run 1½–4 hours after the 1.0 M sucrose feeding.

The fed members of each pair were unresponsive in all cases during the water posttest. The unfed members of each pair were as responsive as the individual unfed nonparabiotic flies run at 60-70 hours of deprivation (see Fig. 204). As shown in Table 26, the differences between the responsiveness of the fed and unfed members of the parabiotic pairs were striking (percentage of responsiveness: $U = 0$, $p \leq 0.002$; mean net excursion: $U = 0$, $p \leq 0.002$; mean maximum extension: $U = 0$, $p \leq 0.002$). The unfed parabiotics and unfed normals did not differ significantly in posttest responsiveness. This result demonstrates the good condition of the parabiotic flies.

Because of the rapid and vigorous extension of the proboscis during the sugar stimulation, it was often impossible to prevent the ingestion of minute amounts of sucrose. Two kinds of evidence, however, indicate that the increased responsiveness to water during the decay

Table 26. Comparison of Responsiveness during Water Posttest.

Experimental condition	No. fly-pairs	Percentage of proboscis extension	Mean net excursion	Mean maximum extension
Fed parabiotic	11	8	0.01^a	1.04^a
Unfed parabiotic	11	95	2.62^a	4.85^a
Unfed normal	12	87	2.65^b	5.11^b
With salt stimulation to legs	13	7	0.00^c	0.47^c
Without salt stimulation to legs	13	69	1.86^c	3.32^c
With water stimulation to legs	4	10	0.10^c	1.00^c
Without water stimulation to legs	4	75	1.90^c	4.50^c

From Dethier, Solomon and Turner, 1965.

[a] Means based on 10 observations per fly for 5 parabiotic pairs and 5 observations per fly for 6 parabiotic pairs.

[b] Means based on at least 4 observations per fly.

[c] Means based on 5 observations per fly, per condition, with each fly as its own control.

time cannot be attributed to the act of ingestion or to continuing stimulation by ingested sucrose of unknown receptors in the alimentary canal. The first evidence comes from experiments in which proboscis extension was elicited by stimulating the sucrose receptors on the feet with 1.0 M sucrose. The second line of evidence is electrophysiological.

The technique used to stimulate the sucrose receptors on the feet was essentially the same as that employed in stimulating the labellar hairs. Two groups of flies were run: one group after 60-70 hours and a second after more than 90 hours of food deprivation. For the water pretest a drop of water was touched to one of the longest anterior labellar hairs for 5 seconds. Because the flies were water-satiated, proboscis extension usually was absent. Then the feet of the fly were touched to the surface of a dish of 1.0 M sucrose for 5 seconds. This stimulation produced proboscis extension, but none of the sucrose was allowed to touch the proboscis; hence, there was no ingestion. Now a decay period of 15, 30, 45, or 60 seconds was allowed to pass; decay periods were randomly varied from trial to trial. The water posttest was then given to the same labellar hair that had received the water pretest.

As Figure 205 illustrates, a CES could be established by supplying sensory input via the leg (tarsal) receptors sensitive to sucrose. The

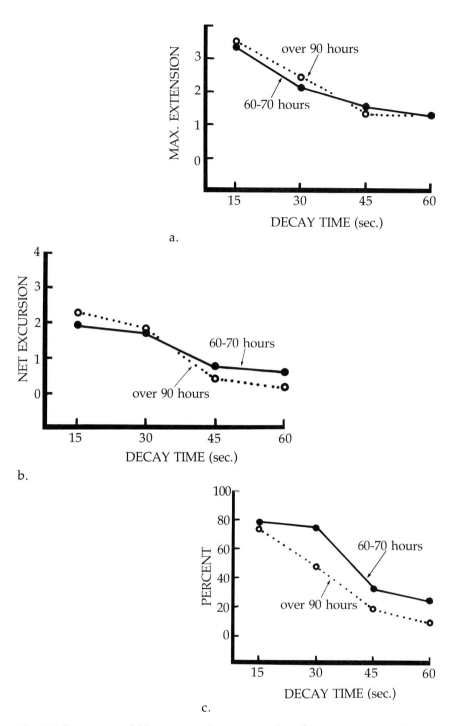

Fig. 205. Percentage of (a) mean maximum extension, (b) mean net excursion, and (c) extension of proboscis during water posttest as a function of decay time following stimulation of the legs with 1.0 M sucrose (from Dethier, Solomon, and Turner, 1965).

decay of responsiveness in the water posttest, after the CES had been aroused by sucrose stimulation of the legs, was more rapid than that obtained after the CES had been aroused by a 1.0 M sucrose stimulation of a labellar hair. A comparison of the 1.0 M sucrose responsiveness curves shown in Figure 203 with the curves in Figure 205 shows, in addition, that the initial magnitude of responsiveness after a 15-second decay time was somewhat less when the legs were stimulated (percentage of responsiveness: $U = 4$, $p = 0.07$; mean net excursion: $U = 8$, $p = 0.25$; mean maximum extension: $U = 2$, $p = 0.02$) than when the proboscis was stimulated.

The second line of evidence in support of the idea that sucrose receptor input alone is sufficient to establish a CES is electrophysiological. The evidence was obtained by making use of a technique of recording developed by Hanson. An Ag-AgCl electrode was attached externally to the neck of a fly fixed to a paraffin block. A pad of cotton soaked in saline lay between the electrode and the neck of the fly. This was the reference electrode. A recording electrode in the form of a glass pipette, 10 μ in diameter at the tip, was placed over one of the long, anterior labellar hairs. This pipette contained 0.05 M NaCl. Sodium chloride at this concentration provided the necessary conductivity for the electrode, yet was so dilute that it stimulated only the water receptor in the hair and not the salt receptor. Placing this electrode on the hair constituted the water pretest. As long as the electrode was in position the hair was being stimulated with water, and the action potentials generated by the water receptor were picked up by the electrode and recorded in the usual fashion (Fig. 206). After 5 seconds the electrode was removed, and a different electrode containing 1.0 M sucrose, in addition to 0.10 M NaCl, was placed over the tip of a long, posterior labellar hair. This constituted the sucrose stimulation. The first sign of a response was a trembling of the proboscis and muscle potentials seen as large, rapid fluctuations in the record (P). When the fly finally extended its proboscis, contact with this electrode was broken. The record obtained while the electrode was in contact shows the period (80 msec.) of activity from the sugar receptor required to elicit proboscis extension. The actual spikes were not seen because the amplifier blocked. After this response a 15-sec. decay time was allowed to elapse. Then the water (i.e., 0.05 M NaCl) electrode was again placed on the anterior labellar hair which had been previously stimulated in the pretest. This was the water posttest. The fly responded by muscle activity and then by extending its proboscis, thus breaking contact with the water electrode. The record obtained during this period of contact shows the number of impulses from the water receptor required to produce the proboscis response.

As Figure 206a shows, the first one second of stimulation (water pre-

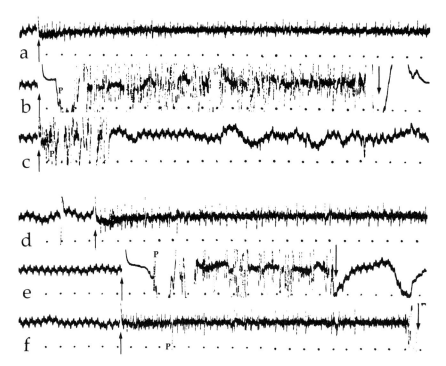

Fig. 206. Two sets of records showing the number of action potentials required from a water receptor to elicit proboscis extension in a water posttest as compared to the number that failed to elicit extension in a water pretest (↑ , onset of stimulation; P, beginning of proboscis activity; ↓ , proboscis extension; 10 mm = 40 msec.). (a) Water pretest; (b) sucrose stimulation; (c) water posttest; (d) water pretest; (e) sucrose stimulation; (f) water posttest (from Dethier, Solomon, and Turner, 1965). For full explanation, see text, Chapter 13.

test) of the water receptor yielded 55 action potentials, and even though output continued at approximately this rate for 15 seconds, there was no proboscis extension. In contrast (Fig. 206c), eight action potentials from the same water receptor during the posttest did elicit proboscis extension from position 2 to 4 after 200 milliseconds. The rate of firing in both cases was the same. The introduction of 80 milliseconds of activity from the sugar receptor into the central nervous system (Fig. 206b) (causing the proboscis to extend from position 3 to 6) had already augmented the responsiveness of the fly to subsequent stimulation by water even though the peripheral firing from the water receptor was unchanged. Another set of records (Fig. 206d-f) from a different preparation shows a similar augmentation of responsiveness to water. In the water pretest (Fig. 206d) there were action potentials from the water receptor but no extension of the proboscis. Stimulation with sugar (Fig. 206e) resulted in receptor activity after 100 milliseconds and extension of the proboscis from position 0 to 6 after 600 milliseconds. In the water posttest (Fig. 206f) there was activity of the muscles of the proboscis after 13 impulses from the water receptor and extension of the proboscis from positions 0 to 5 after 44 impulses.

The results of these two very different experimental approaches prove conclusively that ingestion is not essential to the establishment of the perseverating excitatory state. Taken together with the fact that the sugar receptor ceases to fire when the pipette containing sucrose is removed from its tip, these results prove that the excitatory state is set up by receptor input but is not an artifact of continuing receptor input after the removal of sucrose stimulation. Whatever the nature of the perseverating excitation may be, it does not reside in the receptor units. Because the axons from most of the receptors proceed into the CNS without synapsing, the conclusion is inescapable that the excitatory state is centrally located.

The observation that stimulation of a salt receptor (which normally prevents or reverses proboscis extension) raises the threshold to subsequent sucrose stimulation (Arab, 1959) suggested that there might be an inhibitory counterpart of the CES. Many procedures were tried in attempt to discover such a phenomenon. For example, there were trials composed of a sucrose pretest, which always elicited a proboscis extension, salt stimulation, and, following a decay time, a sucrose posttest; labellar hairs were employed for all stimuli. The sucrose posttest revealed a perseverating inhibitory effect only in very few cases, and only with decay times of durations less than 5 seconds. This rapid decay challenged the dexterity of experimenters in carrying out sequential tests.

More success in demonstrating apparent inhibitory effects was obtained by utilizing the receptors on the legs. The procedure employed was the same as that described for establishing CES by leg stimulation, except for the intercalation of salt stimulation between sucrose stimulation of the legs and the water posttest of the labellar hair. The order of presentation, therefore, was water pretest on the long anterior labellar hair, sucrose stimulation of the leg receptors, salt stimulation on the leg receptors, and water posttest on the same anterior labellar hair. The concentration of sucrose was high enough (1.0 M) to guarantee a high frequency of proboscis extension to the usual water posttest. The concentration of NaCl was sufficient (1.0 M) to produce complete retraction of the fully extended proboscis. It was reasoned that sucrose would set up an excitatory state. If an inhibitory state were established, and if it perseverated in time, the responsiveness to the water posttest would be decreased. Each fly was given five trials without and five trials with intercalated NaCl, each its own control.

The results, presented in Table 26, demonstrate clearly that stimulation by NaCl markedly decreased responsiveness to the water posttest given 10 seconds later (percentage of responsiveness: $U = 14.5$, $p \leq 0.002$; mean net excursion: $U = 6$, $p \leq 0.002$; mean maximum extension: $U = 5$, $p \leq 0.002$).

Perhaps, however, the decrease in responsiveness was an artifact of an intercalated stimulation event rather than a special inhibitory effect of salt. In other words, it is possible that stimulation per se merely discharged or dissipated the CES established by sucrose stimulation. If this effect were masquerading as an inhibitory effect, then the intercalation of an additional water stimulation between the sucrose stimulation and the water posttest should decrease stimulation to the water posttest. The control that was designed used the usual procedures for testing. Two sequences were compared: (a) water pretest, sucrose stimulation, water posttest; and (b) water pretest, sucrose stimulation, water stimulation, water posttest. Each fly served as its own control.

Table 26 shows the results of the control experiment. Intercalated H_2O reduced responsiveness in the water posttest significantly (percentage of responsiveness: $U = 1$, $p = 0.03$; mean net excursion: $U = 0$, $p = 0.01$). Table 26 shows that intercalated water stimulation decreased responsiveness in the posttest almost as markedly as did the intercalated NaCl stimulation. It has been demonstrated that water stimulation, used in all posttests in previous experiments, was doing more than measuring the preseverating CES; it was effectively decreasing subsequent responsiveness.

These findings explain an apparent paradox in the first experiment on the temporal decay of CES. On the one hand, with long decay times, responsiveness to water could last as long as 5 minutes; on the other hand, responsiveness to the water pretest, which came after a 2-minute intertrial interval, was well below 1%. The obvious conclusion from the experiments on intercalated salt and water stimulation is that the water posttests were serving to discharge the CES so that 2 minutes later, at the time of the water pretest, the fly was unresponsive. The water posttest thus was serving as an intercalated stimulus between the sucrose stimulation and the water pretest for the subsequent trial. The fact that the responsiveness after an intercalated water stimulus was 10%, contrasted with the usual pretest responsiveness of less than 1%, indicates that there was indeed a small decrement in CES over the 2-minute intertrial interval. That is, the intercalated water stimulation did not completely dissipate the CES level aroused by the sucrose stimulation.

The discharge of CES by water stimulation posed a difficult methodological problem for the study of central inhibitory processes. The measurement procedure modifies the process to be measured.

All of the experiments thus far described have demonstrated that the stimulation of a sucrose receptor either in a labellar hair or in a tarsal hair of a hungry blowfly will increase the subsequent responsiveness of the fly to water stimulation, even though the fly is thoroughly water-satiated, and even though it will not normally show a proboscis extension to water. The increased responsiveness to water stimulation fades in time. The amount of increased responsiveness to water stimulation depends upon the sucrose concentration previously applied to the labellar or tarsal sucrose receptors. With sucrose concentrations as low as 0.08 M, responsiveness to water disappears within 30 seconds. With concentrations as high as 1.0 M the effect is still easily measurable 120 seconds after sucrose stimulation.

In addition to the passage of time and the concentration of the stimulating sucrose solution, the degree to which the fly has been deprived of food greatly affects its responsiveness to water stimulation. If a fly is food-deprived but a few hours, the heightened responsiveness to water as a consequence of prior sucrose stimulation is difficult to detect. As food deprivation increases to about 60 hours, responsiveness to water stimulation increases in a monotonic fashion and then drops off as the fly becomes moribund from lack of food. Neither the distension of the crop nor blood sugar level can account for increased responsiveness to water as a function of food deprivation.

The increased responsiveness to water, induced by sucrose stimulation, does not depend on actual ingestion of sucrose is evidenced by the fact that sucrose stimulation of the tarsi is a sufficient condition for producing subsequent proboscis extension in response to water. Additional evidence of an electrophysiological nature shows that the responsiveness to water is enhanced simply by delivery of impulses from a sugar receptor to the CNS for as short a period as 80 milliseconds. The peripheral neural activity of the water receptor is not altered as a consequence of this previous input. The occurrence of a previous sucrose stimulation merely changes the probability of occurrence of proboscis extension in response to constant neural input from the water receptor.

Also demonstrated was what appeared to be a perseverating decreased responsiveness of the proboscis to water stimulation induced by intercalated salt stimulation. That is, the expected increase in responsiveness to water as a consequence of prior sucrose stimulation did not occur when salt stimulation intervened between the sucrose stimulation and the water posttest. This apparent perseverating inhibitory effect of salt was shown to be an artifact of the testing procedure because substitution of water stimulation for the salt stimulation produced almost the same decremental effect. Such a phenomenon implied that some new procedures had to be developed in order to study the arousal of and the decay of inhibitory effects.

It is difficult to put in precise theoretical terms the explanation for the experimental findings. It seems likely that sucrose stimulation applied to a hungry fly produces a perseverating excitatory state which is not specific to the sucrose ingestion system alone. Instead, this state seems to prime the proboscis response system so that water stimulation, normally incapable of arousing it, now becomes capable of doing so. Water deprivation is not necessary for this effect to occur; food deprivation definitely is necessary, and yet the response that indexes the excitatory state is aroused by water stimulation. Since the excitatory state cannot be peripheral in nature, it has been referred to as CES (Sherrington's term). Sherrington employed the term primarily in reference to reflex responses in the spinal cord and to changes in excitability outlasting by a few milliseconds stimulation of an afferent nerve.

It is puzzling that one can produce a sudden decrease in the CES by intercalating either salt or water stimulation. Casual observations suggest that almost any stimulation applied to the food and water detection system of the fly, including ordinary mechanical stimulation, can suddenly decrease the CES as long as the stimulation is not arousing the sucrose receptors. Taken in conjunction with experiments on the effect of decay time, such speculations lead to the conclusion that the CES aroused by sucrose stimulation will either decay slowly as a function of time or will discharge quickly as a function of stimulation by something other than sucrose.

Viewed from an entirely different theoretical framework, a fly stimulated 15 seconds previously by sucrose, and then subsequently extending its proboscis in response to water, looks as though it had made an error. If the sucrose had not been presented, the fly would not have extended its proboscis to water. The fly appears to be responding to water as though it were sugar. This is the type of statement sometimes made by conditioning theorists in pointing out the similarity between responses to the conditioned stimulus (CS) and to the unconditioned stimulus (US). Yet the phenomenon described cannot be considered to be a conditioning phenomenon because associative contingencies between stimuli are not necessary for its occurrence. The CES phenomenon may be more like sensitization or pseudoconditioning, but even here the analogy is strained. If one looks on sucrose as the US and water as the CS, pseudoconditioning should be characterized by heightened responsiveness to the CS which gradually declines with repeated stimulation. However, a single water stimulation is sufficient to eliminate the CES.

Although it is not yet possible to locate the postulated CES anatomi-

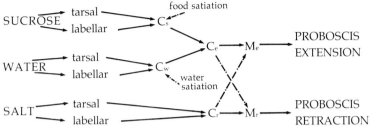

Fig. 207. Schematic flow diagram of hypothetical neural pathways to account for excitatory and inhibitory influences on proboscis response. Solid arrows are inferred excitatory processes; dashed arrows, inferred inhibitory processes. C and N are inferred neural centers (from Dethier, Solomon, and Turner, 1965).

cally, enough behavioral information relative to the proboscis response has been accumulated to permit construction of a tentative flow diagram model of the neural pathways involved (Fig. 207). The reasoning behind the schema is as follows: because summation between tarsal and labellar sugar receptors has been demonstrated (Arab, 1959; Dethier, 1955) there must be a point of junction for the sensory pathways from the legs and the mouthparts (C_s). The same must be true for the pathways from the water receptors (C_w). The two centers must be distinct because there are central events that affect input from the water receptors and from the sugar receptors independently. For example, flies can distinguish between water and sugar, and the ingestion of each is independently regulated (Dethier and Evans, 1961; Evans, 1961a). Taste thresholds to sugar vary with feeding and starvation, and all evidence suggests that it is the presence of sugar in the foregut (exclusive of the crop) that sets the level of sugar threshold (Dethier and Bodenstein, 1958; Dethier, Evans, and Rhoades, 1958; Evans and Barton Browne, 1960; Evans and Dethier, 1957). Water responsiveness, on the other hand, is regulated by blood volume (Dethier and Evans, 1961). Consequently, there must be an independent center into which both tarsal and labellar sugar receptor fibers run and which also receives input from mechanisms related to food satiation. Similarly there must be an independent center for input from all water receptors, and this must have pathways connected to the water regulating mechanism. These two intermediate centers are designated on the schema as C_s and C_w, respectively. But because water and sugar do cause the same behavioral response under appropriate conditions and because sugar stimulation raises the CES for water, it is clear that eventually there must be a junction (C_e), the output of which activates the motor center for extension (M_e).

It is also possible, of course, that the junction of C_s and C_w is at the motor center itself (M_e); however, the integrative events in a system as complicated as that concerned with feeding seem less likely to take place directly at the motor center than at some intermediate point (C_e). The same reasoning applies to C_r.

At the present time it is fruitless to speculate on the location of these theoretical centers in the CNS, but nothing now known about the CNS vitiates the schema. According to the evidence presented, the excitatory state that has been measured can probably be assigned to the center for extension (C_e).

Because of the methodological problems encountered in searching for a central inhibitory state (CIS), that Dethier, Solomon, and Turner (1968) evolved another strategy: to induce CIS before induction of CES rather than afterwards. Tests could still be made for residual CES. There are two time intervals during which CIS could decay: (1) the time between the onset of induced CIS and the onset of induced CES, and (2) the time between the onset of CES and the water posttest. The effects of varying both of these intervals were studied.

Tests were run with three groups of flies. With the first group the course of decay of CES initiated by 0.125 M sucrose was established with one hundred flies and that initiated by 0.25 M sucrose established for another hundred. These represented control groups, and the curves describing their responses were identical to those obtained in the earlier study of CES. The first experimental group was given an inhibitory stimulus immediately following the water pretest. The sucrose stimulation was presented 5 seconds later. Decay times between sucrose stimulation and water posttest of 15, 30, 45, or 60 seconds were varied randomly, each subject being its own control. Two subgroups of one hundred flies each received 0.125 M or 0.25 M sucrose as the excitatory stimulus. In the second experimental group the delay of 5, 15, 30 or 60 seconds was interposed between the establishment of inhibition and the presentation of sugar. Although some inhibitory effect could be produced by NaCl as the only stimulus, more dramatic results were obtained by a composite tactile, chemical, and thermal stimulus. It consisted of totally immersing the fly in a beaker of distilled water for 3 seconds after which it was shaken vigorously three times to remove adhering water. As with the studies of CES extension of the proboscis was counted and evaluated quantitatively.

The results are summarized in Figures 208-213. The top two curves in Figure 208 show the excitatory decay function for 0.125 M and 0.25 M sucrose for the control flies (Group 1). The findings agree closely with the earlier experiments on CES. Both percentage of response and mean net excursion indicate that the responsiveness to the water posttest was enhanced for at least as long as 60 seconds after sucrose stim-

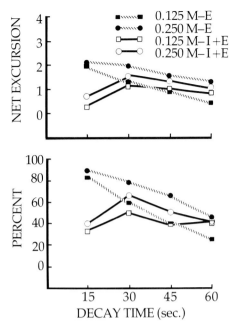

Fig. 208. Percentage of extension and mean net excursion of the proboscis during the water posttest as a function of decay time. E curves represent the response after priming stimulation by 0.125 M sucrose and 0.25 M sucrose. I + E curves represent response when the sucrose stimulation was preceded by immersion, the inhibitory stimulus. Each point represents the mean for 100 flies (from Dethier, Solomon, and Turner, 1968).

ulation. This elevation of responsiveness to water was significant even at 60 seconds because the responsiveness in the water pretest was less than 1%. The decline in responsiveness with increased delay was orderly but steeper in slope with 0.125 M than with 0.25 M sucrose. The decline in responsiveness was also evident when the mean retracted position was plotted as a function of time after stimulation with sucrose (Fig. 209); however, there was no significant difference between the curves for 0.125 M and 0.25 M sucrose.

The delay function resulting from combining excitatory and inhibitory stimulation was markedly different as the bottom two curves in Figure 209 illustrate. Responsiveness to the water postest as measured both by percentage of response and mean net excursion was significantly depressed at 15 seconds after stimulation by sucrose, but not at 30, 45 and 60 seconds. By this time the inhibitory effects of immersion appeared to have dissipated after stimulation both by 0.125 M and by 0.25 M sucrose. Also the proboscis was always retracted farther when the fly had been subjected to immersion (Fig. 209).

The perseverating effect of combining immersion and stimulation with NaCl was greater than that produced by immersion alone (Fig.

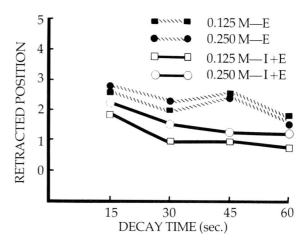

Fig. 209. The mean retracted position as a function of time following removal of sucrose stimulation. Other details as in Figure 208 (from Dethier, Solomon, and Turner, 1968).

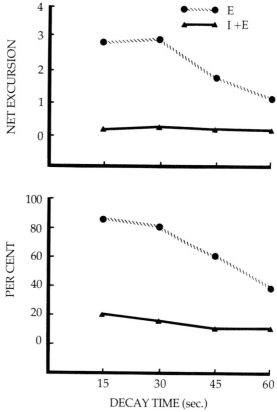

Fig. 210. Percentage of extension and mean net excursion of the proboscis during the water posttest. E curves represent the response after priming stimulation by 0.5 M sucrose. I + E curves represent response when the sucrose stimulation was preceded by NaCl and immersion. Each point represents the mean for 27 flies (from Dethier, Solomon, and Turner, 1968).

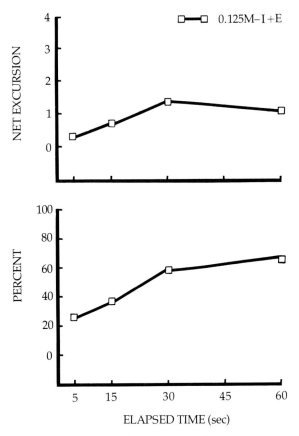

Fig. 211. Percentage of extension and mean net excursion during the water posttest as a function of elapsed time between immersion and 0.125 M sucrose stimulation. Decay time was held constant at 15 sec. Each point represents the mean for 40 flies (from Dethier, Solomon, and Turner, 1968).

210). The effect lasted for 45 seconds and was not measurable at 60 seconds because of "floor effect."

When the delay was introduced between immersion and the water posttest, the results were essentially the same as before (Figs. 211-213). The three indices of responsiveness to the water posttest (always given 15 seconds after stimulation by sucrose) as a function of the elapsed time between immersion and excitation by sucrose show again that the effects of inhibition were dissipated between 15 and 30 seconds.

All of the experiments agreed in showing that stimuli that normally inhibit proboscis extension can induce a central inhibitory state that has a measurable duration in time. Its existence is indicated by three phenomena: (1) failure to give normal maximal proboscis extension to an acceptable stimulus if this was closely preceded by an inhibitory stimulus; (2) faster rate of retraction following termination of sugar stimulation if it was preceded by an inhibitory stimulus than if it was

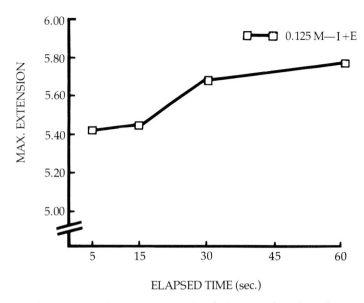

Fig. 212. Mean maximum extension to sucrose stimulation as a function of elapsed time between immersion and 0.125 M sucrose stimulation. Each point represents the mean for 40 flies (from Dethier, Solomon, and Turner, 1968).

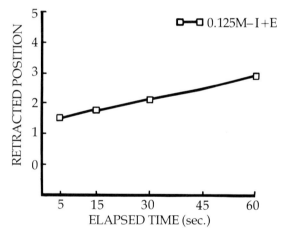

Fig. 213. Mean retracted position of the proboscis for a constant 15-sec. decay time as a function of the elapsed time between immersion and 0.125 M sucrose stimulation. Each point represents the mean for 40 flies (from Dethier, Solomon, and Turner, 1968).

not; (3) failure to respond maximally to a behaviorally subliminal stimulus (e.g., water) superimposed on an induced CES when the CES was preceded by an inhibitory stimulus. The time course of CIS can be inferred by measuring the change in response decrement in each case. All results agreed in showing that CIS was induced by immersion.

Although the decay rates for CES induced by 0.125 M and 0.25 M sucrose were significantly different, there was no corresponding demonstrable difference at 15 seconds between the responsiveness induced by the two sucrose solutions followed by immersion (see Fig. 208). Thus at the moment in decay time (15 seconds) when CIS was readily detectable, no differential effect on CIS on the two CESs could be noted. After 30 seconds, when the differences between CESs were significant, the CIS had dissipated.

Although comparison of the relative rates of decay of CES and CIS is of considerable theoretical interest, the comparison is actually difficult to make. Ideally, at decay time zero CES and CIS should be equal. Given that immersion induces a CIS of a certain magnitude it must be followed instantly by a sucrose stimulus that will produce a CES of equal magnitude. If this condition were achieved, a water posttest given immediately would elicit no response because the two states would be balanced. There would also be no response if CIS were stronger than CES. It was technically impossible to test close to zero decay time; however, at 5 seconds decay time responsiveness was very low. This suggests either that an approximate balance between CIS and CES was obtained or that CIS slightly exceeded CES. The shape of the curves favors the former interpretation.

The best estimate of relative rates of decay of CIS and CES is derived from the nonmonotonic functions plotted in Figure 208. The rise in responsiveness to the water posttest between 15 and 30 seconds decay time, when CIS and CES were interacting (I and E curves), as compared with the declining CES during the same decay time (E curves), can most easily be understood as a faster decay of CIS than of CES.

If a deliberate attempt is made to unbalance CIS and CES at decay time zero, a condition can be established in which the total duration of CIS exceeds that of CES. Apparently this condition was achieved when several inhibitory stimuli, NaCl and immersion (tactile and thermal), were combined. The total duration of CIS was at least 45 seconds and probably 60 seconds (Fig. 210). This is as long as some of the CESs produced by low concentrations of sucrose. In the original experiments on CES, stimulation by 0.016 M and 0.032 M sucrose induced a CES that dissipated completely by 45 seconds. Employing different intensities of induced CES and CIS, one can set up conditions in which either can outlast the other; nevertheless, the ease with which CES can be established, as compared with the difficulty of establishing a measurable CIS, suggests that perseverating states in the central nervous system of the blowfly are biased toward the excitatory side.

CES and CIS are terms of often used to interpret both behavioral and neurophysiological events. CES is generally understood as an increase of spontaneous activity in the central nervous system or an increased tendency to respond to stimuli. CIS has the phenomenological meaning of preventing something from happening (Bullock,

1965). Immersion in water is a rather violent treatment. The choice of the term "inhibition" rather than terms such as "distraction", "depression due to trauma," or "attempts to escape interfering with proboscis extension" was deliberate. When NaCl, which is known to have an inhibitory effect arising in a specific receptor (Dethier, 1955), is combined with immersion, the negative effect on proboscis extension is greater than with immersion alone. It seems reasonable, therefore, to characterize the effects of immersion as inhibitory.

At the neurological level several kinds of central excitation and inhibition have been demonstrated by electrophysiologists. Bullock (1965) has described six different kinds of inhibition. Because the experiments with *Phormia* were behavioral they gave no indication of the kind of neural activity occurring at the time. Yet there are some marked similarities in the time-course of CIS in the blowfly and inhibitory hyperpolarization in the snail *Helix*. Tauc (1960) showed that this hyperpolarization lasts for as long as 20–30 seconds. Thus the interpretation of the behavioral phenomena in *Phormia* is not invalidated by known physiological evidence on perseverating inhibitory neural events.

Fredman and Steinhardt (1972) discounted the action of a salt-induced central inhibitory mechanism affecting response to sugar receptor input. Their conclusion was based on the observation that the simultaneous application of salt to one labellar hair and sugar to another did not cause a decrease (as compared with sugar stimulation alone) of motor output to the extensors of the labellum. They argued further that total immersion is too nonspecific to admit of any involvement of salt receptors. However, the difference in perseverating effect between immersion alone and immersion combined with salt stimulation clearly implicates salt. A subsequent study by Fredman (1975) using the same preparation does not confirm the conclusions of the first study regarding CES and CIS. It shows instead that a CES can be established and that stimulation by salt can inhibit the CES.

Once the prominent role played by central excitatory states in the fly was recognized it became possible to design more critical experiments on conditioning. The first irrefutable evidence for the existence of classical conditioning in *Diptera* was provided by Nelson (1971). She studied the same motor pattern, extension of the proboscis, as did Dethier, Solomon, and Turner (1965, 1968).

Flies were attached by the wings to tackiwax-coated paraffin cylinders on applicator sticks and allowed to rest generally for 2 hours before being tested. Some flies had only 20-minute rest periods; however, the longer rest yielded higher levels of responsiveness and greater uniformity within groups. No anaesthetics were employed.

Since sensitivity to water arising from dehydration can be a serious confounding element in behavioral experiments involving the chemical senses, rigorous control of satiety must be exercised. After the rest period flies were offered distilled water in a watch glass and permitted to drink until the proboscis was completely and stably retracted. Water was again offered 5 minutes later. Flies that drank this time were given a final presentation of water 5 minutes later. If they still responded to water, they were discarded. Despite these precautions a few flies did respond to water on their first experimental trial. Many investigators have observed that tethered insects become excessively thirsty; consequently, constant alertness must be maintained to guard against thirst. Nelson was able to evaluate responses to water because she noted a difference in the character of response to water evoked by dehydration and that given to conditioned stimuli. In the latter situation the labellar lobes remain closed even when the proboscis is fully extended and brought forcibly into contact with the solution. Dehydrated flies, on the other hand, spread the lobes when the proboscis is extended, and drink.

In all experiments the same basic procedure was followed. Three stimuli were used. Distilled water and 1.0 M NaCl applied to the tarsi were the neutral or conditioned stimuli (CS) and 0.5 M sucrose applied to the labellum, the unconditioned stimulus (US). Application of the unconditioned stimuli to the tarsi was made by lowering the fly to the solution in a watch glass so that the distal tarsomeres touched for a period of 4 seconds. In this position the fully extended proboscis could not reach the solution. When two successive stimuli were to be applied to the tarsi, the fly was lifted clear of the first and immediately lowered onto the second for 4 seconds. When the unconditioned stimulus was presented contingent on tarsal stimulation, a droplet (approximately 0.3 ml) of sucrose was touched to the labellum during the fourth (last) second of tarsal stimulation. Almost invariably this elicited complete extension of the proboscis and ingestion of the sugar. The various criteria for evaluating proboscis extension were those developed by Dethier, Solomon, and Turner (1965) in their studies of CES.

Nelson carried out seven classes of experiments. The first was designed to detect CES and/or learning with relatively long intervals between stimuli. Experiments of this class employed a compound conditioned stimulus. The intention was to separate excitatory and associative elements. The compound stimulus consisted of two successive stimuli, salt and water, and water and salt applied to the tarsi according to the paradigm already outlined. Each fly received 15 trials with ten-minute intertrial intervals. Controls received the two stimuli

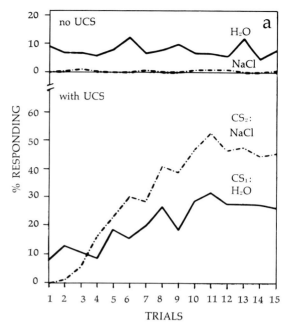

Fig. 214. Proboscis extension elicited by 1.0 M NaCl and water in a compound conditioning paradigm. The ordinate represents the percentage of flies out of all those tested which responded on a given trial. Water and salt were applied successively to the tarsi for 4 sec. each. The order of presentation was water-salt (from Nelson, 1971).

without reinforcement as a control for thirst and sensitization. As Figures 214 and 215 illustrate, increasing *numbers* of flies receiving sugar reward (with US) extended the proboscis after the first three to five trials to tarsal stimulation alone as well as to the sucrose reward that followed. They consistently responded more frequently to the second CS than to the first regardless of whether this was salt or water. The control flies that received no sucrose reward only rarely responded to NaCl and gave only a low level of response to water. These low levels resulted from the sporadic response of individual flies, none of which responded more than three times; 76% of the responses included drinking.

In order to ascertain how individual flies were behaving, criteria were established to categorize flies as good, fair, and poor learners. To qualify as good a fly had to give six or more conditioned responses to CS_2 in the last eight trials. Fair required from three to five; poor, less than three. The performances of three representative flies are diagrammed in Figure 216. In the group that was stimulated with NaCl first, 39% were good, 24% were fair and 37% were poor. In the group stimulated first with water 31% were good, 18% were fair, and 51% were poor.

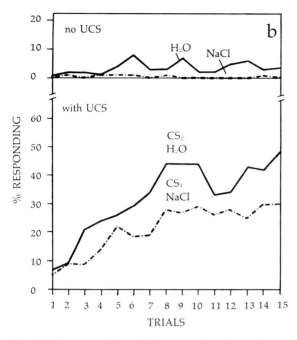

Fig. 215. Same as Figure 214 except that the order of presentation was saltwater (from Nelson, 1971).

The evidence for associative learning is impressive. Two conditioned stimuli were used in the hope that the first would fully discharge any CES set up by the preceding reward, so that any responses to the second would clearly be attributable to conditioning. In fact, it only partially dissipated CES; nevertheless, a comparison between control and experimental flies suggests learning. The order in which the controls received tarsal stimulation had little effect on their responses, and the fact that 76% of the responses included drinking suggests that thirst was the dominant factor. The chemical nature of the stimulus (i.e., whether it was water or salt) was a prime determinant of the flies' reactions. In the control groups, on the other hand, taste cues were subordinate to temporal cues. Even though some discrimination on the basis of taste was occurring, the position of the CS relative to the US was of paramount importance. If CES were primarily responsible for the response, one would have expected more response to the first stimulus since the CES had less time to decay. On the other hand, if conditioning were of prime importance greater response would be expected from the second stimulus since this was closer in time to the reward. A critical test to separate the effects of CES and those of conditioning would be one in which contiguity was varied, because an associative process requires a consistent temporal relationship between stimulus and reward while CES does not. This was the rationale of the next set of experiments that Nelson performed.

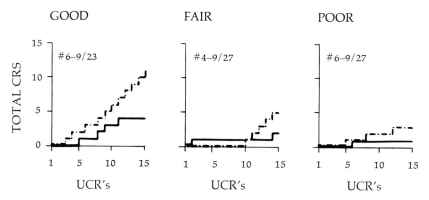

Fig. 216. Cumulative CRs for three individual flies representative of three classes of performance. Criteria for classification were: good = six or more CRs to CS_2 in the last eight trials; fair = three to five CRs to CS_2; poor = less than three CRs to CS_2 (from Nelson, 1971).

Random presentation was accomplished by presenting sucrose every 10 minutes (the intertrial interval had been 10 minutes in previous experiments) and designating four time slots around the sugar (US) stimulation; namely, 45 and 5 seconds before and 45 and 5 seconds after. Each CS was randomly, and independently assigned to a slot in each trial so that on successive trials flies received salt, water, and sugar in randomly varying order. In general, responses were numerous to CSs that followed sugar immediately; those that followed later, less so (i.e., at 45 seconds) (Fig. 217). The fact that there were responses to the second of two stimuli following sugar indicated that the first postexcitatory stimulus did not completely discharge CES although it reduced it greatly. Responses to stimuli that preceded sucrose are interesting because they show how much CES remains after a long period without stimulation; that is, the fact that there was any response at all can be attributed to the presence of CES, remaining from the sucrose stimulation of the previous trial 10 minutes earlier. This perseveration greatly exceeded the maximum observed by Dethier, Solomon, and Turner (1965). The difference might be accounted for on the basis of different testing situations (i.e., tarsi instead of labellum or several hundred chemoreceptive hairs instead of one) or might be indicative of two different levels of CES. The fact that the response was higher to salt (with no drinking) than to water rules out thirst as a factor. The fact that there was no progressive increase in

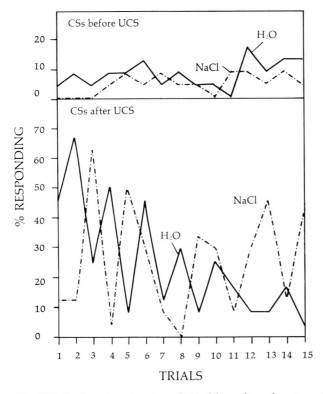

Fig. 217. Proboscis extension elicited by salt and water when those stimuli were presented randomly with respect to the unconditioned stimulus (US). (top) Responses to stimuli which preceded the US on a given trial. (bottom) Responses to stimuli which followed the US on a given trial (from Nelson, 1971).

response (compare Figs. 214 and 217) indicates that the increase previously observed involved some associative process.

In order to determine just how great a role nonassociative factors (i.e., CES) played in the behavior so far observed, Nelson designed a further experiment, a pseudoconditioning control, in which flies received an uninterrupted series of USs (sugar) followed by a series of CS-US pairs. This was the missing control in Frings's (1941) experiment. If CES is the major determinant in shaping the behavior described in the first experiment (Fig. 214), the first trial after a series of sugar stimulations alone should give the same results as the equivalently numbered trial in the conditioning paradigm, where a compound conditioned stimulus (water first followed by salt) is also given. The results are depicted in Figure 218. On trial 11 the unconditioned flies responded vigorously to water and negligibly to salt, whereas the conditioned flies responded more to salt (the second of the two conditioned stimuli and the one most closely associated in time with the reward). By trial 14 the pseudoconditioned flies began to respond

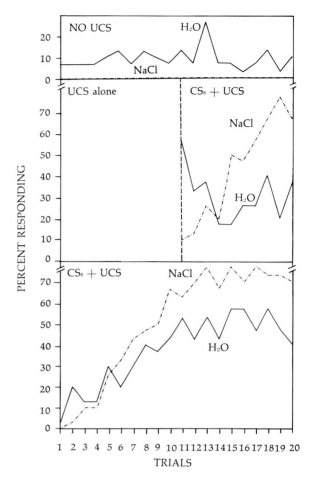

Fig. 218. Effects on proboscis extension of a pseudoconditioning vs. a conditioning paradigm when a compound conditioned stimulus (CS) was used. Middle portion of the figure represents responses of the pseudoconditioning group; lower portion represents responses of the compound conditioning group (from Nelson, 1971).

more to salt. In short, the response to water was due to CES and the reversal due to conditioning.

Once the fact of conditioning had been established, it became possible with additional experiments to characterize it more fully. Tests were conducted to ascertain whether or not the fly could discriminate between two CSs on the basis of their differing temporal relations to the US (the reward). The plan involved presenting a tarsal stimulus every 5 minutes. Salt and water were the stimuli. One (CS+) was always rewarded; the other (CS−) never. Figure 219 reveals that responding consistently increased with CS+ and never with CS−. Obviously the flies could discriminate. Responses to NaCl were always

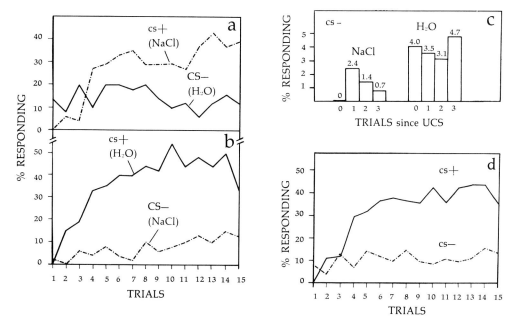

Fig. 219. Proboscis extension elicited by salt and water in a discriminative conditioning paradigm. (a) CS+ was water, CS− was salt; (b) CS+ was salt, CS− was water; (c) number of responses to CS− as a function of number of trials since the last US; (d) grouped data of (a) and (b) (from Nelson, 1971).

less than to water, probably because salt sets up a CIS (cf. Dethier, Solomon, and Turner 1968). It can be inferred that the behavior of the flies in this discriminative conditioning experiment was compounded of two processes—one associative, involving conditioning, and the other nonassociative, involving central excitation and central inhibition. Salt presumably set up a CES acting against excitation; hence the response to salt was always lower, irrespective of whether it was serving as CS+ or CS−.

This explanation of differential effects raised the possibility that the greater response observed to the second stimulus when compound conditions were used (cf. Fig. 214) was due to nonassociative factors, that is, to salt and water, which were exerting their own inhibitory (e.g., with salt) or excitatory (e.g., with water) influence on responses. Or perhaps each exerted an excitatory influence. These possibilities were examined by again conducting experiments with compound stimuli, with the difference that both stimuli were the same in each case; that is, two presentations with salt in one series and two presentations with water in the other. There were also salt-salt and water-water controls in which no reward (sugar US) was given. In every case the response curves of the salt-salt groups were lower than those of the water-water groups, indicating that salt was indeed exerting an inhib-

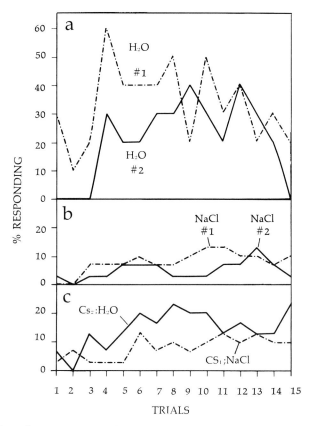

Fig. 220. "Compound" conditioning. (a) Both CS_1 and CS_2 were water; (b) both CS_1 and CS_2 were salt; (c) CS_1 was salt, CS_2 was water (from Nelson, 1971).

itory effect. In contrast to the first experiments with compound stimuli (e.g., water-salt and salt-water), it was the first stimulus here that evoked the greater number of responses (Fig. 220). Nelson interpreted this as indicating that the flies perceived the double homogeneous CS as a single stimulus and responded immediately upon encountering it. In any case this experiment negates the idea that the CSs were producing a CES that mimicked conditioning.

Two final experiments were run to eliminate any final doubts about the reality of conditioning. In one, the intertrial intervals were varied. The results were essentially the same as with regular intertrial intervals (Fig. 221). In the other, flies were given an opportunity to walk on small styrofoam balls between trials in an attempt to discharge CES by a stimulus (tactile) not related to conditioning. As Figure 222 shows, exercise had a profound effect. It diminished the number of responses to water and to salt. Either of two alternatives would explain these results: exercise discharges CES and the presence of excitation in-

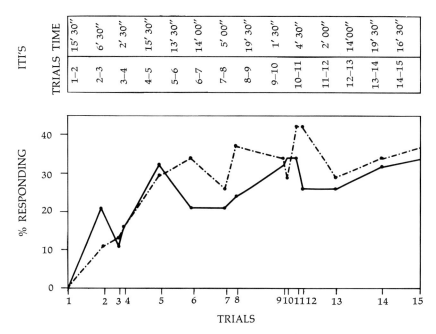

Fig. 221. Effect of variable intertrial intervals on proboscis extension elicited by a compound CS (water-salt). Relative length of the intertrial intervals is indicated on the abscissa (total time, 149 min.). Average ITI was 10 min. Dashed line, salt; solid line, water (from Nelson, 1971).

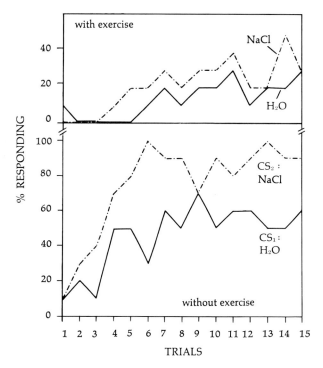

Fig. 222. Effects of exercise (from Nelson, 1971).

creases the likelihood that the flies will condition; or exercise may interfere with the learning process, as Minami and Dallenbach (1946) found for cockroaches forced to remain active during trials.

Aside from these experiments of Nelson's, which clearly prove that the blowfly is capable of acquiring a conditioned response, there are only two other strictly comparable investigations on classical conditioning in insects, that of Takeda (1961) with honeybees, and a more recent study of *Musca* (Fukushi, 1973). The validity of the conclusions drawn from the experiments with bees is difficult to assess because there were no controls for CES, and the previous experiences of the bees were not recorded.

The data relating to *Musca* are more detailed. Houseflies were fastened to a bed of clay, stored at a temperature of 20°C for one day, then fed to repletion on 0.5 M sucrose. On the third day after fastening a glass rod soaked in a mixture of sucrose and one of several odorous compounds was brought within 5 mm of the head and held there for 1 to 2 seconds. Then the rod was touched to the labellum and the fly allowed to drink for 2 seconds when the labellum was extended. This procedure was repeated eight to twelve times at intervals of 15 seconds. After a 2-hour period of rest, the fly was tested for conditioning. Extension of the proboscis upon presentation of the odor and before contact with the sugar was taken as evidence of conditioning. Conditioning was expressed as percentage of flies responding. Chemicals used as conditioned stimuli were: ethanol. *n*-propanol. *iso*-propanol, *n*-butanol, *tert*-butanol, *n*-butyraldehyde, and acetic acid. Ninety-two percent of a sample of 101 flies (51 males and 50 females) conditioned to *n*-butanol responded positively the next day. Flies conditioned to *n*-butanol responded to that compound and not to *n*-butyraldehyde or acetic acid. Flies conditioned to either of the other compounds responded only to the particular compound. This discrimination was taken as evidence that the responses observed represented true conditioning and not sensitization, as was probable with Frings's flies. After eleven presentations at 1-minute intervals of odor unrewarded with sugar, response was extinguished. Two to four hours after extinction there was an 85% recovery of responsiveness. Isolated heads removed from flies that had been conditioned to the odor of acetic acid responded to the odor for a few minutes.

All other data concerning conditioning in insects relate to instrumental learning. It was in this area that so many failures were encountered with Diptera. It is puzzling in this connection that although attempts to train blowflies to avoid electrical shock failed, decapitated flies and isolated thoraces "learned" to flex a leg to avoid shock exactly as cockroaches had been shown to do by Horridge (1962, 1964; Hoyle, 1965; Pritchatt, 1968). Some success in demonstrating avoidance conditioning in Diptera has recently been achieved with *Drosophila* (Quinn, Harris, and Benzer, 1974). Flies were trained by exposing them alternately to two odors, one of which was associated with a shock provided by 90 volts A.C., 60 Hz, or with crystalline quinine sulfate. Three training sessions of 15 seconds with intertrial intervals of 60 seconds were sufficient to establish conditioning. Without reinforcement the effect lasted 24 hours. Exacting controls ruled our pseudoconditioning, excitatory states, odor preference, sensitization, and habituation.

Considering the apparently minor role played by conditioning in the life of the fly, it is unlikely that more complex forms of learning exist. Latent learning and insight learning have been demonstrated in some Hymenoptera (Thorpe, 1963) but thus far not in Diptera (with the one questionable exception of preimaginal conditioning). Nor is there any evidence for behavioral changes resembling imprinting.

Some attempts were made to investigate imprinting in *Phormia* after gustatory behavior resembling imprinting was observed in caterpillars. The experiments with caterpillars showed that the first brief exposure of a caterpillar that had never experienced a plant food to a particular plant established a dietary preference for that plant (Jermy, Hanson, and Dethier, 1968). A preference can be established by feeding for one day only, and it is not eliminated by two larval molts and subsequent feeding on an artificial diet. Experiments designed to test the reversibility of the preference were inconclusive. There is strong presumptive evidence that the modification in behavior is neural in basis (as opposed to metabolic influences) (Jermy, Hanson, and Dethier, 1968; Hanson and Dethier, 1973), but whether the neural change is peripheral or central is unknown (cf. Schoonhoven, 1969). Comparable experiments were conducted with adults of *Phormia*. Flies were fed from emergence on specific sugars and their subsequent behavior with respect to these sugars was tested. No preferences or changes in acceptance thresholds were detected. It is possible, however, that the failure can be attributed to an inability on the part of the flies to discriminate among sugars and that in future experiments individual substances with more distinctive characteristics should be chosen. The hypothesis that early experience in the *adult* may have an influence on subsequent behavior is thus unproven.

Early experience in the larval stage, on the other hand, does influence later behavior.

Experiments with gustatory stimuli given to larvae of *Phormia* have conclusively demonstrated a transference from larval experience to adult behavior; however, here the nature of the phenomenon is extremely confusing and can in no way be construed as evidence for habituation, imprinting, or any kind of associative learning. Evans (1961b) described experiments in which the addition of specific sugars to a basal diet lacking all sugars except lactose reduced the gustatory sensitivity of adult flies to the particular supplementary sugar. For example, flies reared in the presence of fructose gave a mean behavioral response to tarsal stimulation that was 46 times higher than that of control flies reared in the absence of fructose. The sucrose threshold of the experimental flies was also elevated. Evans suggested that the effective sugars acted by depressing sensitivity specifically to themselves and to other sugars that act at the same site on the receptor because of a decrease in the number or affinity of the combining sites.

Certain critical controls were missing from Evans' experiments. One major flaw was that the larval medium was not sterile and supported a luxuriant growth of bacteria and fungi; consequently, there was no assurance that the diet actually contained the supplementary sugar throughout larval development. Another cause of concern was the high mortality rate encountered; therefore the possibility that differentially sensitive strains of flies were being selected was not eliminated. The method of testing also introduced potential sources of error. The ascending technique of measuring thresholds is not the most rigorous because each fly is exposed to more than one test solution and sensory adaptation becomes a confounding factor. Finally, the use of a relative measure of sensitivity is less desirable than an absolute measure.

The experiments of Evans were repeated and extended with more rigorous controls by Dethier and Goldrich (1971). The flies used were from the same stock as those used by Evans. They had been maintained in continuous culture since 1947 and since 1955 had been reared on a diet consisting of yeast, powdered whole milk, agar, and Tegosept. Eggs were sterilized within 24 hours after oviposition by being soaked in 1% NaOH for 10 minutes and rinsed twice in 70% ethanol. They were then placed in a sterile medium consisting of 200 grams of powdered yeast, 200 grams of powdered whole milk, 8.5 grams of Tegosept, 40 grams of agar, and 3 liters of water, all of which had been poured while hot into sterilized flasks, plugged with cotton, and autoclaved for 7 minutes at 121°C. Up to 900 flies were used for each threshold determination, which was made by the random technique. The results are summarized in Table 27.

Significant differences appeared in the median acceptance thresholds in some cases; there were no differences in the slopes of the dosage-response curves in any case. This means that when changes in sensitivity did occur in a population they were not changes in the distribution of sensitivities within the population but changes that affected all members. The changes that did occur were not in agreement with Evans' data. For one thing, they were very small. Whereas he found differences between control and experimental groups as great as 46 times, the greatest difference found in the new study was only 2.5

Table 27. The Tarsal Sensitivity to Sugars of Adult Blowflies Raised on Media Supplemented with Sugars. The Sensitivity Is Expressed as Median Acceptance Threshold.

Larval medium	Median acceptance threshold (mole/liter) to					
	Arabinose	Fructose	Galactose	Glucose	Mannose	Sorbose
Control	0.105	0.114	0.506	0.263	>1.0	0.479
Arabinose	.531	.0108	.586	.247	>1.0	.374−[a]
Fructose	.175	.011−[a]	.760+[a]	.179−[a]	>1.0	.535
Galactose	.090	.007−[a]	.418	.177−[a]	>1.0	.242−[a]
Glucose	.129	.054+[a]	.622+[a]	.330	>1.0	.859+[a]
Mannose	.131	.005−[a]	.513	.244	>1.0	.460
Sorbose	.132	.009	.690	.323	>1.0	.360

From Dethier and Goldrich, 1971.
[a] Value significantly different at $P = 0.01$.
+ or −: the direction of the difference.

times. Furthermore, the directions of the differences were not in complete agreement. Supplementing the diet with a sugar did not always induce a decrease in sensitivity of the adults to that sugar. Specifically, supplementing the diet with fructose slightly increased adult sensitivity to that sugar. Supplementing with glucose slightly decreased sensitivity to fructose and had no effect on the response to glucose.

The data do not correlate with metabolic phenomena nor are they consistent with ideas concerning glucose repression of inducible enzyme synthesis (cf. Evans, 1961b). Additionally, analysis of mortality data collected over five generations of breeding on the respective fortified diets, plus the fact that there were no differences in the slopes of the dosage-response curves, argues against the occurrence of selection. Clearly the composition of the food consumed during development affects the behavioral gustatory sensitivity of the adults, but the nature of the reaction is unresolved. The data are not consistent with ideas of preimaginal habituation or latent learning.

One other brief report is available indicating that the addition of 0.01–0.05 M NaCl or KCl to diets of larval and adult flies causes adults to show an increased preference for salt. The preference becomes increasingly stronger with each succeeding generation (Sinitsina and Elizarov, 1971).

Higher learning in some form or other has been found in insects representing several different orders but clearly reaches its highest attainment in Hymenoptera. Not only the social species such as ants, bees, and wasps display considerable learning powers but also the subsocial and solitary species, so that many writers of an earlier age could speak admiringly of the "intelligence" of these creatures. Diptera are evolutionarily about as advanced as the Hymenoptera although they have never become social. They might be expected, therefore, to have some comparable learning capabilities. For over twenty years persistent attempts were made in a number of laboratories, but especially our own, to demonstrate some form of learning. All attempts either failed outright or were disturbingly equivocal. The attempts included conditioning flies to extend the proboscis to light, sound, vibration, and odor associated with sugar; training them to associate a particular color or particular location in a cage or time of day with food, to learn how to escape from a container, to cease trying to fly after the wings had been clipped, to run a T maze in order to obtain food or water, to avoid or escape shock or lethal temperatures, etc., etc. Experiments were even run with miniature Skinner boxes equipped with bars which the flies would be physically capable of depressing in order to obtain food. This recalcitrance and weak performance is puzzling, especially since many other insects learn so readily. Bees (*Melipona rufiventris*), for example, can learn to press levers in order to obtain food (Pessotti, 1972). It could be argued with justification that experimenters have just not been clever enough to design the appropriate tests, but this hardly seems likely in view of the successes reported with other insects, including the more primitive species such as cockroaches (cf. Thorpe, 1963, p. 241). Many instances of operant conditioning have been reported, and success or failure bears no obvious relation to the phylogenetic position of the species.

In all those insects in which learning has been demonstrated, in cockroaches (Szymanski, 1912), in mayfly nymphs (Wodsedalek, 1912), in ants (Schneirla, 1962), in *Drosophila* (Quinn, Harris, and Benzer, 1974), and in *Phormia* (Nelson, 1971), a great variation in ability among different individuals has been observed. This finding crops up again and again. Furthermore, the effect of learning is seldom strong. This may be indicative of a mode of learning in insects that is qualitatively different from that in vertebrates (Schneirla, 1962). Perhaps, as suggested earlier (Dethier, 1966, 1969), CES and other central and reflex mechanisms obviate the necessity for long-term plasticity. Nelson has suggested that the classical conditioning that does exist is essentially an extension of a central excitatory state.

The demonstrably weak learning capacity of many insects and the repeated failures experienced with Diptera led to a hypothesis that some organisms may have lost the ability to learn; that is, may have evolved away from that capacity in the interests of parsimony (Dethier, 1966). Flies are not social insects nor do they have nests or fixed sources of food (e.g., a patch of flowers) or other points of reference the location of which they must ascertain. Consequently, one might question the adaptive value of learning, especially since their life-span is so short (approximately 60 days) and their reproductive potential so high. We tend to think of learning in some form or other as a universal phenomenon. It is not implausible that its occurence in nature is sporadic, and that where it has little adaptive value it has been eliminated so that the very small brain can be put to better uses. Two dipterous species that might find learning a good evolutionary strategy are the bee flies (Bomyliidae), which suck nectar and have much the same relationship with flowers as do honeybees, even though they have no nest; and robber flies (Asilidae), which dart off for the insectan prey from favorite vantage points much as do hawks. A search for forms of higher or stronger learning in these species, correlated with a comparative study of their brains, might be unusually rewarding.

Literature Cited

Arab, Y. M. 1959. Some chemosensory mechanisms in the blowfly *Phormia regina*. *Bull. Coll. Arts Sci. Baghdad*, 4:77–85.

Bullock, T. H. 1965. Mechanisms of integration. In *Structure and Function of the Nervous Systems of Invertebrates*. Vol. I, T. H. Bullock and G. A. Horridge, eds., pp. 257–351. W. H. Freeman, San Francisco.

Crombie, A. C. 1944. On the measurement and modification of the olfactory responses of blowflies. *J. Exp. Biol.* 20:159–166.

Cushing, J. E. 1941. An experiment on olfactory conditioning in Drosophila guttifera. *Proc. Nat. Acad. Sci. U.S.A.* 27:496–499.

Dethier, V. G. 1952. Adaptation to chemical stimulation of the tarsal receptors of the blowfly. *Biol. Bull.* 103:179–189.

Dethier, V. G. 1955. The physiology and histology of the contact chemoreceptors of the blowfly. *Quart. Rev. Biol.* 30:348–371.

Dethier, V. G. 1966. Insects and the concept of motivation. In *Nebraska Symposium on Motivation, 1966,* D. Levine, ed., pp. 105–136. University of Nebraska Press, Lincoln.

Dethier, V. G. 1969. Feeding behavior of the blowfly. *Adv. Study Behavior* 2:111–266.

Dethier, V. G., and Bodenstein, D. 1958. Hunger in the blowfly. *Zeit. Tierpsychol.* 15:129–140.

Dethier, V. G., and Evans, D. R. 1961. Physiological control of water ingestion in the blowfly. *Biol. Bull.* 121:108–116.

Dethier, V. G., Evans, D. R., and Rhoades, M. V. 1958. Some factors controlling the ingestion of carbohydrates by the blowfly. *Biol. Bull.* 111:204–222.

Dethier, V. G., and Goldrich, N. 1971. Blowflies: alteration of adult taste responses by chemicals present during development. *Science* 173:242–244.

Dethier, V. G., Solomon, R. L., and Turner, L. H. 1965. Sensory input and central excitation and inhibition in the blowfly. *J. Comp. Physiol. Psychol.* 60:303–313.

Dethier, V. G., Solomon, R. L., and Turner, L. H., 1968. Central inhibition in the blowfly. *J. Comp. Physiol. Psychol.* 60:144–150.

Dethier, V. G., and Stellar, E. 1970. *Animal Behavior,* 3rd ed. Prentice-Hall, Englewood Cliffs, N.J.

Evans, D. R. 1961a. Control of the responsiveness of the blowfly to water. *Nature* 190:1132.

Evans, D. R. 1961b. Depression of taste sensitivity to specific sugars by their presence during development. *Science* 133:327–328.

Evans, D. R., and Barton Browne, L. 1960. The physiology of hunger in the blowfly. *Amer. Midl. Nat.* 64:282–300.

Evans, D. R., and Dethier, V. G. 1957. The regulation of taste thresholds for sugars in the blowfly. *J. Insect Physiol.* 1:3–17.

Fredman, S. M. 1975. Peripheral and central interactions between sugar, water, and salt receptors of the blowfly, *Phormia regina. J. Insect Physiol.* 21:265–280.

Fredman, S. M., and Steinhardt, R. A. 1972. Mechanism of inhibitory action by salts in the feeding behavior of the blowfly, *Phormia regina. J. Insect Physiol.* 19:781–790.

Frings, H. 1941. The loci of olfactory end-organs in the blowfly, *Cynomya cadaverina* Desvoidy. *J. Exp. Zool.* 88:65–93.

Fukushi, T. 1973. Olfactory conditioning in the housefly, *Musca domestica. Annotationes Zoologicae Japonenses,* 46:135–143.

Getting, P. A. 1971. The sensory control of motor output in fly proboscis extension. *Zeit. vergl. Physiol.* 74:103–120.

Hanson, F. E., and Dethier, V. G. 1973. Role of gustation and olfaction in food plant discrimination in the tobacco hornworm, *Manduca sexta. J. Insect Physiol.* 19:1019–1034.

Hershberger, W. A., and Smith, M. P. 1967. Conditioning in *Drosophila melanogaster. Animal Behav.* 15:259–262.

Hinde, R. A. 1970. *Animal Behaviour,* 2nd ed. McGraw-Hill, New York.

Horn, G., and Rowell, C. H. F. 1968. Medium and long-term changes in the behaviour of visceral neurones in the tritocerebrum of locusts. *J. Exp. Biol.* 49:143–169.

Horridge, G. A. 1962. Learning of lep position by the ventral nerve cord in headless insects. *Proc. Roy. Soc.* B 157:33–52.

Horridge, G. A. 1964. The electrophysiological approach to learning in isolatable ganglia. *Animal Behav. Suppl.* 1:163–182.

Hoyle, G. 1965. Neurophysiological studies on "learning" in headless insects. In *Physiology of the Insect Central Nervous System,* J. Treherne and J. W. Beament, eds., pp. 203–232. Academic Press, N.Y.

Ilse, D. 1949. Color discrimination in the dronefly, *Eristalis taenax. Nature* 163:255–256.

Jermy, T., Hanson, F. E., and Dethier, V. G. 1968. Induction of specific food preferences in lepidopterous larvae. *Ent. Exp. Appl.* 11:211–230.

Lukowiak, K., and Jacklet, J. W. 1973. Habituation and dishabituation: Interactions between peripheral and central nervous systems in Aplysia. *Science* 178:1306–1308.

Manning, A. 1967. Preimaginal conditioning in Drosophila using geraniol. *Nature* 216:338–340.

Manning, A. 1972. *An Introduction to Animal Behaviour,* 2d ed. Edward Arnold, London.

Minami, H., and Dallenbach, K. M. 1946. The effect of activity upon learning

and retention in the cockroach, *Periplaneta americana. Amer. J. Psychol.* 59:1–58.
Mourier, H. 1965. The behaviour of house flies (*Musca domestica* L.) towards "new objects." *Vidensk. Medd. fra Dansk Naturh. Foren.* 128:221–231.
Murphy, R. M. 1967. Instrumental conditioning of the fruit fly, *Drosophila melanogaster. Animal Behav.* 15:153–161.
Nelson, M. C. 1971. Classical conditioning in the blowfly (*Phormia regina*). *J. Comp. Physiol. Psychol.* 77:353–368.
Pessotti, I. 1972. Discrimination with light stimuli and a lever-pressing response in *Melipona rufiventris. J. Apicult. Res.* 11:89–93.
Pritchatt, D. 1968. Avoidance of electric shock by the cockroach, *Periplaneta americana. Animal Behav.* 16:178–185.
Quinn, W. G., Harris, W. A., and Benzer, S. 1974. Conditioned behavior in *Drosophila melanogaster. Proc. Nat. Acad. Sci. U.S.A.* 71:708–712.
Rowell, C. H. F., and Horn, G. 1968. Dishabituation and arousal in the response of single nerve cells. *J. Exp. Biol.* 49:171–183.
Schneirla, T. C. 1962. Psychological comparison of insect and mammal. *Psychol. Beiträge* 6:509–520.
Schoonhoven, L. M. 1969. Sensitivity changes in some insect chemoreceptors and their effect on food selection behaviour. *Proc. Koninkl. Ned. Akad. Wet.* (C) 72:491–498.
Sinitsina, E. E., and Elizarov, Yu. A. 1971. Some properties of contact chemosensory sensilla excitation to sodium chloride solutions according to food reactions change of insects. Proc. Ist All Union Symp. on Insect Chemoreception, Vilnius, 8-10 Sept. 1971. Inst. Zool. Parasit. Akad. Sci., Lithuania SSR.
Szymanski, J. S. 1912. Modification of the innate behavior of cockroaches. *J. Animal Behav.* 2:81–90.
Takeda, K. 1961. Classical conditioned response in the honeybee. *J. Insect Physiol.* 6:168–179.
Tauc, L. 1960. Evidence of synaptic inhibitory actions not conveyed by inhibitory post-synaptic potentials. In *Inhibition in the Nervous System and Gamma-aminobutyric Acid,* E. Roberts, ed. Pergamon Press, New York.
Thompson, R. F., and Spencer, W. A. 1966. Habituation. *Psychol. Rev.* 73:16–43.
Thorpe, W. H. 1938. Further experiments on olfactory conditioning in a parasitic insect. The nature of the conditioning process. *Proc. Roy. Soc.* B 126:370–397.
Thorpe, W. H. 1939. Further experiments on pre-imaginal conditioning in insects. *Proc. Roy. Soc.* B 127:424–433.
Thorpe, W. H. 1963. *Learning and Instinct in Animals,* 2d ed., Methuen, London.
Wodsedalek, J. E. 1912. Formation of associations in the may-fly nymphs *Heptagenia interpunctata* (Say). *J. Animal Behav.* 2:1–19.
Wright, R. H. 1974. Housefly memory. *Canad. Ent.* 106:223–224.
Yeatman, F. R., and Hirsch, J. 1971. Attempted replication of, and selective breeding for, instrumental conditioning of *Drosophila melanogaster. Animal Behav.* 19:454–462.

14
The Fly and the Concept of Motivation

And appetite, a universal wolf,
So doubly seconded with will and power,
Must make perforce a universal prey,
And last eat up himself.

Shakespeare, *Troilus and Cressida*

This book began with a descriptive account of a day in the life of a fly as it might have been narrated by a naturalist. The time could be today or some bright day sixty million years ago. In its completed form the story continues like this. The sun rises. As the rays become more vertical they penetrate the interstices of the vegetation. The ambient temperature rises. The increasing heat begins to warm the resting fly, which thereupon begins to groom. Thirty minutes later the fly launches itself into the air and commences flying a random course. This flight pattern bears no orderly relation to wind direction, light source, or visual landmarks.

In another area there is a pine tree. Here, too, among the needles the temperature has risen. The millions of aphids that have been sitting with their mouth stylets inserted deep among the phleom cells of the needles begin to suck sap. It contains an excess of carbohydrates, so much so that the aphids secrete drop after drop of concentrated sugar until the needles become sticky with honeydew. This mixes with the droplets left on earlier days, which the ubiquitous yeasts of the air have already begun to ferment. A faint alchoholic, yeasty odor escapes to the breeze.

Eventually in its erratic flight the fly encounters a cloud of odor. Immediately the flight pattern changes, becomes less random, resolves itself into a crude zigzag course until the fly is within a meter or less of the source of fermentation. From this point the flight path is direct to the honeydew. When the fly lands, it begins a random walk over and among the pine needles. At some point one of the front feet touches the honeydew. The fly stops at once, turns in the direction of the stimulated foot, and quickly and forcefully plunges its extensible proboscis into the liquid.

As soon as the mouthparts touch the sugary fluid the fly sucks vigorously. It continues for nearly 70 seconds. At the end of this time sucking stops, the proboscis relaxes, and finally is slowly retracted. The fly takes several steps away from the honeydew. For 15 or 20 seconds it grooms, regurgitates a drop of fluid, sucks it back, defecates, and then resumes walking. After five minutes or so of alternate walking and stopping, the fly encounters another drop of honeydew. The proboscis extends slowly, touches the fluid briefly. One or two sucks ensue; then the proboscis is retracted sharply, and the fly wanders away. A third drop is encountered, the proboscis is partially extended, then retracted before having touched the fluid. More walking follows. A fourth drop is encountered, and a fifth and a sixth. At none of these does the fly stop nor is the proboscis extended. Arriving at the end of a needle, the fly stops and remains motionless. It spends two or three hours here before launching itself into random flight. The following day the performance is repeated—with only minor variations. If the fly is of a late summer generation and fortunate enough to escape predators and disease, it eats less as days shorten and eventually crawls into a crevice to spend the winter in dormancy.

In the pages that followed the introductory description the various sequences constituting the total pattern of the fly's behavior were correlated with demonstrable imbalances in energy states. Let it be emphasized that these were correlations. Bit by bit the behavior was dissected experimentally. The identity, characteristics, and fluctuations of various physiological mechanisms were established and assessed. Finally the parts were reassembled to produce a coherent, if still incomplete, description of the fly's behavior in terms of physiological mechanisms. Although the picture is still flawed by many omissions and uncertainties, and will require perfecting as more facts are assembled and fitted into place, the venture has been reasonably successful. The results are logically satisfying enough to encourage one in the hope that with more time, more sophisticated techniques, and more laborers a finished canvas can be produced.

A simple stimulus-response model satisfies many of the requirements imposed by our observations. Feedback loops, also envisioned as stimulus-response systems (the stimuli this time being within the fly), nearly complete the model: one feedback loop to cause the crop to deliver fluid, via the mid-gut, to the blood when required metabolically; and another feedback loop to raise the behavioral taste threshold after feeding and lower it progressively as energy requirements become more demanding. Additional feedback loops can be added to maintain water balance and to provide protein for egg development. One can even envision the extension of the stimulus-response model to regulate antagonistic feeding and locomotory sequences by means of complex series of reflexes, all deriving from the neural properties and circuitry that are genetically programmed into the fly.

What the reflex model really does, however, is to supply logical explanations for the *average* behavior of the fly. It does not have as high a predictive value as one desires nor does it explain every facet of behavior. It does not explain the selective recruitment of organized temporal spatial sequences of motor patterns—how, for example, one set of proboscis muscles causes extension for feeding at one time and aversive action at another. It does not explain the apparent spontaneity of some behavior, and it does not explain fluctuations in responsiveness. When, for example, an experimenter controls all of the variables of which he is aware and then applies sugar to the oral chemoreceptors, the fly *usually* extends its proboscis; it does not respond 100% of the time. Electrophysiological evidence shows that the receptors do respond to the stimulus 100% of the time but the nature of the response is highly variable, as we have seen. The sensing elements are not constant. Here is a source of variability within the organism that must be taken into account. But even if we are content to say that noise and inconstancy are structural characteristics of the sensing elements, the fact remains that the receiving portion of the nervous system is also variable. Somehow or other the organism is not the same from one moment to the next and this inconstancy affects output, that is, behavior. There are great and numerous fluctuations in response.

The first two explanations of inconstancy that come to mind are that there are changes in central nervous states brought about by antecedent events and/or that there are endogenous oscillations. Ample evidence in support of the first explanation has been advanced in the preceding pages. The existence of central excitatory and inhibitory states and of classical conditioning are examples. The best example of endogenous oscillating systems in the fly is circadial rhythm. The existence of other systems can be inferred from the observation that the locomotor center in the thoracic ganglion is slave to constantly-operating, antagonistic, excitatory and inhibitory centers in the brain. These have not actually been demonstrated in the fly but have been studied in other insects, notably the flight centers of the locust (Wilson, 1967) and copulatory behavior in the mantis. Roeder (1937) has shown that once the abdominal ganglia are released from the inhibitory control of ganglia in the head of a mantis by cutting every connective, all movements associated with copulation are executed even in the absence of external stimuli. Endogenous activity in receptor and central neurons is widespread among invertebrates, having first been detected in caterpillars by Adrian (1930). Its characteristics and neuronal bases have been discussed in detail by Kennedy (1962), Van der Kloot (1962), Bullock (1961, 1962), and Bullock and Horridge (1965). The need for invoking spontaneous activity in the central nervous system as a supplement to reflex systems to explain behavior has been eloquently stated and documented by Roeder (1955). It comes as no surprise, therefore, that the reflex model satisfies only some of the requirements imposed by the naturalistic description of the fly's behavior. The concept of continuously firing centers in the central nervous system, operating together with reflexes responsive to internal as well as external stimuli, offers a more satisfactory framework for explaining behavior. Add to this an intrinsic structural inconstancy and a lability such that antecedent events impinging on central neurons have perseverting effects of longer or shorter duration, and add further the ebb and flow of hormonal tides; then it would appear that the model is adequate. This conclusion is not to be construed as an assertion that the behavior of the fly is now completely explained; rather, it is to be interpreted as indicating a belief that this model or one resembling it is totally adequate for understanding the behavior. And generally speaking, for behaviorists studying invertebrates, it is.

Students of vertebrate behavior come to a different conclusion regarding their animals. Does this mean that there is a quantum difference between vertebrates and invertebrates, insofar as behavior is concerned, or does it mean that respective students of the two groups ask fundamentally different questions? Does it mean that our model of feeding behavior in the fly is so elegant simply because we have been naive and have not asked the same difficult questions that the students of vertebrate animals ask? One way of analyzing these uncertainties is to compare the feeding behaviors of the blowfly and the rat, the vertebrate about which most is known in the realm of feeding. Here, too, there are gaps in our knowledge because students of the rat's behavior have been more concerned with experimental manipulation than with observing the normal animal in a natural setting. Some of the following description is an extrapolation from field observations of wild Norway rats and field and wood mice.

The rat has a circadial rhythm of activity. It normally sleeps during the day and forages at night. Upon waking it is active. It moves about partly at random but also in patterns established by previous experience. It tends to return to places where it had previously found food and responds to stimuli usually associated with food. It is guided to food by odor. Having come close to food, it is stimulated to ingest by odor and/or by visual stimuli. Once the food has entered the mouth, taste buds are stimulated, and the food is either accepted or rejected. Taste discrimination is based presumably upon the across-fiber pattern generated by the taste receptors and is probably influenced by odor (although the involvement of odor except in the case of alcohol is not well established). After a certain amount of food is consumed, eating stops. The amount eaten at a given meal is related to the taste, to the time since the last meal, to the volume of the last meal, to learned preferences, and to the caloric value of that meal. Under normal circumstances monitoring and regulatory mechanisms ensure that a rat maintains a constant body weight.

Up to this point the parallel between the two animals is striking: they both exhibit endogenous circadial rhythms of activity, are active before feeding, are inactive after feeding (locomotion and feeding are in opposition), exhibit qualitative and quantitative control of ingestion, and regulate ingestion in relation to energy demands (the fly less efficiently than the rat). The most obvious difference is that the rat profits by experience. Initially one asks the same questions of the two animals: what starts and what terminates locomotion? what starts and what terminates eating? Let us examine these questions one at a time.

Little work has been done on the central nervous system of the fly. By analogy with locusts, whose flight mechanism is fairly well understood (Wilson, 1967), and with mantids (Roeder, 1955), a picture emerges of spontaneously active excitatory centers and spontaneously active inhibitory centers. A decapitated fly can be placed on a sheet of paper, a ring drawn around it, and left until the next day. The drawn circle apparently has all the power of the magical circles of mythology, because the next day the fly is still standing within it. If the headless trunk is stimulated mechanically, it will walk briefly, indicating that the locomotory mechanisms are functional. If instead of decapitation only the supraoesphageal ganglion is destroyed, the fly will walk interminably, indicating that there is a spontaneously active center that energizes walking in the absence of external stimuli. Elegant experiments by Huber (1960, 1967) involving local brain lesions and point stimulation in the cricket have shown that there are two centers in the brain concerned with locomotion, one in the corpora pendunculata, which inhibits the suboesophageal ganglion, and one in the corpus centrale, which excites the suboesophageal ganglion. The suboesophageal ganglion regulates the degree of excitation of the thoracic ganglion. The thoracic ganglion together with proprioceptive input from the legs actually promotes locomotion.

Thus, although the head determines the onset and duration of locomotion and, in conjunction with cephalic sensory input, the direction, the regulation of locomotion is much more complicated than these facts suggest. The internal signals that inform the locomotor command neurons in the brain of the physiological state of the fly are unknown. Although all states of deprivation generally engender activity, they do not engender the same kind. A fly that has been deprived of all food flies in a random pattern until the odor of nectar or alcohol is encountered; then the flight is oriented upwind. A gravid female fly that is deprived of protein ignores odors that might attract it when it is hungry and its locomotion is directed only by odors of putrefaction. A fly that is primed for copulation will regulate its activity with respect to pheromones but not with respect to the odors of food. In short, the internal physiological state or imbalance influences not only the degree of locomotor activity, it influences also the responses to stimuli, selectively.

When appropriate gustatory stimuli activate the external chemoreceptors the sensory input to the central nervous system not only initiates feeding, it also inhibits locomotion. When feeding has ceased, internal stimuli inhibit any further incoming chemosensory input, which in turn no longer holds locomotory centers in check. These various checks and counterchecks are not merely on and off switches; they operate over a continuous scale such that there are all gradations of locomotion and feeding and all degrees of eating (i.e., various segments of the feeding sequence may be modified—accelerated, slowed, or eliminated).

What starts and what terminates locomotion in the rat? Normally sleep is terminated when the pattern of neural activity in the brain changes. What initiates the change is not known. When the rat awakes, its central nervous system is the recipient of incoming sensory signals. In addition to the specific messages they convey, the sense organs increase the general activity of the cortex via the reticular formation that "arouses" it; that is, the central thresholds to many stimuli, related and unrelated to the first, are lowered. Whether at this point locomotion is driven by spontaneously active centers released from inhibition, or activated from without by environmental stimuli, is not known. In any case, appropriate stimuli when encountered now direct locomotion and elicit sequential responses in the chain of feeding behavior; each response places the animal in a situation where new stimuli are encountered: smell and/or sight of food, food in the mouth, food in the stomach, etc. Feeding is shut off when internal stimuli raise the threshold of the brain to incoming sensory stimuli.

The situation is complicated, however, by the experimental discovery that there are regions in the hypothalamus that profoundly influence feeding behavior. The classical story is as follows. If a small area (feeding center) in the lateral hypothalamus is stimulated, feeding behavior is initiated whether the animal is already satiated or not. A stimulated rat will arouse itself, walk over to food, and eat. It this area is damaged, the rat, even if starving, will ignore normal stimuli associated with food (sight, odor, taste, food in mouth, etc.), will not explore, and will not eat. In acute stages, food placed in the mouth will be allowed to drop to the ground. Later, however, rats with lateral lesions can be coaxed back to eating by weaning them with a highly attractive diet (e.g., milk, chocolate) (Teitlebaum and Epstein, 1962). Even the ability to regulate returns eventually; however, the recovered animal is always a finicky eater and never again works very hard to obtain food. Other areas of the brain apparently compensate for part but not all of the neuronal loss. This is another indication that it is risky to assign specific brain functions to rigidly localized areas (compare Valenstein, Cox, and Kakolewski, 1969, 1970).

If another area in the hypothalamus, the ventromedial region (the satiety center), is stimulated, the starving rat will ignore food. If this area is damaged, eating is greatly increased; that is, rats respond positively to all the normal food stimuli, they explore, they will work very hard for food, they overeat to the point of extreme obesity. Once fat, however, they will not explore, they are finicky eaters, and they will not work very hard for food.

There really is no evidence that cells here actually initiate and drive. It is equally plausible to say that these two centers working together determine to what extent stimuli from the outside (food and from the inside (gastrointestinal and blood) are heeded. Indeed there are cells in the hypothalamus that are normally responsive to physiological imbalance, but it is going beyond the data to say that they *initiate* anything. Even if these cells were spontaneously active groups driving motor output, it is difficult to envision how their activity would drive eating. At best it can be said only that external stimuli that were formerly ineffective in evoking a response are now effective.

A number of investigators are now beginning to question the idea of "feeding" and "satiation" centers. Many feel that deficits in feeding behavior in mammals after hypothalamic lesions result in part from the disruption of afferent and efferent fibers of passage rather than from destruction of cells and synaptic networks within the hypothalamus itself. Gold (1973) has shown in rats that overeating, once associated specifically with destruction of the cells known as the ventromedial nucleus, actually occurs as a result of damage to the nearby ventral noradrenergic bundle, a group of fibers that ascend from the brainstem to innervate limbic areas including several hypothalamic loci. Few of these fibers actually terminate in the ventromedial nucleus.

Since almost every effect produced by manipulating the lateral hypothalamus can also be obtained by interfering with other structures in the brain, it is possible that there is nothing unique about the hypothalamus, insofar as feeding is concerned, except that it is packed with fibers of passage. Interruption of these anywhere in the brain obviously will have profound effects on behavior. The importance of oral-pharyngeal sensory input to feeding behavior in the rat is well established (Epstein, 1967). There is also evidence of sensory input to hypothalamic and other "feeding centers." In the pigeon it has been possible to study selectively the effect of very localized lesions on feeding behavior and to show that some of these affect feeding by influencing sensorimotor processes (Zeigler and Karten, 1973). These lesions also affect the pigeon's *responsiveness* to food. The two types of deficit can be experimentally dissociated. One interpretation is that decrease in responsiveness ("motivational" deficits) may be an indirect effect of the deafferentation of other regions of the brain involved in the integration of the somato-motor response mechanisms underlying feeding behavior (Zeigler and Karten, 1973). A similar hypothesis has been advanced to explain alterations in the feeding behavior of mammals following brain lesions (Wyrwicka, 1969; Marshall, Turner, and Teitlebaum, 1971).

So far in this account no profound differences have appeared between the fly and the rat. Some of the more obvious ones reflect differences in the experimental approaches of those who study rats and those who study invertebrates. Feeding and satiety centers have not been found in the brain of the fly (they have not been sought); but even if they were found, their existence would do little to change the original model of feeding. But the students of rat behavior, and vertebrate behavior in general, are impressed by the fluctuating nature of behavior, the spontaneity of behavior, and the way in which patterns of behavior are grouped together. They ask such questions as: why does an animal respond to a particular stimulus at one time and not at another? what determines the beginning and ending of a particular behavior? why is an animal active at all? They are impressed by the fact that specific behavior occurs when particular physiological imbalances occur and that the behavior normally results in restoration of the balance. Now all of these facts and questions are as relevant to the behavior of the fly as to that of the rat, but the students of vertebrates speak of "need" and "goal" and postulate something inside the animal that makes it different from time to time so that stimuli do not invariably elicit the same response or even any response. They also postulate something inside the animal that starts behavior (mood, Stimmung, motivational state, drive state, etc.) From human subjective and emotional experience, and from the traditional rationalistic doctrine the man's behavior results from his mental processes because he has free will, was conscripted the concept of motivation, a hypothetical internal agency or force that causes behavior (cf. Bolles, 1967). However, not all students of vertebrate behavior think that motivation causes or energizes behavior. Hebb (1949) proposed that motivation was the *organization* of behavior rather than the *production* of behavior because even at the neural level "unmotivated" animals showed the same level of activity as "motivated" animals. The flaw in that argument is that one does not know precisely where to look for activity or what kind of activity to search for in the first place. If indeed it exists, it can be tucked away in odd places or be so subtle as to pass unnoticed in the kaleidoscope of activity emanating from the vertebrate brain. Gallistel (1974) views motivation in terms of processes that potentiate and inhibit the lower-level mechanisms of sensorimotor coordination in order to ensure an overall coherence and direction to behavior.

It did not take long to extend the idea of internal driving forces in man to driving forces in nonhuman vertebrates. Postulated internal forces energizing the various patterns of behavior were termed drives (motives, tensions). Drives are also defined as mathematical intervening variables, hypothetical biogenic states (related to internal conditions), hypothetical psychogenic states (unrelated to internal conditions), etc.

Implicit in the concept of drive are the ideas of need and goal. The achievement of a goal reduces the need. These ideas have so permeated the field of behavior that without further ado certain kinds of behavior, notably drinking, feeding, and sexual behavior, are saddled with the adjective "motivated" (see, for example, Dethier and Stellar, 1970; Morgenson and Huang, 1973). Whereas psychologists might speak of feeding behavior as characterized by a drive leading to goal-directed behavior which results in satiation (drive reduction) when the goal is achieved, physiologists would describe the same behavior as characterized by an internal state (hormonal, neural, metabolic, etc.) that initiates locomotor activity or releases inhibition of locomotor activity, lowers central threshold to specific stimuli (proximate and ultimate goals) that were until then ineffective, and releases these effects when the initiating internal states are altered.

One of the several difficulties erected by the drive concept was the tendency to think of drive in unitary terms. For example, in this book the adjective "hungry" has been employed as a convenient way of referring to a fly that has been deprived of food and will eat it upon presentation. That has been its only connotation here. "Hungry" as employed by motivation theorists refers to an intervening variable, a "hunger drive." As Hinde (1970) has emphasized, however, the postulation of *one* intervening variable is too simple a hypothesis to account for the behavioral facts. The changes of behavior observed (in the rat, for example) — bar pressing, number of licks, amount of food eaten, amount of shock or quinine tolerated, etc. — all bear a different relation to the period of deprivation or state of metabolic imbalance. Deprivation induces many diverse neural and extraneural changes and these influence to different degrees a whole constellation of behaviors.

Once the concept of drive was recruited to explain behavior, it was deemed desirable to measure the strength of the drive state (motivation). Feeding behavior lends itself very well to this game. Of the several possible measures the following are most commonly used: the intensity of adverse stimuli that an animal will tolerate to get food; the amount of work (general activity, frequency, speed, quantity) that an animal will perform to the same end; the amount of general activity that a rat will indulge in; the amount of physical work it will perform; the degree of bitterness of food it will accept; and the intensity of electric shock that it will tolerate. All increase as the length of deprivation of food increases, that is, as the metabolic energy imbalance becomes more acute.

No attempts have been made thus far to measure the intensity of electric shock that a fly will tolerate in order to obtain food. We know that the fly will tolerate more and more adverse mechanical stimulation as the period of deprivation increases. Thus it is almost impossible to prevent an acutely starved fly from feeding by roughly pushing the proboscis aside, whereas a fly that has recently fed is easily discouraged from extending its proboscis. More accurate measurements have been made of the amount of gustatory adulterants that will be tolerated as a function of duration of deprivation, but, as will be seen, the data cannot be interpreted as usefully as one would like. At first glance, a starved fly does indeed seem to tolerate more salt in its food than a fed fly, but this finding is deceptive. Feeding represents a favorable balance between acceptable and unacceptable sensory input. For example, if a small amount of salt is added to sugar, ingestion continues unabated; if more salt is added, sucking stops. Now if the sugar concentration is increased, intake resumes even though the high concentration of salt remains. In other words, the sensory input from sugar receptors must exceed the input from salt receptors if feeding is to result. It has been pointed out, however, that the sugar threshold drops with deprivation. That is to say, fewer impulses from the sugar receptor are required to trigger the event because there is less inhibition from gut receptors to balance the chemosensory input. When a deprived fly tolerates more salt in a *standard* sugar solution, it means simply that the lower threshold to sugar is, in effect, an increase in sugar concentration; the net result is a sugar/salt balance in the central nervous system still in favor of sugar. It would be highly instructive to reapply the test of salt tolerance by adjusting the sugar concentration with each stage of deprivation to a threshold criterion. A few experiments of this sort were conducted by Haslinger (1935) with the related fly *Calliphora erythrocephala*. The rejection threshold for hydrochloric acid during starvation was measured by presenting the acid in a fructose solution, the concentration of which was varied so as to be just three times the threshold of fructose on each day of the test. Under these conditions no change in the rejection threshold for acid was observed. Similar results were obtained with unacceptable sugar alcohols, salts, and quinine. In other words, the fly did not tolerate more adversity. The experimental results are further complicated by the fact that mixed solutions cause complicated interactions to occur in the receptors. For comparison it would be interesting to know more about the mechanism of aversion and tolerance of adverse gustatory stimuli in the rat.

Another test of the fly's propensity for tolerating adverse stimuli with increasing deprivation was made by adulterating water with sodium chloride and observing drinking responses as desiccation due to water loss increased. As in the case with food deprivation, there is no evidence of any change in the threshold of the receptors themselves; the change is central. That change, as in the case of ingestion of sugar, appears to be brought about by interaction between a standard sensory input (from the water receptors) and a variable inhibitory feedback initiated by a mechanism that detects changes in blood volume or pressure or ionic balance.

Turning to general activity, evidence was presented showing a positive correlation between general bodily activity of the fly and its state of deprivation; however, the fly does not move faster, it moves more often. This type of activity pattern suggests an on/off switching mechanism controlling spontaneous activity. The experiments described in Chapter 2 point to the involvement of a blood-borne regulatory factor. It has been postulated that the foregut receptors when stimulated by food release a factor from the corpus cardiacum. The probable relationships are summarized in Table 28. Experiments with a locust (*Locusta migratoria*) have produced evidence that the corpus cardiacum does indeed release a factor during feeding (see also Bernays, Blaney, and Chapman, 1972) and that the act of feeding causes neurosecretory cells of the pars intercerebralis to become active (Clark and Langley, 1962). In a close relative of the blowfly, Thomsen (1952) has shown that the corpus cardiacum acts as a storage organ for neurosecretory materials formed in the pars intercerebralis. These neurosecretory products are transported via the nervi corporis cardiaci I to the corpus cardiacum to produce a factor of their own. In addition there is evidence (Davey, 1962) that feeding causes release of a pharmacologically active factor from the corpus cardiacum of the cockroach *Periplaneta americana*. If such processes occur in *Phormia* and if some of these blood-borne factors suppress activity in the ganglia of the central nerve cord (Ozbas and Hodgson, 1958; Milburn, Weiant, and Roeder, 1960) or

Table 28. Possible Mechanism for the Control of Spontaneous Locomotor Activity in Adults of *P. regina* in Relation to Feeding and Deprivation. Activity Curves Shown in Figures 163 and 165b Are Explained on the Basis of a Neurohumoral Mechanism Involving Foregut Receptors, Neurosecretory Cells in the Protocerebrum, and the Corpus Cardiacum.

Condition of fly	Theory of mechanism	Behaviour of fly
Starved, crop empty	No sugar in foregut; foregut receptors not stimulated; corpus cardiacum not induced to release active factor; locomotor activity not depressed. Periods 16–17, Figures 163 and 165b	Continuously active
Fed to repletion on 0.5 M sucrose	Foregut full of sugar; foregut receptors stimulated; corpus cardiacum induced to release active factor; median neurosecretory cells induced to produce active factors; locomotor activity depressed. This is the immediate condition caused by feeding and shown in most of the activity patterns presented in Green (1964).	Inactive
Fed to repletion, fly begins to regurgitate	Foregut has been full of sugar for some time; foregut receptors either firing continuously, or beginning to adapt; corpus cardiacum stimulated to release stored factor continuously or, if foregut receptors adapt, only when they disadapt; corpus cardiacum becomes depleted of active factor faster than it can be manufactured; or releases it only if receptors disadapt; locomotor activity depressed or released in relation to availability of corpus cardiacum factor. This explains the low but persistent activity shown during Periods 1 and 2, Figures 163 and 165b.	Inactive at first, but shows periods of activity in relation to reduction in blood titer of corpus cardiacum factor
Regurgitation has ceased but rate of removal of sugar from crop still very rapid.	Foregut periodically filled with sugar; foregut receptors firing frequently, or only as they disadapt; corpus cardiacum stimulated to release factor frequently; corpus cardiacum begins to increase store of factor as production catches up; corpus cardiacum factor present continuously in high titer in blood; locomotor activity depressed. Periods 3–5, Figures 163 and 165b.	Overall decrease in activity, inactive most of the time

Table 28. (continued)

Condition of fly	Theory of mechanism	Behaviour of fly
Removal of sugar from crop slower	Foregut periodically filled with sugar; foregut receptors only periodically stimulated; corpus cardiacum releasing factor periodically; corpus cardiacum storing additional factor produced. This is a more or less progressive stage; corpus cardiacum factor in blood is depleted before more released; blood titer of factor showing overall decline. Periods 5–15, Figures 163 and 165b.	Periodically active and inactive, with amount of inactivity decreasing regularly
Removal of sugar almost complete	Foregut infrequently filled with sugar; foregut receptors stimulated infrequently; corpus cardiacum releasing factor infrequently; much factor stored in corpus cardiacum; production of factor decreases; locomotor activity not depressed much. Periods 15–16, Figures 163 and 165b.	Active most of time
Crop empty	This represents a starved fly, plenty of stored factor in corpus cardiacum, and the situation is the same as shown in the first condition in this table. Activity reaches a peak between Periods 16 and 17 and then begins a steady decline as metabolic reserves are depleted.	

From Green, 1964.

depress neural or neuromuscular transmission in peripheral regions, a logical explanation of fluctuations in locomotor activity patterns is available.

Will a starved fly suck faster than a satiated fly? Will it eat longer? Clearly, for any given concentration the answer to the second question is yes. But this phenomenon is explicable in terms of interaction between a standard sensory input and a variable inhibitory feedback from the foregut. A satiated fly receives maximum inhibitory feedback so that the sensory input is behaviorally ineffective. As deprivation increases, inhibition wanes and sensory input becomes increasingly effective in initiating feeding and regulating its rate. Since the duration of sucking increases as the frequency of sensory impulses increases, and since the time to adaptation similarly increases, it is to be expected that deprivation would cause changes in the duration of eating. Waning inhibitory input is equivalent to increasing sensory input.

Will a fly work harder to obtain food? Some attempts were made to detect an increase in flying effort as measured by frequency of wingbeat when a deprived fly is exposed to the odor of food (Schoettle, 1963). In this case the fly was *Drosophila* and the food, bananas. No increase was observed. In this connection it is of interest that hungry flies are more "persistent" in their efforts to come to food (as are hungry mosquitoes) and can be discouraged only with great effort, but this behavior is explainable on the basis of increased general activity with deprivation and with greater tolerance for mechanical disturbance. One must take heed, however, of the experiments with rats. Different measures of activity do not relate in the same way to internal states. Correlations with activity differ, depending upon whether one measures activity in an exercise wheel or in a stabilimeter. With the fly one should measure, for example, rate of walking, rate of turning, force exerted in flying, etc.

Will flies work harder to obtain food in a learning situation? Here the fly clearly differs from the rat. As Chapter 13 indicated, all attempts thus far to set up operant conditioning in *Phormia* have failed. In *Drosophila* operant conditioning is weak; nevertheless, its very existence should be a spur to renewed efforts.

None of the measures described (except for the last to which we shall return presently) support the need for a concept of motivated behavior as usually defined because all are amenable to explanation in terms of known physiological processes as far as the fly is concerned, and probably also with respect to the rat. If one continues to think of motivation as an internal agency or force whose existence is hypothesized, it is reasonable to expect to find neural evidence. One might then formulate the problem in the following terms (Dethier, 1964): "Motivation is a specific state of endogenous activity in the brain which under the modifying influence of internal conditions [metabolic] and sensory input leads to behavior resulting in sensory feedback or changes in internal milieu which then causes a change (reduction, inhibition, or another) in the initial endogenous activity". The essence of motivation would then be endogenous activity. Failure to demonstrate it (cf. Hebb, 1949) would not constitute evidence against it. Furthermore, it would be applicable only in the case of a unitary drive concept, and we have already indicated that a unitary concept is too simplistic. It is further complicated by the fact that animals can have acquired motivations. Finally, the definition does not exclude the activities of centers such as the respiratory center of vertebrates.

If we can reduce motivation to these mechanical terms (centers, oscillators, and organizing hierarchies), the concept, along with the drive, seems to be superfluous. Broader definitions offer little cause for reinstating the concept. If motivated behavior is characterized simply as goal-directed behavior with a drive component (change in activity, or work performed, or aversions tolerated), then too many patterns of behavior that are never conceived of as motivational would have to be included in the category. For example, the phototactic behavior of a moth with respect to a candle could be construed as motivated, the light being the goal, compulsive flight representing the drive component, and the moth's coming to rest in the vicinity of the light representing drive reduction. Similarly the chemotactic responses of male moths to the sex attractants of the female would come under the category of motivated behavior because the female would be the goal, flight would be the drive component, and cessation of flight and failure to fly again after copulation would represent drive reduction.

Even broader is the designation of all random, unoriented, restless behavior, the appetitive behavior of Lorenz, as the outward manifestation of a mounting drive. Definitions so broad as to encompass within their bounds at one and the same time tactic behavior of moths and feeding by the rat would seem to raise the danger of obscuring meaningful differences of a fundamenatl nature.

Let us now return to the matter of operant conditioning. Teitlebaum (1966) suggested that the most unequivocal measure of what is meant by motivation is operant conditioning. He argued that the operant is essentially a voluntary act, not dependent upon specific afferent input, which the animal can use to obtain reinforcement. Since in this situation the animal exerts control over the occurrence of its response, the behavior is distinct from reflexes and from complex fixed motor patterns. Bolles (1967) in his penetrating account of the history and development of the concept of motivation came finally to the thought that perhaps what we call motivated behavior is no more and no less than learned behavior. If behavior is unlearned it is unmotivated. Perhaps, he continues, we do not need the concept of motivation and drive to explain behavior.

Certainly the concept adds nothing to our understanding of the fly's feeding behavior. And the foregoing comparison between the two animals strongly suggests that it adds little to our understanding of the rat's behavior. The salient difference is that the rat can easily be operantly conditioned, and the fly with difficulty, if at all. Perhaps if research relating to this concept were to be shifted to an insect such as the honeybee that can readily be operantly conditioned there would be no qualitative differences between the behavior of the insect and that of the rat.

Literature Cited

Adrian, E. D. 1930. The activity of the nervous system of the caterpillar. *J. Physiol.* 70:34–35.

Bernays, E. A., Blaney, W. M., and Chapman, R. F., 1972. Changes in chemoreceptor sensilla in the maxillary palps of *Locusta migratoria* in relation to feeding. *J. Exp. Biol.* 57:745–753.

Bolles, R. C. 1967. *The Theory of Motivation.* Harper and Row, New York.

Bullock, T. H. 1961. The origins of patterned nervous discharge. *Behaviour* 17:48–59.

Bullock, T. H. 1962. Integration and rhythmicity in neural systems. *Amer. Zool.* 2:97–104.

Bullock, T. H., and Horridge, G. A. 1965. *Structure and Function of the Nervous Systems of Invertebrates,* Vols. 1 and 2. W. H. Freeman, San Francisco.

Clark, U. K., and Langley, P. 1962. Factors concerned in the initiation of growth and moulting in *Locusta migratoria* L. *Nature* 194:160–162.

Davy, K. G. 1962. The release by feeding of a pharmacologically active factor from the corpus cardiacum of *Periplaneta americana. J. Insect Physiol.* 8:205–208.

Dethier, V. G. 1964. Microscopic brains. *Science* 143:1138–1145.

Dethier, V. G., and Stellar, E. 1970. *Animal Behavior,* 3rd ed. Prentice-Hall, Englewood Cliffs, N.J.

Epstein, A. N. 1967. Oropharyngeal factors in feeding and drinking. In *Handbook of Physiology,* Sec. 6, Alimentary Canal, C. E. Code, ed., pp. 197–218. Amer. Physiol. Soc., Washington, D.C.

Gallistel, C. R. 1974. Motivation as central organizing process: The psychophysical approach to its functional and neurophysiological analysis. In *Nebraska Symposium on Motivation,* 1974, University of Nebraska Press (in press).

Gold R. M. 1973. Hypothalamic obesity: the myth of the ventromedial nucleus. *Science* 182:488–490.

Green, G. W. 1964. The control of spontaneous locomotor activity patterns in *Phormia regina* Meigen. II. Experiments to determine the mechanism involved. *J. Insect Physiol.* 10:727–752.

Haslinger, F. 1935. Über den Geschmackssinn von *Calliphora erythrocephala* Meigen und über die Verwertung von Zucker und Zuckeralkoholen durch diese Fleige. *Zeit. vergl. Physiol.* 22:614–640.

Hebb, D. O. 1949. *The Organization of Behavior.* Wiley, New York.

Hinde, R. A. 1970. *Animal Behavior,* 2d ed. McGraw-Hill, New York.

Huber, F. 1960. Untersuchungen über die Funktion des Zentralnervensystems und insbesondere der Gehirnes bei der Fortbewegung und der Lauterzeugung der Grillen. *Zeit vergl. Physiol.* 44:60–132.

Huber, F. 1967. Central control of movements and behavior in invertebrates. In *Invertebrate Nervous Systems,* C. A. G. Wiersma, ed., pp. 333–351. University of Chicago Press.

Kennedy, D. 1962. The initiation of impulses in receptors. *Amer. Zoologist* 2:27–43.

Marshall, J. F., Turner, B. H., and Teitlebaum, P. 1971. Sensory neglect produced by lateral hypothalamic damage. *Science* 174:523–525.

Milburn, N. S., Weiant, E. A., and Roeder, K. D. 1960. The release of efferent nerve activity in the roach *Periplaneta americana* by extracts of the corpus cardiacum. *Biol. Bull.* 118:111–119.

Morgenson, G. J., and Huang, Y. H. 1973. The neurobiology of motivated behavior. *Progress in Neurobiol* 1:55–83.

Ozbas, S., and Hodgson, E. S. 1958. Action of insect neurosecretion upon central nervous system *in vitro* and upon behaviour. *Proc. Nat. Acad. Sci. U.S.A.* 44:825–830.

Roeder, K. D. 1937. The control of tonus and locomotor activity in the praying mantis (*Mantis religiosa,* L.). *J. Exp. Zool.* 76:353–374.

Roeder, K. D. 1955. Spontaneous activity and behavior. *Sci. Monthly* 80:362–370.

Schoettle, H. E. T. 1963. Master's thesis, University of Pennsylvania, Philadelphia.

Teitlebaum, P. 1966. The use of operant methods in the assessment and control of motivational states. In *Operant Behavior: Areas of Research and Application,* W. K. Honig ed., pp. 565–608. Appleton, Century, and Crofts, New York.

Teitlebaum, P., and Epstein, A. N. 1962. The lateral hypothalamic syndrome; recovery of feeding and drinking after lateral hypothalamic lesions. *Psychol. Rev.* 69:74–90.

Thomsen, E. 1952. Functional significance of the neurosecretory brain cells and the corpus cardiacum in the female blowfly, *Calliphora erythrocephala* Meig. *J. Exp. Biol.* 29:137–172.

Valenstein, E. S., Cox, V. C., and Kakolewski, J. W. 1969. The hypothalamus and motivated behavior. In *Reinforcement,* J. Trapp, ed., pp. 242–285. Academic Press, New York.

Valenstein, E. S., Cox, V. C., and Kakolewski, J. W. 1970. Reexamination of the role of the hypothalamus in motivation. *Psychol. Rev.* 77:16–31.

Van der Kloot, W. G. 1962. Muscle and its neural control. *Amer. Zool.* 2:55–65.

Wilson, D. H. 1967. An approach to the problem of control of rhythmic behavior. In *Invertebrate Nervous Systems,* C. A. G. Wiersma, ed., pp. 219–229. University of Chicago Press.

Wyrwicka, W. 1969. Sensory regulation of food intake. *Physiol. Behav.* 4:853–858.

Zeigler, H. P., and Karten, H. J. 1973. Brain mechanisms and feeding behavior in the pigeon (*Columba livia*). II. Analysis of feeding behavior deficits after lesions of quinto-frontal structures. *J. Comp. Neurol.* 152:83–102.

15
Epilogue

Even as this book is being written people are busy at work in our laboratory and others, designing and conducting experiments to challenge the conclusions drawn here, to correct experimental fact, to add new data. The picture of the hungry fly as presented is certainly an oversimplification. The human mind for all its transcendental endowments can best grasp only one fact at a time. To analyze a complex problem one must first simplify it. Until the final synthesis is achieved the explanation must represent an oversimplification. If oversimplification is a hazard in studies of such a "simple" organism as the fly, how much more of a hazard it is in studies of "higher" animals. At the most, analyses of simple organisms can illuminate the tortuous path toward understanding ourselves; at the least, they can give heightened awareness of the extreme complexity of higher organisms.

*Lord,
shall I always go in black
for this life?
Fugitive from its tumult
on my transparent wings,
humming my prayers
and pausing weightless
on my thin legs,
I,
whom the world finds such a burden?
You have made me
stick to what lures me.
Yet, if I am caught
clinging there,
don't let me die
like the poor useless
thing that I am.*

Amen.

Carmen Bernos de Gasztold, *The Creatures' Choir*

Index

Abbott, C. E., 74
Abdominal innervation, 316–320
Acceptance, 188, 191, 193, 223, 343
Acids, response to, 203, 206–209
Across-fiber patterning, 211–219
Action potential, 139, 140, 142
Activity: measuring, 5–9; patterns, 10, 290–295
Actograph, 6
Adams, J. R., 82, 83, 90, 197
Adams, T. S., 314
Adaptation, 98–101, 109, 111–113, 121, 143, 190, 219, 230, 231, 257, 284, 341, 413
Adrian, E. D., 463
Aedes aegypti, 18, 84
Age, 11, 283, 357, 358
Agranoff, B. W., 164
Alcohol, 56–63; rejection thresholds, 63, 149–151
Aldehydes, 57
Alkaloids, 202
Alkyl group, 150
Altmann, G., 348
Amakawa, T., 176
Amino acids, 177–178
Anautogeny, 300
Anemotaxis, 18, 19
Angiotensin, 350
Anion receptor, 132–136, 146
Anions, 146, 205
Anisotarsus cupripennis, 337
Anodal stimulation, 139
Anopheles gambiae, 10
Antennal sensilla, 90
Anwyl, R., 280
Aplysia, 412
Appetitive behavior, 5, 290, 300
Arab, Y. M., 88, 95, 105, 111, 138, 139, 417, 431, 434
Arevad, K., 15, 16
Arnett, D. W., 15
Arnica alpina, 37
Arousal, 418–442, 462, 464, 465
Associative learning, 411, 444–457
Autogeny, 300
Autrum, H., 15, 403
Aversive behavior, 206, 208, 223

Bailey, S. F., 353

Balboni, E. R., 299
Bárány, E., 107
Barber, G. W., 13, 16
Bare, J. K., 309
Barelare, B., 303
Barlow, H. B., 406
Barness, L. A., 12
Barnhard, C. S., 44, 173
Barrows, W. M., 72
Barton Browne, L., 5, 142, 210, 235, 257–260, 293, 307, 330, 347–349, 435
Beck, S. D., 52, 55, 353
Begg, M., 339
Behavioral thresholds, 234–237
Beidler, L. M., 143–144, 162, 173, 174, 178, 187
Belzer, W. R., 39, 42, 300, 301, 304, 320, 321, 324, 325, 329
Bennett-Clark, H. C., 103
Bennettova-Řezábová, B., 312, 315, 316, 320
Benzer, S., 453, 456
Bernays, E. A., 234, 281, 282, 470
Berridge, M. J., 316
Berry, L. B., 15
Beulter, G., 289
Biological clocks, 5, 13
Bishop, L. G., 14, 377, 405
Bissar, A., 112
Blake, G. M., 300
Blaney, W. M., 197, 217, 234, 470
Block, B., 417
Blood: osmotic pressure, 275, 287; sugar, 240–244, 426
Blue, S. G., 82, 83, 84
Bodenstein, G., 247, 251, 254, 257, 259, 260, 300, 311, 435
Boeckh, J., 88, 89
Boettcherisca peregrina, 83, 164, 171, 172, 176
Boistel, J., 126
Bolles, R. C., 467, 474
Bolwig, N., 240
Bombyx mori, 17
Bracken, G. K., 300
Bradley, R. M., 309
Brady, J., 10, 292, 293, 295
Brain: frontal sections, 369–375; dorso-ventral sections, 376–393; association centers, 393–407

Braitenberg, V., 16
Bretschneider, F., 377, 403
Brink, F., 152
Broadbent, A. H., 368
Brobeck, J. R., 289
Bullock, T. H., 189, 190, 441, 442, 463
Burk, D., 165
Burkhardt, D., 15
Bursell, E., 336, 339

Cafferty, D., 4
Cagan, R. H., 161
Cajal, S. R., 14, 377, 403
Calabrese, E. J., 309, 355
Calcium, effects of, 146–148
Calcium chloride, 202, 212
Calliphora, 15, 132, 162, 235, 268, 285, 300, 301, 312, 322, 325, 349, 368, 369, 392, 414
Calliphora erythrocephala, 72, 88, 469
Calliphora phaenicia, 14
Calliphora vicina, 145, 349
Cameron, A. T., 49
Campbell, K. H., 39, 47
Carausius morosus, 316
Carbohydrates, utilization of, 11, 52, 53
Case, J., 177
Cathodal stimulation, 139
Caterpillar, 188, 214, 217–219, 285, 463
Cations, stimulation by, 145, 172
Celerio euphorbiae, 285
Central excitatory state, 100, 418–422, 462, 464
Central inhibitory state, 418–422, 462, 464
Chadwick, L. E., 12, 44, 60, 72, 110, 152, 173, 181, 205, 234
Chapman, R. F., 197, 234, 281, 282, 470
Chemosensory hairs, structure of, 73–85
Chemoreceptors, 138, 139, 214; duplication, 217
Chen, S. S., 356
Choice, definition of, 37, 63, 64
Choice situation, 344
Chorda tympani, 193
Cibarial pump, 69, 94–96, 101–104, 404
Ciliary structure, 83, 94
Circadial rhythms, 5, 7–9, 210, 462–464
Clarke, U. K., 287, 470
Cleaning behavior, 13, 20, 58, 206
Clegg, J. S., 12
Clements, A. N., 12
Cockroach: sensilla, 84; gustatory organs, 210
Cole, P., 339
Color preferences, 15, 16
Colorado potato beetle, 202, 355
Concentration-response curves, 57, 168, 170
Conditioned reflex, 411
Consummatory behavior, 290

Contaminants, 342, 344
Cook, A. G., 197
Coraboeuf, E., 126
Corn borer, 52, 55
Cornus sericea, 37
Corpora pedunculata, 375, 376, 406, 464
Corpus allatum, 254, 261, 295, 303, 311, 312, 315, 320, 324, 348
Corpus cardiacum, 254, 261, 295, 348, 370
Coumarin, 416, 417
Cousin, G., 355
Cox, V. C., 466
Cragg, J. B., 339
Crickets, 407, 464
Crombie, A. C., 414
Crop emptying, 258, 268, 269, 272–278
Crow, S., 74
Cubbin, C. M., 10
Cuccati, J., 368, 376, 389, 392
Culex pipiens, 300
Cushing, J. E., 414
Cynomyia, 416

Dahlberg, A. C., 49, 52
Danilevskii, A. S., 5, 354, 355
Davey, K. G., 275, 470
Davies, D. M., 300
Davies, J. T., 152
Davis, R. A., 12
Day, M. F., 348
De Buck, A., 355
Deficiency: protein, 320; salt, 309; thiamine, 304
Deficit: hypothesis, 323, 324; motivational, 467
Den Otter, C. J., 132, 133, 145, 146, 147, 148, 153, 181
Dendrites, 82–84
Denton, D. A., 303
Deomier, C. C., 74, 146, 176
Depolarization, 143–148
Depression of receptor activity, 176, 177
Deprivation, 235, 268, 292–295, 340, 426, 433
Desiccation, 336, 340, 341
Desmosomes, 83, 186, 210
Deterrence, 123, 344
Deuterium, 205
Diamond, J. M., 148, 153
Diapause, 353–356
Digestive System, 236, 238–240
Dill, J. C., 14
Disadaptation, 257, 258
Discrimination, 190, 212, 306; definition of, 37, 63, 64; factor, 52; of sugars, 50–52
Diuretic hormone, 348
Divalent salts, 146, 147
Dodt, E., 15
Doetsch, G. S., 212

Dondero, L., 355
Drinking, by insects, 338; prandial, 341
Drinks, 231, 232
Drives, 467–469
Drosophila, 15, 17, 251, 254, 364, 368, 376, 377, 381, 392, 473; orientation to odor, 18, 19, 61; tarsal sensitivity of, 72
Drosophila guttifera, 414
Drosophila melanogaster, 11, 413, 417
Drosophila repleta, 12
Dudzinski, A., 348
Duplication, in chemoreceptor field, 217

Edney, E. B., 337, 338
Egg maturation, 311
Eichner, J. T., 88
Electrokinetic streaming, 178, 179
Electrotonic coupling, 210, 211
Elizarov, Yu. A., 455
Eltringham, H., 72
Endogenous activity, 5, 462–465
Engelmann, F., 275, 322, 325, 329
Epipharyngeal region, 95
Epstein, A. N., 341, 350, 466, 467
Erickson, R. P., 211
Eriogonum ovalifolium, 37
Eristalis tenax, 368, 417
Ernst, K-D., 94
Evans, D. R., 5, 12, 53, 58, 62, 123, 125, 130, 131, 142, 144, 146, 162, 175, 178, 190, 191, 194, 210, 230, 231, 235, 239, 241, 254, 258, 259, 260, 307, 312, 340, 342, 345, 347, 435
Experience, role of, 104, 453
Eyes, neural network, 14, 15

Face fly, 355, 359
Falk, D., 103
Farkas, S. R., 18, 19
Fat body, 11, 12, 312, 315, 320, 325, 356, 359
Fatty acid salts, 133–136
Faull, J. R., 145
Fay, R. W., 16
Feedback loops, 461
Feeding: habits, 34–36; patterns, 36, 94–101, 228, 229; stimuli, 104; effects on blood sugar, 241–247; process, 257, 258; model for, 266, 267; centers, 465, 466
Fender, D. M., 15
Ferguson, J., 152
Fermi, G., 15
Fernandez, H. R., 15
Fernandez Perez de Talens, A., 16
Filtering, central, 187
Finlayson, L. H., 101, 316
Fitzsimons, J. T., 350
Flavor, distinguishing, 223
Flight, 243, 246; energy requirements, 11–13; factors controlling, 17–19; patterns, 17–19, 465; centers, 462, 464
Flügge, C., 17
"Fly dance," 27–30
"Fly factor," 44, 173
Food: selection of, 36; of wild blowflies, 36; searching, 407
Forgash, A. J., 82
Fraenkel, G., 12
Frank, M., 194, 211, 212
Frank, S., 406
Fredman, S. M., 100, 442
Freeborn, S. B., 15
French, R. A., 338
Frings, H., 145, 193, 342, 416, 417, 447
Von Frisch, K., 52, 58, 109, 161, 234
Frizel, D. E., 18
Frontal ganglion, 404
Fructofuranoid site, 175
Fukushi, T., 452
Functional unit of hair, 215, 217

Galun, R., 125
Gans, J., 37, 53, 162, 168, 235, 285
Gelperin, A., 231, 260, 268, 269, 272, 281, 282, 285, 288, 293, 316, 330, 347
Gelvin, D. E., 88
Geotrupes stercorarius, 17
Getting, P. A., 96, 97, 101, 142, 144, 190, 193, 268, 413
Giacchino, J., 234
Gieke, P. A., 355
Gieryng, R., 368
Gillary, H. L., 132, 133, 138, 145, 146, 194, 210, 230, 359
Gilmour, D., 12
Glossina austeni, 95
Glossina morsitans, 9, 292, 293–295, 339
Glycogen synthesis, 244
Glucosidases, properties of, 174–177
Goal, 467
Gold, R. M., 466
Goldrich, N. R., 178, 454
Goldsmith, T. H., 15
Goodmann, L. J., 16
Goryshin, N. I., 354
Götz, K. G., 15
Grabowski, C. T., 74, 77, 82, 180
Graham-Smith, G. S., 106
Green, G. W., 6, 9, 238, 247, 257–259, 269, 283, 290–295, 312
Greenberg, B., 324
Greenberg, S., 355
Grooming, 13, 20, 58, 206
Grubb, T. C., 18
Guilliam, G. F., 177
Gupta, B. L., 316
Gut innervation, 253–256
Gwadz, R. W., 280

Gymnemic acid, 178

Habituation, 100, 111, 411–414
Hackley, B. E., 88
Hahn, M., 112
Hall, D. G., 354
Hall, D. W., 360
Halpern, B. P., 144, 145
Hamster, 212
Hanamori, T., 164, 176
Hansen, K., 171, 174, 175
Hanson, F. E., 132, 139, 140, 146, 153, 154, 177, 180, 181, 202, 205, 211, 215, 222, 230, 285, 429, 453
Harlow, P. M., 312
Harris, W. A., 453, 456
Hartline, H. K., 193, 199
Haskell, P. T., 202
Haslinger, F., 74, 162, 234, 469
Hassett, C. C., 37, 53, 162, 168, 235, 285
Hayes, W. P., 74
Heavy metal salts, 146, 176
Hebb, D. O., 467, 473
Hecker, E., 126
Helix, 442
Hellekant, G., 199
Heracleum lanatum, 37
"Herd instinct," 15, 44
Herold, R. C., 299
Herschberger, W. A., 414
Hertwick, M., 74, 368
Hewitt, C. G., 316
Hidaka, I., 165, 174, 179, 211
Hierodula crassa, 284
Highnam, K. C., 278
Hill, L., 278
Hill, R. M., 406
Hinde, R. A., 411, 468
Hinton, H. E., 5, 337
Hintz, A. M., 314
Hirsch, J., 417
Hocking, B., 12
Hodgkin, A. L., 145
Hodgson, E. S., 126, 131, 138, 139, 142, 153, 164, 165, 174, 180, 194, 202, 210, 211, 470
Hogben, L., 339
Holbert, P. E., 82, 83, 197
Holling, C. S., 232, 284
Honeybee, 58, 161, 324, 338, 348; sugar consumption by, 46, 125; discrimination factor for, 52
Hori, N., 145, 146
Horn, G., 416
Horning, D. S., 37
Horridge, G. A., 452, 463
Hoskins, W. M., 61
Housefly, 14, 15, 74, 125, 177, 240, 324, 414, 452

Hoyle, G., 452
Huang, Y. H., 468
Hubel, D. H., 406
Huber, F., 406, 407, 464
Hudson, A., 11, 12, 243, 246, 247, 250
Humidity, 339; receptors, 338, 339
Hygroreceptors, 339
Hymenoptera, 376, 453, 455, 456
Hyperphagia, 257–264, 280, 329
Hyperpolarization, 146
Hypocerebral ganglion, 254
Hypothalamus, 465–467

Ilse, D., 417
Imprinting, 453
Information processing, 188
Ingestion: motor response in, 94–104; sensory control in, 95, 104; pattern of, 231; process of, 238
Inhibition, 125, 133–138, 211, 330, 347; by mannose, 123; competitive, 162, 168, 175; metabolic, 172; noncompetitive, 175
Insight learning, 411, 453
Instrumental conditioning, 411
Interpseudotracheal papillae, 85, 95, 222, 230
Interspike interval, 194
Intrinsic variability, 218
Ions, stimulation by, 145
Irregular firing, 208
Irwin, F. W., 37, 63, 64
"Isosweetness," 49

Jacklet, J. W., 412
Jakinovich, W., 164
Jander, R., 15
Jarnicka, H., 368, 376
Jermy, T., 453
Johansson, A. S., 300, 311
Jones, M. D. R., 10
Junction body, 83

Kabuta, H., 83, 88
Kahn, M., 63
Kaib, M., 88, 89
Kaissling, K-E., 88, 89
Kakolewski, J. W., 466
Kalf, G. F., 240
Karten, H. J., 467
Katz, B., 145
Kawabata, K., 176
Kay, R. E., 88
Keehn, D. G., 14, 377, 405
Keely, L. L., 299
Keiding, J., 16
Kellogg, F. E., 18
Kennedy, D., 463
Kennedy, J. S., 19

Kijima, H., 163, 176
Klinokinesis, 18, 19
Knight, B. W., 113, 200
Knight, M. R., 238, 239
Koizumi, O., 176
Konishi, J., 179
Koyama, N., 176
Krijgsman, B. J., 74
Kuckulies, G., 112
Kühner, J., 171, 174, 175
Künckel, D'Herculais, J., 85
Kunze, G., 109
Kurihara, K., 176
Kuwabara, A., 83, 88, 139, 177
Kuwabara, M., 211

Labeled Lines, 189, 190, 193, 194, 200
Labellar hairs: distribution of, 76–77; structure of, 72–85, 140, 141; number of neurons in, 78–85; specificity of receptors in, 136, 181, 182; identity of receptors in, 180, 181; response to acids, 206–209
Labellum, attitudes of, 106–107
Landing place, factors determining, 14–16
Lange, D., 193, 199
Langer, H., 15
Langley, P., 278, 404, 470
Larimer, J. L., 178
Larsen, J., 78, 84, 85, 88, 90, 300, 365
Larsen, W. J., 299
Latency of response, 142, 190, 219
Latent learning, 411, 453
Laverack, M. S., 177
Lea, A. O., 312
Learning process, 411–415
Leclerq, J., 338
Lecompte, J., 126
Lemberger, F., 52
Lepkovsky, S., 39
Lettvin, J. Y., 126, 131, 406
Leucophaea maderae, 275, 324, 329
Levenbook, L., 299
Levick, W. R., 406
Lewis, C. T., 74
Licklider, J. C. R., 107, 108
Limulus, 198
Lindquist, D. A., 16
Lineweaver, H., 165
Liu, Y-S., 74
Livingston, R. B., 234
Localization, in receptor fields, 105
Locomotion: spontaneous, 5; patterns, 17–30; relation to feeding, 229, 290–295; as food search, 407; regulation, 464, 465
Locust, 197, 202, 214, 215, 278, 281, 282, 416, 462, 470

Lorenz, K., 474
Lowne, B. T., 316
Lucilia caeser, 139
Lucilia cuprina, 338, 348
Lucilia sericata, 88
Lukowiak, K., 412

McCann, G. D., 14, 15, 377, 405
McCutchan, M. C., 74, 181, 202, 215, 218, 278
MacGinitie, G. F., 15
Maddrell, S. H. P., 280, 316, 348
Mail, G. A., 354
Manduca sexta, 219, 234
Manning, A., 4, 411, 414
Mannose, inhibition by, 123
Mantis, 284, 285, 462, 464
Marsh, D., 10, 19
Marshall, D. A., 212
Marshall, J. F., 467
Maturana, H. R., 406
Mayfly nymphs, 456
Maynard, D. M., 367, 406
Mayer, J., 289
Mazokhin-Porshnyakov, G. A., 15
Meal, defined, 231
Mechanoreceptor, 82, 95, 136
Melipona rufiventris, 456
Mellanby, K., 337, 338, 348
Mellon, De. F., 130, 131, 142, 144, 146, 178, 190, 191, 194, 210, 230
Membranes: sites of, 141; models of, 144; selectivity of, 148, 153
Milburn, N. S., 470
Miller, P. L., 336
Mimura, K., 14
Minnich, D. E., 58, 72, 109, 235, 268
Mistretta, C. M., 208
Miura, T., 355
Mixtures, effects of, 122–126
Models: membrane, 144; sugar stimulation, 173–175; water stimulation, 178–180; feeding, 266–267; stimulus-response, 461
Modes of action, of receptors, 136, 173, 175, 178–180
Møller, I., 322
Moncrieff, R. W., 52
Monosaccharides, receptor sites for, 165, 167; stimulation by, 165
Monovalent salts, 145
Moore, G. P., 201
Mordue, W., 278, 359
Morgenson, G. J., 468
Morita, H., 84, 98, 127, 128, 138, 139, 140, 153, 164, 165, 171, 172, 173, 174, 175, 176, 178, 179, 202, 210, 211

Mosquito: sensilla, 82, 84; protein needs, 300; feeding behavior, 355, 356
Moth "ear," 187
Motivation, 467, 473
Motor output, 94–104
Moulins, M., 84, 210
Mourier, H., 17, 414, 416
Murphy, R. M., 417
Murray, R. G., 186
Musca autumnalis, 355, 359
Musca domestica, 14, 15, 74, 125, 177, 240, 324, 414, 452
Musculature, of proboscis, 69–71
Mushroom bodies, 375, 376, 406, 464

Nair, K. K., 300
Narcosis, 153
Nauphoeta cinerea, 4
Need, 467
Nelson, M. C., 417, 442–450, 456
Neural coding, 188–190
Neuropile, 366, 377
Neurosecretion, 295, 312, 315, 316, 320–322, 348, 406, 470
Nobel-Nesbitt, J., 337
Noirot, C., 210
Nonelectrolytes, stimulation by, 148–154
Norris, K. R., 9, 36
Novel stimulus, response to, 304, 414–416
Nuñez, J. A., 234, 260, 262, 337, 348
Nuorteva, P., 8, 355
Nutritive value of sugars, 37, 38, 52; role in feeding, 287–289

Oakley, B., 178
Odors: effect on flight, 17–19, 57, 58; behavioral response to, 208–216
Odor specialists, 88–90
Ogawa, H., 212
Okasha, A. Y. K., 337, 338
Olfactometer, 54, 59
Olfactory receptors, 88–94; of rabbit, 188; of caterpillars, 188, 218–221
Olfactory thresholds, 55–57
Oligomyrmex, 363
Omand, E., 168, 175, 211, 234, 357
O'Neal, B. R., 145
Operant Conditioning, 411, 473, 474
Optomotor responses, 15, 19
Oral papillae, 222, 223
Orientation: to odors, 18, 19, 57, 61, 300; to humidity, 339
Orr, C. W. M., 309, 312, 324, 328
Osborn, M. P., 316
Otto, E., 17
Ovipositor, chemoreceptors on, 88
Ozbas, S., 470

Painter, R. R., 323
Parabiosis, 247–250, 312–314, 426
Paratenodera sinesis, 285
Patterning, across-fiber, 211, 212, 218, 223, 344
Patterns: discrimination, 15, 16; flight, 17–19, 465; locomotion, 20–27; feeding, 36, 94–101, 228, 229, 231, 264, 282
Penczek, E. S., 49, 52
Periplaneta americana, 52, 275, 342, 470
Peristalsis, 238, 239
Perkel, D. H., 189, 190, 201
Perttunen, V., 339
Pessotti, I., 456
Peters, W., 76, 78, 82, 84, 88
Peterson, B. V., 300
Pfaffmann, C., 39, 49, 194, 211, 212, 309
Pflumm, W. W., 164, 171, 175
pH, effect of, 143, 172, 205, 206
Pharyngeal receptors, 88
Phasic response, 141, 144, 190, 212, 217, 230
Pheromones, 18, 19
Phormia terraenovae, 8, 37, 164, 171, 174, 312, 355, 357
Photoperiods, 354
Pieris rapae, 58, 72, 109
Pigeon, 467
Pink bollworm, 18
Pirenne, H., 107
Plodia interpunctella, 19
Polar groups, 149
Pollack, G., 96, 100, 101
Polydipsia, 346, 348
Polypedilum vanderplanki, 5
Pomonis, J. G., 314
Populations of receptors, 105–113, 200, 214, 215
Populus trichocarpa, 37
Pores, 234; in taste hair, 197
Pospíšil, J., 15, 300
Posternak, J. M., 152
Postexcitatory depression, 199
Power, M. E., 316, 364, 368, 375, 377, 389, 392, 394
Precision of response, 193, 194, 200, 201
Preference: color, 15, 16; studies of, 37–53; defined, 37, 63, 64; of sugars, 285–287; in feeding, 307; dietary, 453
Preference-aversion, 46–55, 289, 342
Price, G. M., 325
Pritchatt, D., 452
Proboscis extension: to odors, 58; types of, 70; control of, 94–101
Proboscis musculature, 69–71
Protein: insect's need for, 299–302; deficit, 320; synthesis, 355
Psilopsa petrolei, 34
Pyrameis, 72, 235

Pyrausta nubilalis, 55
Pyrrhocoris apterus, 355

Quinn, W. G., 453, 456

Raabe, M., 316
Rabbit, olfactory receptors of, 188
Ramade, F., 316
Rat: feeding behavior, 289, 304, 341, 463–467; locomotion, 465
Ratliff, F., 193, 199
Rau, I., 325, 329
Receptor fields, 105–113
Receptor Proteins, 176–177
Receptors: olfactory, 88–94; multiple, 105–113, 188, 200, 214, 215; specificity of, 136, 180–182
Recurrent nerve, 253–257; effects of cutting, 260–264; effects on feeding, 346–348
Reed, M. R., 61
Rees, C. J. C., 105, 130, 140, 141, 145, 146, 147, 164, 178, 181, 190, 210, 211, 230, 344, 359
Reichardt, W. E., 15
Rejection, behavioral, 121–125, 136, 188–191, 223, 341–344
Reproductive system, 310–312
Respiratory rate, 355
Resting potential, 141
Reynierse, J. H., 4
Rhoades, M. V., 39, 44, 53, 55, 62, 123, 125, 162, 175, 211, 232, 435
Rhodnius, 103, 228, 229, 231, 264, 311, 338
Rhythmic sucking pattern, 104, 112
Rhythms: circadial, 5; endogenous, 5
Rice, M. J., 95, 101, 102
Richardson, C. H., 74
Richter, C., 39, 47, 303
Richter, S., 76, 78, 82, 88
Robbins, W. E., 300
Roberts, R. A., 354
Rockstein, M., 299
Roeder, K. D., 126, 131, 138, 187, 194, 462, 463, 464, 470
Rolls, B. J., 350
Roubaud, E., 355
Rowell, C. H. F., 416
Roys, C., 126
Rozin, P., 303, 304

Saccharin, 161
Salt, R. W., 359
Salt receptor, 101, 136; response to salts, 143–148
Sanchez, D., 377, 403
Sarcophaga, 178, 368
Sarcophaga argyrostoma, 90

Sarcophaga bullata, 171, 172, 312, 322
Satiety center, 466
Satija, R. C., 369
Sato, M., 212
Saturation deficit, 339
Savage, C. P., 359
Sawyer, W. H., 12
Saxena, K. N., 58
Saxifraga oppositifolia, 37
Scharrer, B., 407
Scharrer, E., 407
Schistocerca gregaria, 202, 278
Schneider, D., 88, 126, 218
Schneider, G., 15
Schneirla, T. C., 456
Schoettle, H. E. T., 470
Schoof, H. F., 355, 359
Schoonhoven, L. M., 202, 217, 218, 219, 234, 453
Schweder, M., 15
Schwinck, I., 17
Scolopoid sheath, 82–84, 94
Segundo, J. P., 201
Sekhon, S. S., 90, 94
Shaefer, C. H., 355
Shaw, I. R., 355
Shimada, I., 171, 176
Shiraishi, A., 74, 164, 165, 171, 172, 174, 175, 176, 177, 181, 211
Shorey, H. H., 18, 19
Shortino, T. J., 300
Side-wall recording, 128–131
Silk moth, 300
Silkworm, 228
Simonds, B. J., 350
Sinitsina, E. E., 455
Siverly, R. E., 355
Slama, K., 355
Slifer, E. H., 84, 90, 94
Slow potential, 138, 139
Smallman, R. L., 187
Smith, D. S., 366
Smith, M. H., 107, 108
Smith, M. P., 414
Smyth, T., 126
Snodgrass, R. E., 364, 377, 388, 389
Solomon, R. L., 142, 312, 417, 436, 442, 443, 446, 449
Solubility of compounds, in water, 150
Somomya erythrocephala, 368
Soulairac, A., 39
Specific hungers, 303–306
Specificity, 200, 211; of receptors, 136, 181–182
Spencer, W. A., 412
Spike generation, 139, 140, 142
Spontaneous activity, 130, 133, 218, 294, 295, 462, 463, 464, 465

Spontaneous recovery, of receptors, 176
Squirrel monkey, 49, 412
Starnes, E. B., 13, 16
Starvation, 234, 235, 293
Stegwee, D., 355
Steiner, G., 17
Steinhardt, R. A., 96, 100, 132, 133, 153, 174, 180, 202, 211, 268, 442
Stellar, E., 63, 411
Sterilization, 329
Stimulating effectiveness, order of, 145–147, 150, 172; of sugars, 161
Stimulating power, 286, 287, 288, 307
Stimulus fractionation, 214
Stimulus-response, model, 461
Stoffolano, J. G., 105, 309, 324, 357
Stomatogastric system, 253, 277
Stomoxys calcitrans, 84
Strangways-Dixon, J., 39, 42, 300, 301, 303, 312, 321, 322
Stretch receptors, 268, 278, 281, 356, 359
Stumpf, H., 15
Stürckow, B., 78, 82, 83, 84, 130, 132, 140, 197, 198, 202
Sucking, 67, 95, 96, 101–104, 341
Sugar receptor, 98, 100, 136; nature of, 171–178; models for, 173, 175; activity of, 417
Sugars, as energy source, 11–13; stimulating effects of, 122–126, 161–163, 175
Sulphhydryl reagents, 170
Summation, 98–101, 105, 417
Supplementation, 105–109
Swellengrebel, N. H., 355
"Sweetness," 49, 160, 161
Sweeteners: natural, 161; synthetic, 161, 177
Synapses, 366
Synergism, 162, 165, 168, 170, 211
Syrjämäki, J., 339
Sytshevskaya, V. I., 9
Szymanski, J. S., 456

Tabanus sulcifrons, 145
Taddei Ferretti, C., 16
Takeda, K., 84, 139, 210, 452
Tanabe, Y., 74, 181
Tapper, D. N., 144
Tarsal receptors, 74–77, 180–181, 202, 208; life of, 187; mammalian, 199, 211, 212
Tarsal threshold, 234, 235, 242, 250, 359
Taste: behavioral response to, 211–216; modalities, 223
Tate, P., 355
Tateda, H., 98, 139
Tauc, L., 442
Taylor, F. H., 152
Teitlebaum, P., 341, 466, 467, 474

Telea polyphemus, 240
Telfer, W. H., 324, 325
Temperature, effect on receptors, 138, 143, 172
Temporal patterning, 221
Texture: preference, 16; response to, 107
Thermobia domestica, 337
Thermodynamic analysis, 152
Thiamine deficiency, 304
Thompson, R. F., 412
Thomsen, E., 309, 312, 322, 325, 348, 470
Thomsen, M., 309, 325
Thomson, A. J., 232
Thorell, B., 15
Thorpe, W. H., 411, 412, 413, 414, 453, 456
Thorsteinson, A. J., 58, 177
Thresholds: acceptance, 46, 57; measuring techniques, 55–57, 109–111; olfactory, 55–57; rejection, 57, 134; of alcohol, 152; changes in, 243–247; regulatory mechanisms, 250–253, 259
Thurm, U., 83
Tight junctions, 83
Tip recording, 126–128
Tominga, Y., 83, 88
Tonic response, 141, 144, 190, 212, 217
Tormogen, 82, 83
Trail following, 22–27
Traps, sugar-baited, 173
Trehalose, 240
Treherne, J. E., 275
Trichogen, 82, 83
Trichogenius, 34
Tsetse fly, 292, 293, 295, 339
Tubifera, 368
Turner, B. H., 467
Turner, L. H., 142, 312, 417, 436, 442, 443, 446, 449
Tyshchenko, V. P., 5, 354

Urine production, 348

Valenstein, E. S., 466
Van der Kloot, W. G., 280, 463
Van der Poel, A. M., 132, 133
Van der Starre, H., 181, 194, 217, 218, 349
Van der Wel, H., 161
Vanessa indica, 139
Vapors on taste receptors, 208
Variability, 193, 194, 197–199, 214, 215, 218
Verlaine, L., 58, 72
Vincent, M., 355
Vitellogenesis, 309–311, 312, 315
Volleying, of receptors, 133–135, 202
Vowles, D. M., 375, 376, 406

Wagner-Jauregg, T., 88
Walking, 407, 464; patterns of, 20–27

Wallis, D. I., 88
Wallis, R. C., 355
Walther, J. B., 15
Waning responses, 412
Washino, R. K., 355
Water, stimulation by, 172
Water balance, 345, 350
Water receptor, 97, 136; model, 178–180; of fish, 179
Waterhouse, D. F., 15
Wehner, R., 15
Weiant, E. A., 470
Weis, I., 109
Wells, P. H., 234
Wenking, H., 15
Wiersma, C. A. G., 406
Wiesel, T. N., 406
Wiesmann, R., 15, 44, 174
Wieting, J. O. B., 61
Wigglesworth, V. B., 11, 12, 311
Wilczek, M., 76, 84, 85, 105, 365, 368
De Wilde, J., 359
Wilkens, J. L., 312, 322

Williams, C. M., 12, 299
Wilson, D. H., 462, 464
Wodsedalek, J. E., 456
Wolbarsht, M. L., 88, 131, 138, 139, 140, 141, 177, 180, 210
Wright, E. M., 148, 153
Wright, R. H., 18, 416
Wyatt, G. R., 240, 355
Wykes, G. R., 46, 125, 126
Wyman, R. J., 17
Wyrwicka, W., 467

Yamashita, S., 139, 140, 194, 212
Yeandle, S., 199
Yeatman, F. R., 417
Yin, L. R., 82, 83
Yolk formation, 311–315, 320, 321
Yost, M. T., 55, 58, 59, 61
Young, P. T., 39

Zacharuk, R. Y., 82, 83, 84, 210
Zawarzin, A., 403
Zeigler, H. P., 467

THE LIBRARY
ST. MARY'S COLLEGE OF MARYLAND
ST. MARY'S CITY, MARYLAND 20686

079116